"If you're interested in the world of game sound, you need this b[...] more excited that this book exists and wish I had had such a comp[...] organized handbook when I was starting out as a sound designer/composer. Gina and Spencer take us through a complete overview of the craft and the skills involved in this growing field where sound is an essential component of game development. They shed light on the creative process, the technical skills, and the practicalities, from sound and music creation to game audio implementation.

This book is a culmination of Gina and Spencer's many years of professional experience wrapped into an accessible and effective guide. Not to mention, they are seasoned educators, so as you can imagine this book is not only bursting with invaluable tips from professionals, but it's also instructive and even includes assignments and additional practical information on a companion website to make sure the concepts really hit home. That combination is invaluable and an enormous help! A must-read for up-and-coming game audio folks, in my opinion."

Michele Darling, professional sound designer and assistant chair of
Electronic Production and Design at Berklee College of Music Boston, USA

"This is an astonishing achievement. It is the definitive book of reference that the industry has needed for years. It will help guide a whole new generation of creative technical audio professionals into this immensely challenging and rewarding area of the video games industry. If you have any serious interest in video games audio or music, you must read this book."

Guy Michelmore, composer

"A delightful and useful addition to the Game Audio Toolbox of Your Mind (and bookshelf!)."

George Alistair Sanger, "The Fat Man"

"*The Game Audio Strategy Guide* is a comprehensive text for anyone wanting to better understand the nuts and bolts of music composition and sound design for games. In their information-packed treatise, Gina and Spencer provide an excellent roadmap for understanding the complex and not-so-straightforward art of game music and sound."

Brian Schmidt, executive director, GameSoundCon and
Brian Schmidt Studios, LLC, USA

"This exciting book is a comprehensive 'walkthrough' of the creative processes of game audio. It uses the experience of industry professionals in tandem with theoretical discussion to present practical wisdom in an accessible manner. The book encourages readers to become reflective, innovative composers and sound designers. A strategy guide, yes, but absolutely no cheating!"

Dr Tim Summers, lecturer in Music, Royal Holloway University of London, UK,
author of Understanding Video Game Music *and co-founder,*
Ludomusicology Research Group

The Game Audio Strategy Guide

The Game Audio Strategy Guide is a comprehensive text designed to turn both novices and experienced audio designers into technical game audio pros. Providing both a theoretical foundation and practical insights, *The Game Audio Strategy Guide* offers a thorough look at the tools and methods needed to create industry-quality music and sound design for games. The text is supported by an extensive companion website, featuring numerous practical tutorials and exercises, which allows the reader to gain hands-on experience creating and implementing audio assets for games.

The Game Audio Strategy Guide is the essential manual for anyone interested in creating audio for games, inside or outside the classroom.

Gina Zdanowicz is an accomplished audio designer for games, film, TV, new realities, and podcasts. A graduate from Berklee College of Music with a degree in Music Synthesis, Gina honed her skills in game audio by working in house with game developers and founding her own independent sound studio, Serial Lab Sound. Gina has worked on over 100 video game and interactive titles and her work with Nickelodeon (TV) has been nominated for two Daytime Emmy awards. In addition to her career as an audio designer, Gina is a game audio course author and instructor at Berklee Online, a tutor at Thinkspace Education, and a lecturer at universities such as NYU and conferences around the world like GDC. Follow Gina at: www.seriallab.com

Spencer Bambrick is a composer for games, film, and the concert stage. His recent work includes *BestLuck* (2018, PC, Mobile), *Block Party* (Emmy-Nominated Nick Jr. Series of Animated Shorts), and additional music for *Paladins* (Hi-Rez Studios). Spencer is the Game Audio Network Guild Northeast Regional Director and has presented at GameSoundCon (2018, 2017), PAX Dev (2016), and MAGFest (2015). He is currently pursuing his doctorate in Music Composition at the University of Hartford (The Hartt School), studying interactive music, game technology, and social activism.

The Game Audio Strategy Guide
A Practical Course

Gina Zdanowicz and
Spencer Bambrick

Routledge
Taylor & Francis Group

NEW YORK AND LONDON

Visit the companion website: www.routledge.com/cw/zdanowicz

First published 2020
by Routledge
52 Vanderbilt Avenue, New York, NY 10017

and by Routledge
2 Park Square, Milton Park, Abingdon, Oxon, OX14 4RN

Routledge is an imprint of the Taylor & Francis Group, an informa business

Library of Congress Cataloging-in-Publication Data
Names: Zdanowicz, Gina, author. | Bambrick, Spencer, author.
Title: The game audio strategy guide : a practical course / Gina
Zdanowicz, Spencer Bambrick.
Description: New York : Routledge, 2019. | Includes
bibliographical references and index.
Identifiers: LCCN 2019023938 (print) | LCCN 2019023939
(ebook) | ISBN 9781138498334 (hardback) |
ISBN 9781138498341 (paperback) | ISBN 9781351016438
(ebook) | ISBN 9781351016421 (adobe pdf) |
ISBN 9781351016407 (mobi) | ISBN 9781351016414 (epub)
Subjects: LCSH: Computer sound processing. |
Video games–Sound effects. | Computer game music.
Classification: LCC TK7881.4 .Z424 2019 (print) | LCC
TK7881.4 (ebook) | DDC 794.8/165–dc23
LC record available at https://lccn.loc.gov/2019023938
LC ebook record available at https://lccn.loc.gov/2019023939

ISBN: 978-1-138-49833-4 (hbk)
ISBN: 978-1-138-49834-1 (pbk)
ISBN: 978-1-351-01643-8 (ebk)

Typeset in Sabon
by Integra Software Services Pvt. Ltd.

Somewhere along the way I learned there is nothing greater than appreciation for those who feed your soul and propel you forward on this amazing journey called life. I dedicate this book out of gratitude for...

Mom and Dad (for always believing in me); Jaimee Figueras (for all the support on this crazy ride); Paul (my first creative partner), Danielle, Ava, and Clive Zdanowicz; Julie Chase (for my first gig in the industry); Jason Kanter (for being an amazing mentor and ally); Mike Santasieri; Ron, Vickie, and Arnold Figueras; Rob Khor; Geneviève; Vel; Dr Richard Boulanger (for teaching me to think outside the box); Ellen Early, MD (for all these extra years); Vera Evenson and Allan Wasserman (for sculpting my journey in music early on).

To all the students and mentees over the years who have taken my classes and sat in on workshops and talks, you help me continue to strive to be better at what I do. And all those who have touched my life in one way or another – thank you!

Gina

For Natasha.

You will always be the music in my heart.

Spencer

Contents

Foreword

One of the key differentiators between game audio artists and our cousins in linear media is the implementation of our sounds into the game engine. In a linear workflow, all sounds go onto a DAW timeline, which is then mixed to complete the finished product. In game audio, however, we have no DAW timeline that matches picture. Instead we're creating sounds, putting them into a package that gets loaded by the game, and then building and tuning systems to control their playback. This stage is crucial; not only can we control the mix within the game engine, we can control the playback of layers, pitch modulation, and other creative effects.

I will even go as far as to say that implementation is half of the creative process. If anything is written on my tombstone, I want it to be that. My wife insists it should be something about family, but no: implementation. It's important stuff. Let it be written in stone forever. And I suppose you can put beloved husband and father on there too.

In my time in the industry, I've seen implementation grow in importance quite a bit, and I've heard the benefits of that growth in the games I've played. Certainly, there were stunning examples of implementation long before I started paying attention, some of my favorites being the interactive scores in LucasArts games. But two magical things have happened in the last 15 years: audio tools and game engines have been made widely available, and social media has become an enabler of broader discussion throughout our industry. These two factors have been massive in their impact on the way we as an industry think about, create, and implement game audio today.

The availability of tools has allowed more brains to think about how sounds are hooked up in games. More brains mean more ideas. And publicly available, shared toolsets mean those brains can talk to each other without violating NDAs (for the most part). Now that we have Twitter, Facebook, Slack, and other communication methods, those brains can very *easily* talk to each other. A sound designer in Singapore can start up a chat with a sound designer in London to discuss how to assemble a dynamic vehicle system in Wwise. That sort of conversation was the stuff of fairy tales 15 years ago. We've come a long way.

In addition to the openness of tools and discussion, when we hit the current console generation, many resource constraints fell away. We're six years into developing for the current generation consoles as I write this, and I'm still not thinking very much about RAM. Occasionally, disc footprint/download size becomes an issue. We have more polyphony than I really want present in a mix. Sure, there are definitely trails we still need to blaze, both technically and creatively, but this generation feels like the first where we're not making do with the tiny scrap of resources we have.

It's a magical time to be a game sound person. Truly. I feel fortunate to be doing this right now, and you should too.

So, about implementation – I have some thoughts. The first is that we should not be so focused on the implementation that we lose sight of our role – counter intuitive, but important. I do not like to think of myself as a sound person, but as a game developer that makes sound. It's semantic on the surface, but it's deeper than that. First and foremost, the work we do should not be done for the sake of audio itself, but in service to the game. What that means in practice depends on the individual game, of course. I used to work on *Call of Duty*, where game play is king. Everything I did was in service to the game play. If I was working on a very narrative-focused title, then I would focus my efforts there. In some games that context can shift depending on the current state of the game, and we as sound professionals must consider and accommodate that. This is all to say that we should understand our role in game development and think of ourselves as contributors to the game itself, rather than people who are making sounds. The game is more important than the sound.

Second, while great tools and workflow are very empowering, we must also be mindful not to allow them to limit us. As wonderful as they are, easily accessible interfaces can be a trap. I had a bit of a Twitter rant a while back about my fear of what I call "Middleware Operators," people who understand how to press the buttons and operate a middleware tool, but who don't know how to get deeper into the guts of the game itself or understand how the game communicates with the audio layer. The data that flows to Wwise or FMOD (or any other audio pipeline/toolset) is important, and there's a lot of creativity to be had in managing how that data flows: manipulating it, interpreting it, sometimes throttling it, for creative purposes. You also never know what useful bits of data you might find if you go digging in gameplay code.

On every game I've ever touched, supporting the game with sound goes way beyond just adding assets to the game, handing some sound event names to a designer or coder and walking away. There is tremendous power in controlling how those sounds get triggered, how they are modified based on run-time data, player state, context, and much more. There is tremendous power in understanding all of that and designing content with the implementation context in mind. And, while I would never argue that we should make complex systems for the sake of complexity, we should not shy away from complexity if it allows us to better serve the player experience. This means we should be looking into the space between the game itself and the audio pipeline, whether that space is C# in Unity, Blueprint in Unreal, asset definitions and gameplay code in some other engine, animation keyframing, particle system editors, level editors, you name it. We should discover what systems can interact with sound in a meaningful way, and we should then strive to learn how those systems work so we can hook into them more deeply.

We should not let the challenges of game audio dissuade us. They are constant. Game development, as a process, is one big tangle of problems. We solve one, and then the next, and the next. We solve as many as we can before we ship the game. Well, to be fair, we solve as many as we can before someone above us rips the unfinished game from our hands. Our job is never done.

As you pursue your journey in game sound, it is my hope that you will endeavor to be not a sound designer, but a game developer who makes sound. Be curious and do amazing things. I'll be listening.

Mark Kilborn, audio director and sound designer

Preface

The video game industry is perpetually in flux. Due to rapid growth, changes in the field of technology are constantly challenging game developers to create novel and artistic worlds within which audiences can immerse themselves. Market demands are also a heavy influence on the industry. With the growing prevalence of mobile phones and tablets, and the increased interest in indie and crowd-funded game development, players have more choices than ever before. Added to that are the myriad of VR headsets, which are becoming more affordable and more powerful with every update. The result is an exciting and unprecedented era of game development.

Although discovery and innovation in the game industry are things to be celebrated, it leaves industry professionals with a very difficult task. Developers must not only keep up with the technology of the times, they must also strive to create work that offers something new and interesting to the profession, as well as work that speaks to the hearts and minds of the gaming community. As we will show in this book, audio is one approach to accomplish just that.

To many, audio is "invisible." This is because when the audio is done well, audiences rarely notice it. When it's done poorly however, it is immediately apparent to the listener, even if she cannot articulate why. This often leads developers to underestimate the importance of audio in the development process. But in truth, it is one of the most crucial elements of game design, and it is a powerful tool that can capture the attention of players and unlock the emotional potential of an interactive story. To succeed in this, audio designers (like all game developers – and audio designers *are* game developers) must be intimately familiar with the technology used to develop games as well as the methods used to create audio. A chef cannot expect to create a meal without first learning to use an oven. Likewise a game audio professional cannot expect to implement quality assets before first internalizing the basics of audio production. Experience with the technology and practice with the methods are both necessary for success as an audio designer in games. In this book we will outline these methods and explain the theory behind them. We will also use industry-level software to illustrate our points. Finally, we will provide tutorials, examples, and assignments on our companion website (The Sound Lab) to help you gain experience on your own.

Acknowledgements

A book such as this one takes many minds to bring it to reality. We spent more than a year brainstorming various topics and outlining them. Afterward we spoke to colleagues and friends to figure out the best way to map out our ideas. Thank you to everyone on the Taylor and Francis/Routledge team for your support and for believing in us. We are ever so grateful to Julie Chase, the best cover designer we could imagine. Thank you for taking the time out to create the "face" of our work, and for tackling all the chapter figures. Likewise, Jaimee Figueras – thank you for the expert and artful graphics, and Armina Figueras – a big thank you for the amazing photography.

We would also like to offer our sincerest thanks to UCLA's Game Music Ensemble, and to Jose Daniel Ruiz in particular. Thank you for all of the time and effort you spent making our musical ideas come to life.

Special thanks to our technical and creative contributors from the game audio community for helping us ensure we are delivering a useful text that helps guide the next generation in our industry.

Technical reviewers: Aaron Brown, Adriane Kuzminski, Alexander Brandon, Brian Schmidt, Felix Faassen, Guy Michelmore, George A. Sanger, Jamey Scott, Jason Kanter, Jeanine Cowen, Jose Daniel Ruiz, Lani Minella, Michael Sweet, Michele Darling, Dr. Tim Summers, T. Lin Chase, Thor Bremer, Trevor Kowalsky.

Visiting artists: Alexander Brandon, Ann Kroeber, Suganuma Atsushi, Ayako Yamauchi, Bonnie Bogovich, Brian Schmidt, Damian Kastbauer, Dren McDonald, D. B. Cooper, George A. Sanger, Jason Kanter, Jeanine Cowen, John Robert Matz, Jose Daniel Ruiz, Martin Stig Andersen, Michael Csurics, Penka Kouneva, Rachel Strum, Stephan Schutze, Tamara Ryan, Thomas Rex Beverly, Tom Salta, Watson Wu, Wilbert Roget, II, Xiao'an Li.

Finally we would like to thank Mark Kilborn. Your work and your words have had equal parts in inspiring us to write this book. We hope that we have done justice to your foreword, and that the next generation of game audio professionals will be just a little bit better off thanks to this book.

INTRODUCTION

Getting Started

Welcome to *The Game Audio Strategy Guide: A Practical Course*. This book is the culmination of years of experience as sound designers and composers for games, as well as years of teaching at college and graduate level. At some point we realized that all of the methods and approaches that we use professionally, and teach to our students every day, should be compiled in one place so the budding game audio professional might be able to access them easily. On top of that, we hold a great amount of respect and passion for game audio as well as educational practices. With that, we combined our knowledge into what we hope is a successful dissemination of the theory and practice of game audio. We've also included many tips and tricks from other working professionals in the field, and offer some real-world examples of topics we discuss. Before we get started talking about what game audio *is*, let's cover what game audio *isn't*.

Game audio as a process isn't a particular set of directions that you can learn or copy. Rather it involves understanding games at a critical and technical level. Much like programming and engineering, game audio will require you to think outside the box and problem solve, to arrive at appropriate solutions for your game. You will need a wide breadth of skills and theoretical perspectives to do this effectively. In some ways game audio is more about finding a *variety* of answers than finding the *right* answer. In this book we will not give you a "one size fits all" solution. Instead we will offer our philosophy and theory of game audio, and enough practical knowledge, so that you can come up with your *own* solutions. This, we believe, is more valuable than a single step-by-step manual.

Game development can be creatively rewarding and enjoyable, but is certainly not without its difficulties and challenges. The field of game audio is always evolving and covers a wide variety of defined specialties such as sound design, music composition, field recording, voice-over production, orchestration, implementation, and many others. Whether you are at the start of your career or somewhere in the middle of it, it is always a good time to evaluate your skill set and look at opportunities for improvement. Having multiple skill sets can help push your career forward and make your day-to-day work more interesting. Even in specialized technical positions, learning seemingly disparate skills can come in very handy, and even provide creative solutions that you may never have thought of before. Reading through this book you will find that we consistently advocate for learning as much as you can about the game development process overall, rather than narrowing your view to one part of one field within the game development process. To reiterate: resourcefulness and the ability to think "outside the box" in both creative and technical approaches will greatly help you along in this fast-paced industry.

As teachers, mentors, and game audio advocates we seek to nourish a sense of creative problem solving in those who wish to enter the industry or further their career development. If learning to approach game audio with a creative and analytical mind sounds appealing to you, then let's begin our journey!

OUR GOALS FOR THIS TEXTBOOK

We had three goals in mind when we set out to author this textbook. First, we wanted to present a theoretical and practical framework for teaching game audio that goes beyond the basics. There are fantastic books on the market that outline the basic theory of game audio, and there are equally effective books that offer a more practical guide. This textbook, along with our Sound Lab (companion website), is both. You will start by learning the basics of **nonlinear audio,** and by the end of the textbook you will have experience with practical sound design and music composition as well as implementation into middleware. Along the way there will helpful assignments, thought experiments, and resources for further study.

Our second goal was to bring as much information on game audio as possible into one place. Students can cobble together loads of useful information from various sources on the internet, but we wanted to provide something more. This book is meant to be a cohesive account of game audio from the fundamentals all the way through the more advanced topics. If anything, the subject matter is weighted considerably more toward intermediate concepts because they are harder to find elsewhere. Still, we've worked hard to make sure that the subject matter moves logically and cumulatively so that each topic builds on the last in a way that makes it digestible for students. Using this book, a beginner looking to get into game audio will learn the foundations quickly and move confidently into intermediate and advanced topics.

Our final goal was to begin bridging the gap between academia and industry practice. We are ardent advocates of education and, simply put, we want to see an emphasis on game audio **pedagogy.** We want to see more college- and graduate-level programs specific to game audio around the world, ideally within game development programs themselves. Game audio is *not separate from game development*, it is a specialization *within* game development. Likewise, game audio is not separate from academia. Brilliant minds like William Cheng and Tim Summers[1] have written illuminating papers and books on the topic of **ludomusicology**, the scholarly analysis of music in the context of video games. This writing has done a great service to the field of game audio by reinforcing its validity as an academic topic, paving the way for more research and analysis. In a similar vein we hope to make industry practices more accessible to academic programs, and conversely to make academia more present within industry practice. This manner of "cross-pollination" can only lead to a deeper understanding of game audio practices, which will lead to more pedagogy-focused game audio programs. In turn, the presence of more game audio programs will yield a highly competent and technical field of game audio professionals.

OVERVIEW AND LEARNING OUTCOMES

This book is meant to be viewed as a course that will take you from introductory topics in game audio, to more intermediate asset creation, and then on to more advanced implementation practices. In a very strong sense the textbook itself can be thought of as the *lecture* component of a course, and the Sound Lab (companion) website can be thought of as the *lab* component. We cannot overstate the importance of the companion site. It is not supplementary material; it is a *core component* of this course. Lectures alone can only get you so far without making use of the lab (on the companion site) to internalize the theory. Paying attention to the assignments and coursework here will also get you started on your demo reel and portfolio (see "Demo Reel," Chapter 11, page 358).

We begin this text with an introduction explaining how this book should be utilized and why you should read it. We also outline the roles on a game development team and identify the responsibilities of an audio designer within this paradigm. The size and features of game development studios are broad and diverse, but we will outline some common examples of small-, medium-, and large-scale projects.

In Part I we discuss all aspects of sound design, including techniques and practices from recording to editing and how to address the challenges that stem from working with nonlinear media. We cover the essential tools and skills you will need to succeed as a sound designer on your own and with a team. We also cover the production cycle and planning for various platforms.

Part II is focused on adaptive music. We begin by discussing methods and strategies for generating musical ideas as they relate to real-world game projects. We also discuss techniques and practices for arranging and orchestrating sample-based music and live orchestral cues.

In Part III we explore audio engines, implementation, resource management, optimization, and interactive mixing. We finish by detailing techniques for complex adaptive music systems and dynamic mixing along with a discussion on the creative and aesthetic nuances of experimentation within the game audio framework.

Finally, in Part IV we cover topics on the business end of the game audio field. We will discuss networking, contracts, and marketing yourself in the era of social media. We will also go into some of our personal thoughts on industry ethics and how they fit into a globalized game development industry ethic.

Game audio is an essential component of a vast and ever-expanding industry. It is affecting and artistic in its own right; it breathes life into the worlds that artists and programmers create. It is our hope that this book will help make learning the craft of game audio more accessible and rewarding for students of all levels. Underlying our desire to share our passion for this field is our belief that game audio is special, and that everyone should have the means to appreciate it as we do.

WHO SHOULD READ THIS BOOK?

This book is intended for readers with some experience with digital audio and music composition. Readers should also be familiar with a variety of games. Our rationale for this is simple – there are already tons of helpful sources in books and on the internet that cover these basics. What we want to offer with this textbook is a quick look at the fundamentals *as they apply to game audio*. Most people don't try to score a film if they have never seen one before, and this applies to games to an even greater degree. Games are **nonlinear**, and it's hard to understand nonlinearity if you have never played a game before.

After covering the fundamentals of game audio, we will focus most of our time on intermediate and advanced topics ingrained in the nonlinearity of games. Students looking to gain an intermediate and advanced technical knowledge of game audio are the target audience for this book. However, even if you are a complete beginner, don't worry. The topics can be dense at times, but this book is self-contained. This means in each chapter we will cover everything you need to know to understand subsequent chapters. Beyond that, we have included a slew of other sources and reference for those interested in broadening their game audio studies.

Keep in mind that the terms we define and the opinions we share throughout this book are based on our professional and educational experiences. Some terms may be coined differently across various scenarios, and other audio professionals may have different opinions or techniques. This is why checking out our references and "Further Reading" section on the companion site (see below) can be extremely helpful for a well-rounded view.

To summarize, our intended audience includes everybody who wants an industry-level technical understanding of the theory and practice of game audio. As we have mentioned, this book is organized like a lecture/lab, so game audio teachers and students are prime candidates. As a close second, anyone either looking to get into game audio from other related fields, and anyone teaching themselves game audio will (we hope) find this book indispensable. Likewise, experienced people in the industry can use this book to gain insight into other views and methods in game audio. Finally, those from other disciplines (like film and TV audio) looking to move into game development will find this book useful as well. It will help in understanding the vast amount of technical and creative work that goes into creating and implementing sound effects and music for games. If nothing else, an experienced game developer will finish this book with a vastly greater respect for the game audio process.

How to Use this Book

This textbook can be used in a few different ways. Our recommendation, of course, is to read it all the way through in order. This will give readers the fullest sense of what it takes to produce audio for games, and to develop a sustainable career doing so. Our Introduction sets the foundation for what game audio is and what our approach to teaching it will be. Parts I and II involve sound design and music respectively. We have chosen to cover sound design first because many of the elements in Part I will make the discussions in Part II very easy to understand.

If you are specifically interested in either sound design or music *only*, then feel free to read the Introduction and then jump straight into the relevant chapters. We would encourage you to *read all relevant chapters entirely.* As we mentioned earlier, this course is cumulative. Skipping Chapter 2 may leave you fumbling for answers in Chapters 3 for example. It will be helpful to read through all of Part IV regardless of your speciality. Career development is essential to all areas of game audio.

Readers may also want to use this book simply as a reference for industry techniques and best practices. This is perfectly fine. Just be sure to use the Sound Lab as appropriate to cover the practical aspects of game audio. We've also included a detailed glossary of terms on the companion site, and plenty of references as well. As stated earlier in this chapter, one of our goals was to compile as much information on game audio as possible so that it would be accessible in one place. Here it is!

REQUISITE KNOWLEDGE

Despite the fact that this text dives into some highly advanced topics, we do our best to assume little or no prior knowledge. That said, this book covers so many interdependent fields that some prior knowledge is helpful. For starters, a bit of experience with a DAW (Digital Audio Workstation) is vital. The topics covered here are DAW agnostic, so any particular DAW will do, but it should be industry standard (i.e. **Logic**, Pro Tools, Reaper, but not Garageband). You don't need to be an expert, but familiarize yourself with the process of using plugins and virtual instruments in your DAW so you can get a feel for the basic workflow. It can also be helpful to learn the fundamentals of signal processing since there are some technical mixing concepts discussed.

In terms of prior musical knowledge, the fundamentals of music theory are a must. By this we mean you should be able to read musical notation. This will help you in understanding the composition and orchestration concepts covered in Part II. You should have a basic understanding of diatonic **harmony** as well. We will be using roman numerals at times to describe various chord progressions. Apart from this, we assume no knowledge whatsoever of implementation or game design.

THEORETICAL FRAMEWORK

We like to start each chapter or section by laying out the topical theory. By this we don't mean a hypothesis of an observation. Rather we will lay out as simply and clearly as we can the topic of discussion in a broad sense. We also try to include our personal philosophies on the topic's place within game audio and within the field of game development as a whole. This is to help you formalize a mindset that will suit you when working on your own projects. For instance, before we get into any techniques about designing sounds we will first cover what exactly sound design is, and what it means to create sound design for games. The goal is to give you a wide perspective on each topic so that you can think for yourself and make critical decisions when the time comes.

PRACTICAL FRAMEWORK, A.K.A. COMPANION WEBSITE (SOUND LAB)

The practical side of things is equally important, and is covered almost entirely on the Sound Lab website. We will direct you to the Sound Lab throughout the book where you will read deeper into some practical topics and step-by-step guides and also gain experience with assignments and practice of your own. The Sound Lab will be your portal to additional reading and resources, audio examples, video tutorials, further reading suggestions, musical scores from your favorite games, and much more. TLDR: *don't skip out out the Sound Lab!*

Video Tutorials and *Audio Examples*. It's important not just to understand the concepts we put forward, you need to *see them in action*. This will prove a vital component of your practical experience in game audio. Additionally, in later chapters we will be offering downloadable examples/sessions of *Adaptive Audio Events* and *Implemen-tation Tutorials* among other things. These are crucial to check out because they are your only means of practicing some of the more challenging technical aspects of our lessons.

In that vein, part of the practical component of this book is the smattering of *Exer-cises* you'll find throughout each chapter. These are usually simple thought experi-ments to help you internalize an idea or method. Don't skip these. They can be really helpful when taking audio from theory to practice.

The final aspect of the practical side of this text is the list of *Assignments* at the end of the chapter as found in the Sound Lab. These are more goal oriented and tangible than the Exercises, and you may even have these assigned to you by a professor if you are using this as a course textbook. They are even more important than the Exercises because they closely resemble directives you might receive for an industry project. This is how you will begin to amass actual experience in game audio. Simply put, there is no substitute for this experience. The only way to truly learn game audio is to actually *do it*. Without putting the theory to use, you are only learning about 50 percent of the lessons this textbook has to offer.

In Part II we cover game music. One area that we have found lacking in other sources of game music education is actual *Score Study*. Here we attempt to remedy this by including real-world examples of game scores. In this section we will look at a score and analyze the composition, orchestration, and adaptive recording techniques used. We will share a few bars of the score and detail a few key items to look at. These will also serve as a reference for you when it comes time to prepare sheet music for instrumentalists to record. It's important to note that not all game composers read music or find the need for scores, but it can be a useful resource when incorporating live instruments into the recording.

Finally, the companion site is important to explore because game audio is a fast-moving field. We've included the most contemporary software and techniques in our field to the best of our knowledge, but even a year or two after this text is published there will likely be new techniques and new software to cover. The companion site allows you to keep up to date on industry trends, so make sure you check it out when you see one of the callouts below.

Companion Website Callout Explained

When you see a "Sound Lab" text callout we are letting you know there is add-itional material available on the site. This information isn't critical in under-standing the material to come. You can continue reading through the chapter

(Continued)

and visit the companion site later. There will be reminders at the end of each chapter as well as at the end of each part.

 When you see this icon it signals a callout that is critical, meaning you should stop reading the textbook and visit the Sound Lab. The information presented on the site will be a necessary piece of the puzzle for the forthcoming topic so we highly recommend reviewing the material on the site before moving on in the book.

VISITING ARTIST'S TIPS

Getting advice from industry veterans is a crucial element of "learning the ropes," so to speak. We have included the "Visiting Artist" series as a way for readers to hear from industry professionals. You'll find plenty of topics covered here, ranging from music composition to networking. These sections are written by friends and colleagues speaking about real-world topics that they have had unique experience in, so don't miss out on these helpful tips.

KEY TERMS

The terminology in game audio can be tricky. Whenever you see a word in **bold,** you are looking at a key term. Key terms are essential bits of language that you'll hear often in the industry. They can also clue you in to various techniques and contemporary methods for sound creation and implementation. Check the glossary on the companion site for definitions.

FURTHER READING

Sadly, no book can cover everything of value to any technical or creative field. In fields as demanding as game audio there are always topics that need to be omitted to maintain logical consistency. There are a number of topics that come into play at a foundational level that we have omitted in order to keep our focus on games and their nonlinear characteristics. To counter this we have included a Further Reading section in the Sound Lab (companion site) so that readers have a place to turn for more information on some of these topics. Take note of these sources. They have proven to be valuable assets in our professional lives, and will certainly be the same for you.

Game Audio

Now that we have gotten the logistics out of the way, let's talk about game audio. What exactly is game audio? How is it different from film or television audio? How is game music different from what you hear every day on the radio? The answer to this will be covered in depth in the chapters to follow but, in short, game audio is **nonlinear**. To put it simply, linear audio is arranged sequentially from start to finish and will never change. The sound effects in a movie will all be heard in exactly the same order at exactly the same time every time you watch it. Games are different because *players* are different. One player might use a sword while another uses magic. These two approaches require different sets of sound effects. In fact, this is only scratching the surface of nonlinearity. A single player might play through the same game a thousand times and she will likely *never hear the same exact audio performance twice*.

In practice, game audio is the intersection of many different fields. For a sound designer, game audio is the overlap between sound-effects creation, audio production and post-production, recording, mixing, audio editing, programming, and game design. For a composer game audio is the overlap of music composition, orchestration/arranging, recording, mixing, programming, and game design. And these are only the broad categories. There are still more areas like dialogue recording/editing, voice direction, music supervision, audio direction, and many more areas that overlap into game audio. Our point is that because games are so multidimensional, game audio is multidimensional. There are many subfields within game audio and each of these fields plays an important role.

Finally (and this is our own personal standpoint derived from years of game audio research and experience), game audio is fun! We are creating audio that is unpredictable and tied to a product that you can actually *play*. We aren't just designing sounds and writing music, we are designing *interactive audio systems* for the enjoyment of players. Don't be fooled by this –game audio is technically demanding and takes years of experience to master. It is also a highly competitive industry, which is why we have included Part IV: Business and Networking. However, for those who are patient, and truly enjoy the challenge of designing playable systems of music and sound, this is the field for you.

Game Development Roles Defined

Now that we have defined game audio, let's take a step back and look at some of the important roles in the broader field of game development (Figure 1.1). Keep in mind that learning game audio is really learning a *specialization within game development*. We would encourage readers to start by familiarizing themselves with games in a wide variety of genres. Research the development teams and take a look at who is credited for each role on the development team. Pay particular attention to the audio both stylistically, and in terms of how/when sounds are triggered.

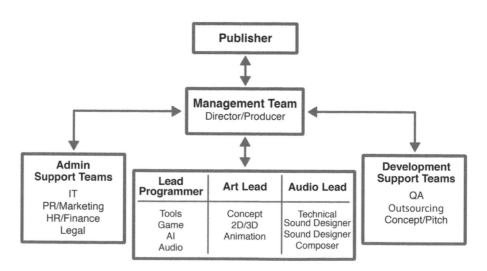

FIGURE 1.1 Game development teams (large and small).

In game development you will find teams of all sizes, which means the roles can be expanded and branched out over multiple people or simplified with one person sustaining all the tasks within multiple roles. Being flexible will help you adapt to different teams throughout your career. In the case of audio, a smaller indie team might mean one person will fill the roles of both composer and sound designer. They may be tasked with asset integration or it may be left with the game programmer to handle. In either case, the audio artists should be involved in testing to ensure the assets are properly implemented. Teams that are somewhere in the middle or what we might label "pro casual" may have a separate composer and sound designer, where the sound designer might be in charge of music editing and implementation and the final mix. This type of team might also have an audio lead or audio director. **AAA** studios (large-budget development studios) and larger development teams might have several sound designers including a senior role, with an audio director overseeing the team. At the end of this chapter we will direct you to the Sound Lab (companion site) to read about game development roles from game design and programming to the various audio roles in the field.

Essential Soft Skills and Tools for Game Audio

Working in game audio can be quite a fulfilling and amazing career. While games are a lot of fun, a career in games requires dedication and hard work, along with patience and practice. There is a variety of essential skills and tools that should be on your radar to master if you are serious about working in the industry. At the end of this chapter we will direct you to the Sound Lab (companion site) for an in-depth look at the basic essential soft skills and tools required by sound designers and composers looking to get started in the industry. These resources include skills such as looping, required technical knowledge, software and hardware to get started. For now, here are a list of skills essential to sound designers and composers interested in a career in game audio.

Job listings are a great place to start pulling together the skills and experience you should acquire to be an audio designer in the game industry. Most listings are detailed enough to be a good resource and reference. Here we present a list of desired traits from various job postings:

- Passion and enthusiasm for music, sound and games
- Knowledge of sound recording, field recording, Foley, synthesis and editing techniques
- Ability to compose music in a variety of styles
- An understanding of acoustics and spatial sound
- Mastery of a DAW and/or audio editing software
- Critical and analytical listening skills
- Imagination and creativity
- Excellent communication, time management, and organizational skills
- Ability to solidify concepts into material assets
- Ability to work well with a team

- Ability to handle feedback and process it into revisions
- Ability to work well under pressure
- Self-sufficient, with problem-solving abilities
- Knowledge of game engines (e.g. Unity, Unreal)
- Knowledge of audio engines (e.g. Wwise, FMOD)
- Knowledge of collaboration software and tools

Production Cycle and Planning

Whether you are working with a large team or a small one it will be important to have your organizational and time management skills sharpened in both the business and creative side of things. The process of game development is a mix of computer science and production, which takes every game from its very basic concept stage to a shipped product. The development process consists of several stages (see Figure 1.2).

The **concept stage** is where the developer is essentially deciding what type of game they will develop. Concept art and/or a prototype will be developed to further share the idea with others.

Pre-production follows the concept idea by allowing the team time to develop resources and design documents that will be used in developing the game. After a **game design document (GDD)** is created, the audio designer can begin to think about what type of tools and resources are needed for the game's audio development. This is a great stage at which to create mock-ups and source material.

Early on in the **production** stage is typically when a **vertical slice** is created. This is a short **playable build** of the game, which acts as a proof of concept. It's typically used to secure funding. Throughout the production stage the game is being refined and further developed.

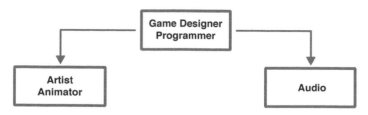

FIGURE 1.2 Typical game development cycle.

As the development matures through the production stage, an **alpha** build of the game is thoroughly reviewed by the **quality assurance (QA)** team. QA is an important phase in the cycle which evaluates the games mechanics, value, and design. As the game runs through QA, testers are looking for **bugs** or missing features through **white-box** and **black-box techniques**. You will notice in Figure 1.2 that pre-production, production, and testing is an iterative cycle as developers may often have to go back to pre-production or production to make changes and fix bugs. It's important to note that an alpha build may not contain all the planned features and may be a very unstable build, which could result in crashes and data loss.

The **beta** phase is entered when a build is generated with complete features. During this phase there may be some instability in the build and it is possible to uncover new bugs, but it's technically in a state in which it can be delivered to publishers, reviewers, and as a limited public release to beta testers.

The **gold master** or release phase is typically the final release prepared for distribution. There may still be some bugs, which can be fixed via **patching**. During this phase the marketing and community management is in full 'go' mode. The development team may still be in the production stage during the release as there is usually additional **DLC** (**downloadable content**) and patches that need to be prepared.

You may be wondering at what stage the audio team is brought onto a project. The answer is – *it depends on the project*. The pre-production stage is an ideal time for the audio team to be brought on. Some teams will even bring in audio during the concept stage for input. Often the audio team is brought on during alpha or later. This is unfortunate because bringing on an audio team early means they have more time to plan, gather source material, and experiment with ideas.

Games are usually on a strict schedule with milestone deliverables along the way. Regardless of when audio is brought in, there will always be necessary planning to keep the team in sync with the production cycle. Publishers want to hit the mark and have the product ready for holiday sales or conferences like E3. Even independent developers (or **indie** devs) need a plan to push their game to the starting gates. For indie developers it's important to choose a time for release that won't be over-shadowed by an AAA release or big industry news.

Whether you get started during the pre-production phase or later, there are a number of questions you can ask the developer to be sure you have all the info you need to complete your tasks. It's important to ask these question *before* contracts are signed (a topic we will cover in Part IV: Business and Networking). Once signed contracts are in place, the next step is to review the game design document (GDD) and (ideally) play an actual build of the game. This will give you a feeling for the game mechanics as well as the overall visual aesthetics. If the project is in the early stages of development however, you may only have concept art and a storyline to work with. This still offers valuable clues toward finding the right source.

During the information-gathering process or pre-production you will want to suggest additional assets that the developer may be overlooking. Once you have enough information and ideally a build of the game to play through, you can start mapping out your plan.[2] If you have a very quick turnaround time you may want to work out which assets are top priority with the developer and see if some lower priority assets can be patched in with an update at a later date.

Milestones are typically set by the developer and their publisher. An asset list may be provided to you by the developer or you may be asked to create your own based on your play through of the build. If this is the case, be sure to review your list with the developer to ensure you didn't miss anything. The development team will utilize **development methodologies** such as Agile, Scrum, Waterfall or Design Sprints to keep the project on track. It would be helpful to be familiar with these methods as well as some of the collaboration tools used in the process.

Reading through the game design document (GDD), if there is one, is a great way to understand all the tiny details about the game. Be aware of the fact that as the game development moves forward deviation from the GDD is not uncommon.

You may have heard of crunch and tight deadlines in game development. Sometimes even the best plans fall short and that leaves a desperate situation in which to turn the ship and get things done in time. Planning from the start and not overplanning can help avoid crunch and a lot of game development companies are doing their part to be mindful of quality of life and limiting crunch. After all, long crunch times can lead to poor work generated from overworked and exhausted individuals.

In the Sound Lab (companion site) we provide some example asset lists, schedules, and additional information regarding development methods and collaboration tools.

RESEARCH

Research is a fundamental part of the pre-production process. Audio references can help determine the sonic direction of your project. Throughout Parts I and II of this book you will be reminded of research opportunities that include critically listening and breaking down a sound in order to reconstruct it.

Developers may already have an idea of the game's sound vision before you are brought on to the project. Often the vision and delivery specifications may be relayed to the sound team in the form of a written brief. Understanding how to interpret the direction and follow through with asset delivery according to spec is an important skill. When you are unsure if your ideas and theirs are on the same page, it's best to open the lines of communication with the developer by asking for further details. Once you have an idea of the direction for your soundscape, careful research will help further flesh out the details. Research isn't copying another game or film sound, but rather it helps inform your own ideas and provides further inspiration.

MAKING USE OF DOWNTIME

Whether you just finished a game audio project or have extra time away from your day job or education, you want to make use of this downtime. When you are playing games make assessments of the audio. Start practicing with a new plugin or try a mic in your collection in a way you haven't used before. If you have a field recorder be sure to do some experiments with all the settings to find the sweet spot. Use this time to increase your skill level and be ready for the next project.

The Sound Lab

 Before moving onto Chapter 2, head over to the Sound Lab for additional reading and practical exercises on the topics discussed in Chapter 1. We will review game development roles in depth, essential skills and tools including looping assets, example asset lists and schedules, and how to make the most of your downtime.

Chapter 2 begins Part I, the sound design portion of our textbook. We encourage you to read through the entire book, but as we mentioned in the Introduction, if you are only interested in music composition for games you can skip Part I and move directly onto Part III, Chapter 8. From there we recommend reading Part IV, Chapters 10 through 12 for the business side of game audio.

NOTES

1 M. Kamp, T. Summers, and M. Sweeney, *Ludomusicology*; W. Cheng, *Just Vibrations*.
2 When working remotely you may not always have the ability to work directly with the game engine or even a build of the game. In this case you would work with animations and screen captures of game play provided by the developer.

BIBLIOGRAPHY

Cheng, W. (2016). *Just Vibrations: The Purpose of Sounding Good*. Ann Arbor, MI: University of Michigan Press,

Kamp, M., Summers, T., and Sweeney, M. (2016). *Ludomusicology: Approaches to Video Game Music*. Sheffield/Bristol: Equinox.

Part I
SOUND DESIGN

The Basics of Nonlinear Sound Design

In this chapter we provide an overview of sound design for game audio as a process. This is a great starting place for those new to the industry or intermediate sound designers looking for a refresher. Our goal is to offer a broad perspective on concepts such as nonlinearity, tools and skills for the game sound designer, as well as foundational components of game sound. We will explore these topics in depth in later chapters.

Many of the concepts discussed in this chapter are entry level and can be understood by novice sound designers. However, we will quickly move on to more intermediate topics in later chapters. If you at any point have trouble with the material in later chapters, we suggest you return to this chapter to refamiliarize yourself with the basics. You can also use the glossary and index as references.

WHAT IS NONLINEAR SOUND DESIGN?

To answer the question "What is nonlinear sound design?" we must first look into how games and other nonlinear media differ from linear productions like film and TV. In the "Game Audio" section of Chapter 1 (page 12) we briefly mentioned some differences between the two processes and here we will discuss further the challenges that come with nonlinear media. In linear media sound and visuals are synced and printed, which means there is no chance for change during viewing. The nonlinearity of games puts the player in control of the gameplay, which means audio must adapt to changes in run-time through pre-planned events.

For some, game audio is an afterthought, but the role of sound in games is much more than ornamental; it sets the mood, adapts to changes in the **game state** (a description of the current contexts and systems in the game, similar to a **snapshot**), provides feedback on player input, and focuses the player's awareness. You can probably recall a moment in your favorite game or film where sound was used to grab your attention – "Hey, look over here!" Sound gets our attention in our everyday lives, and it does so in games as well. In short, nonlinear sound helps shape the direction of gameplay in a very tangible way. Viewers may notice new details about the sound in a film or television show, but the sequence and character of game sound fundamentally changes *every time the game is played*. This begs the question: "*How do we design sound for games?*"

Let's imagine ourselves as the audience of a film. In this scene the main character rushes into traffic on a busy street. The car horns are blaring, drivers are screaming, and general street noise is present underneath everything. In this scenario what is drawing our *focus*? What is the role of the audio when supporting that focus? When we watch films our focus is usually decided for us. The camera makes the choice for us (for the most part) by focusing on important characters, objects, or events. Our focus is on the character in this sequence, and the role of the audio is to then set the scene and fill out details about the environment.

Now let's imagine a new scene, this time we are the developers of a video game. In this scene a huge battle has just occurred and the last few enemies have been defeated. Succeeding in this battle opens up a new area of the map. How do we change the player's *focus* so that attention is drawn to this new area? The challenge is that in a video game the player can choose where to look and what to pay attention to. We don't have the luxury of always controlling specific camera angles – we need a method of drawing the focus of a player to their next task or objective. The answer of course is sound! If we want our players to *choose* to explore a new area without controlling the camera we can plant a sound effect in that location. This is how sound should function in a nonlinear environment. Audio cues should give players enough information to *inform* their choices, but not *choose for them*.

The freedom players have to make their own choices is what sets games apart from linear media. The player's actions affect the game state and in turn the game provides information and feedback to the player. Game sound functions in the same way. Sounds trigger in reaction to the player, and those sounds then influence the player's actions. These sound triggers (user interface [UI], music, dialogue, or sound effects) are critical in guiding the player through the game world.

As sound designers this interactivity between the player and the game creates a fundamental challenge. Imagine a linear scene where a character is walking through the forest. She pauses briefly, listening to some rustling in the brush ahead of her. Then she continues on. As sound designers for this scene we already know exactly what happens and when. We simply have to create a list of necessary **assets**, design or source the sounds, and place them on a timeline in perfect synchronization with the visuals. If this scene were in a game we would have much more to think about as sound designers. First, we would need to create a list of necessary assets and create them, just as we did before. However we can't just place them on a timeline. We need to *implement* them into the game engine so that each footstep triggers in sync with the animations. On top of that, we have to

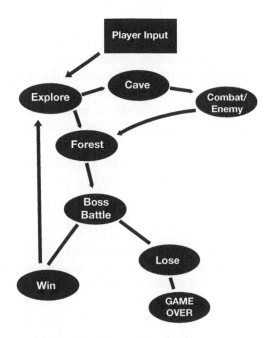

FIGURE 2.1 Player input changing game states.

make sure to include appropriate **aural cues** so that the player can take appropriate actions toward her objectives. In many cases the two scenes would sound *very* different side by side. In linear media, focus should be placed on the sounds that create the mood of the scene without getting in the way of the action. In nonlinear media the sounds need to react to and influence the action.

There are many strategies and tools that we have as sound designers to aid us in creating this kind of nonlinear audio. With regards to the example above we would likely use ambient loops and **random playlists** of sounds to ensure that the audio is detailed and does not cut out to silence. Because player actions are unpredictable we would also have to set this up as an **event** (system of adaptive or interactive sound) rather than on a timeline. This is so that the **game engine** (the framework within which the game is built) can "understand" how the audio is triggered and synchronized to the visuals in real-time (see Figure 2.1).

THE SOUND DESIGN PROCESS

Now that we have a better understanding of how nonlinear sound differs from linear sound let's explore sound design as a process. From a creative perspective, the job of a sound designer is to tell a story through sound that immerses the player in the game environment. The goal is to produce unique sounds that complement or enhance the gameplay and visuals, set the tone, and inform the player of potential actions.

There are three main phases in the sound design process. The first two are **sourcing** (recording, generating, and licensing), and **designing** (**layering** and manipulation). These two phases have been adopted from the tradition of film sound. However, the **implementation** phase is derived from film sound mixing but entails challenges and methods unique to games. As mentioned earlier, this involves organizing sounds into adaptable events which are triggered during **runtime** as the game state changes or based on the player's input. Below we provide a quick explanation of these three phases but in Chapter 3 we discuss the specific details.

Sourcing

Sourcing refers to the process of retrieving the necessary components of a sound effect for use in a game. There are many ways to source sound effects, making this an interesting task. Sounds can be sourced through licensing, designing synthetic sounds from scratch (synthesis), or through custom recordings.

When sound designers license sounds it means that they have paid for the **media license** to a single audio file or a library of audio files. In many cases the audio is in the form of **source recordings** to be used as basic elements of a sound to be combined and designed later on. Other times the audio is more of a complete sound effect that will be shaped or trimmed to fit the needs of the game. In either case licensing sounds is usually a **non-exclusive agreement**, which means that other sound designers can use these same sounds in their projects as well. For this reason licensing sounds is a very quick and easy way to get started, but it can also lead to somewhat generic sounds if you aren't careful.

Let's imagine we have been assigned a project and our first task is to create player character footsteps for a third-person adventure game. It might seem tempting to purchase a sound-effect library and deliver the sounds "as is" to the programmer for implementation. This is a quick and easy way to handle the task, right? The problem with this process is the client is likely assuming they are working with a sound *designer* and not a sound *plucker*. Anyone can choose a sound from a library and place it in game, but games need to be immersive, and sound is a large factor in that immersion. It is the sound designer's responsibility to shape layers of sound into a unique asset that perfectly suits the action in game *and* supports the gameplay. This will not only result in a better fitting sound, but there will also be a slimmer chance that the licensed sound will be recognized in other games.

With library sounds you also have to consider the quality of the source. There are many options for library sounds ranging from no cost to thousands of dollars. No-cost or low-cost libraries run the risk of lesser quality assets, which would need clean-up and restoration before becoming a good candidate for in-game use. It is equally important to consider the type of license, and your rights to use the assets in a commercial project. For this reason we recommend reading the license agreement fully to make sure your project is legally viable for use with the library sound *before* you purchase it.

This isn't to say that you should always avoid licensing sounds. Library sounds are used very commonly in a range of projects from indie to AAA. It's perfectly fine to use a properly licensed library sound effect when time and project budgets don't offer the opportunity to record a new source. It is also beneficial to license sounds when the intention is to edit or process the sound heavily, and layer it with other sounds so that it fits the game more adequately. This doesn't mean you won't run into a situation where a single sound can be pulled from a library, cleaned up and used singularly in game.

Generating sound through synthesis is another way to source sounds. These kinds of **synthesized sounds** can be used on their own or combined with **mechanical sounds**. Plenty of games have retro themes where synthesized sounds are the perfect complement to the art. If you are tasked with creating the sound for a spaceship starting up and taking off, mechanical sounds like power tools or a hair dryer are great sources. But adding synthesized elements can help add movement to the engine and bring it to life. Cyclic sounds such as sub base tones can be generated from scratch, and they will make the engine sound weighty and powerful while the hair dryer or other tonal sources sit up top in a higher **frequency range**. Of course, synthesized sounds (even by themselves) always sound great with retro-styled artwork.

The important thing to remember about synthesized sounds is that they are by definition synthetic. This means they can be tricky to mix in with more organic or natural sound sources. To achieve the best results, be intentional about the layers of synth elements that you add. Adding a synthetic bass layer is often more than enough to add some interesting element to a sound. Don't overdo it, and always pay careful attention to how the event sounds as a whole so that you can properly balance the synthesized elements with the organic ones. We will discuss sound design using synthesis later on in Chapter 3, under the section titled "Sourcing Sound through Synthesis" (page 58).

The process of custom recording is a great way to source unique material to fit your game. It will give you full control over the material that you record and use in the design process, and it offers the greatest flexibility over the recordings themselves. If you are contracted to create sounds for a racing game, you may decide to record custom sounds if the developer is specific about the authenticity of vehicle sounds and the budget allows. By choosing to custom record you could get the *exact* sounds of each model of car present in the game. Not only that, but you could record any aspect of the car that you wanted. You could record **pass-bys** on a racing track, engine revving, mechanical noises, or even put the cars up on blocks and record the engines in isolation. This would come in particularly handy if the game offered players multiple perspectives during a race because there is a vast sonic difference between driving a car in third person and first person. You also have control over the number and types of microphones you use and just about every detail of the recording process is under your control.

Recording custom sounds is in many ways an ideal way to source audio for your game, but it does have its disadvantages as well. Recording custom sounds can be expensive and time consuming. If footsteps in snow are needed, there is a list of items that need to be taken into consideration before a custom recording is even possible; the most obvious item on the list being "Is there snow?" If you are in a location without snow on the ground you would have to consider the costs of traveling to a location with snow. You can of course simulate the sound of footsteps in snow by placing a lot of salt in a bucket and stepping in, or placing a piece of carpet over gravel and walking on top, but it may not yield the sound you are aiming for.

Should you decide to record the sound yourself you also have to consider if you have the proper equipment. In future chapters we will get more specific on microphone techniques, but for now let's just say you will need to think about the microphone, microphone placement, polar pattern, handling noise, room or environment control, clothing noise, breathing, recorder settings, proper walking (gait and heel-to-toe movement), proximity to the microphone, and more.

Regardless of how you source your sounds, you will have to consider how they fit the visuals. The best choice of course is the one that fits appropriately with the game. For example, in a game scene where the snowy terrain is hard and compact it won't make sense to use source sounds of deep footfalls in soft snow. The footfall must sound harder to be true to the animation. Sourcing is the foundation of your game sound, so make sure to search for fitting material before you move on to the design phase.

Sound Design as a Process

The design process is where you put your mark on the assets by editing the source, adding additional textures, layering, and processing. This is the stage where designers add detail and depth to the sound effect as a whole, and continue the process of molding it to fit the needs of the game.

The design process must start with preparing the sound for use in game. This is often referred to as cleaning up or editing the audio. Any source audio that is custom recorded, or comes from a lower quality sound library will likely need to be cleaned up. Try to think of how a game would look if the artist didn't polish up the shading or left the coloring inconsistent. When all the artwork came together, the game would be a mess! Assets would be unclear and ill defined. Colors would blend into each other and players would have a tough time walking around let alone traversing levels. The same effect happens with sound when it's not properly polished or mixed before placing it in game. If the character footsteps were recorded outdoors, and they are left unedited players might hear bird chirps or car engines triggering with each step. These are unwanted artefacts present in the source sound, and they *must* be edited out *before* they are used in the asset. There are a lot of great restoration tools to help aid in this clean-up process. Using de-noise, de-click, or spectral editing tools such iZotope's RX can speed up the workflow and help deliver quality sounds. As we dive further into editing in "Effects Processing as a Sound Design Tool," Chapter 3, page 96, we will further explore **plugin effects** and their use as a sound design tool.

Quality sound effects almost always involve layering of some kind. After you clean up your source, you can begin layering sounds to build an asset that contains all of the sonic qualities necessary to fit the visuals in game. A good example of this type of layering is weapon-fire sound effects. An effective gun-fire sound will typically be built from at least two or three layers. Typically the base layer will be the main firing sound. This layer will be the foundation of the asset, and characterize the sound as a whole. Designers will commonly add a secondary layer, usually in the lower or low-mid frequency range which adds some **punch** to the sound so that players really feel the impact of the weapon. Finally, a mechanical layer can add detail to the sound, and convey information regarding the materials that the weapon is made of.

While layering is an important part of the process, you will probably find yourself needing to use some creative plugin effects to further manipulate the sound. A plugin is software that processes audio in some way. This aspect of design falls under audio editing techniques. After you have the appropriate layers in place, plugins can help make the sound cohesive and powerful. Plugins like EQ, echo, and **compression** can effectively "glue" your layers together, and make sure the asset as a whole is clear and polished. Other plugins like **transient shapers** can add dynamics and punch. Similarly pitch shifters and **modulation** plugins can add some interesting aspects to the sound, and really stretch the sonic boundaries further from the source material. There are many more plugins than these, and this is a crucial step in the design process. As mentioned above, we will dive further into plugin types and usage in later chapters.

Implementation

The implementation phase is the only phase that is entirely unique to games, and it is a highly technical process. Asset implementation plays a huge role in making sure the sound design adapts to (and informs) the gameplay. Effective audio for games is often considered 50 percent creative source and design, and 50 percent implementation.

By definition, implementation is the process of carrying out a design and putting it into action. For our purposes, the "design" is referring to the audio asset and the "action" is the game engine. When it's time to implement, our assets have already been created and now they need to be integrated into the game engine with a set of instructions to control their behavior. There are a variety of proprietary and non-proprietary game engines. Larger development teams may choose to create their own proprietary engine while smaller teams are likely to work with third-party engines such as Unity or Unreal. It is important to understand which engine the game is being built in before you begin the implementation phase. These engines each have their own process for integrating audio and scripted events.

Implementation requires a practical understanding of how sound is hooked up to the game engine. It also requires knowledge of the myriad options for integration. Having this knowledge will allow you to implement audio in a more realistic and immersive way. Let's say that in testing a game you find that only one footstep asset plays regardless of terrain type. In the days of *Super Mario Bros.* this might have been fine, but modern games have advanced technically and visually. As sound designers it is our duty to keep the audio experience up to par. An understanding of scripting and how sound can be implemented will provide opportunities for more effective integration solutions. In this case, delivering six to eight footsteps with each terrain type will offer more variety and help avoid listener's fatigue for your player. The player character walk cycle will sound noticeably more realistic. For additional variety you could write a quick script to randomize volume and pitch on the footstep sounds. It would also be beneficial to add cloth or armor movement sounds to the footsteps, or you could **bake** them into the assets themselves. We will further explore the pros and cons of this in Chapter 8 and discuss how audio middleware offers the sound designer more control through a graphical user interface.

• •

The Sound Design Process Summary

Based on the information presented here we can conclude a few things:

- Game audio design and implementation (nonlinear audio) needs to be approached in a different manner than linear media.
- The recipe for a good sound designer starts with a passion for audio and games, mixed with a well-trained ear, a mastery of the tools, and a solid understanding of implementation and game engines.
- There are three phases of game audio as a sound designer: *sourcing, designing,* and *implementation.*
- Good *source* sounds need to be clean and fit the aesthetic of the game. Source material can be licensed, synthesized, or custom recorded.
- Regardless of how the sound is sourced the assets must be manipulated and processed to uniquely fit its position in the game during the *design phase.*
- *Implementation* requires a broad understanding of nonlinear audio, the specific gameplay mechanics of your project, and a practical understanding of the integration options.

CHALLENGES IN NONLINEAR SOUND DESIGN

Triggering and Game Sequence

Sound design for games as a process wasn't created by completely reinventing the wheel. Many techniques and practices have been adapted from film sound and traditional audio post-production. These post-production techniques gave game audio a bit of a headstart, but interactive experiences come with their own creative and technical challenges that are not present in linear media. Previously we learned that the player's actions in game can change the game state. We can break this down further by examining play styles and preferences. Some players speed through a game like they are in a lightning bonus round while others like to linger and explore, taking in the scenery. Some players will pick up gameplay mechanics quickly, while others might try and fail many times before succeeding. In many games players can even make choices which affect the outcome of the story itself. These are just a few examples of how play styles can create totally different outcomes in a game. As sound designers we have to account for these possibilities. The challenge is how to design sound as the sequence of events triggered in a game varies from player to player and session to session.

Keeping up with the player's unpredictable choices can be likened to working on an assembly line where graphics, sound, and programming do their best to keep up with the changing demands. An example of adaptive sound design is player movement from terrain to terrain in a game environment. The terrain is tagged with a material type which scripting and audio events can adapt to allowing for smooth transitions as the player connects with various floor types while moving about the world. Another example might be reverb zones positioned to change the sense of space as the player moves from room to room or area to area. We address these challenges in Chapter 8 where we discuss perspective, resource limitations, repetition, and the mix.

EXPLORING IMMERSION AND INTERACTIVITY

How do we define **"immersion"** in a game? Generally speaking, immersion means "achieving a state of deep engagement." It's safe to say that in games the goal is to engage the player into the game world so that she feels immersed, as if she is fully and completely part of the experience. To do this, let us first examine the elements of gameplay. Each of the following elements plays a role in immersing the player into the world of the game.

Game Elements

- Story/narrative
- Game mechanics

- Environmental interaction
- Graphics/art style
- Cutscene transitions
- Music and sound design

While sound might not be the first thing that comes to mind when you ask yourself what immerses you in a game, we are going to focus on the importance of sound and how it provides the player with a feeling of being surrounded by the world.

Just like a good film production, good game production will make the player forget about reality and draw them into the game world. This phenomenon called "suspension of disbelief"[1] isn't new or specific to games. It was coined by poet Samuel Taylor Coleridge in 1817. This ability to immerse with sound becomes even more important with virtual reality as we will later discuss in "Audio for New Realities: VR," Chapter 9.

Sound, just like visuals, can enhance the user experience by eliciting emotions, supporting the game's theme, and heightening intense moments. Pairing sound and visuals can enhance the experience and ensure the necessary information is received by the player. Visual cues provide a vast amount of information at any given time in a game. Sound helps the visuals relay information since the player's eyes cannot be in multiple places at once. A human ear can receive numerous sounds at one time from all around the environment. The brain works to dissect the load of aural information and break it down into several recognizable components. In an RPG (role-playing game) collectible card game there are various sounds to provide feedback to the player that don't require them to look away from where they are focused. An example is a timer ticking away to inform the player who might be looking over their deck that time is getting low.

Because sound is such a big part of our everyday life it is an important part of visual media. A truly immersive in-game soundscape provides the player with a feeling of being dropped into the middle of the virtual scene. Sound design done right will add to the experience but when sound is implemented poorly the users really notice. A solid understanding of how sounds interact in real-world spaces will help the designer to sonically recreate them. Unless an abled-hearing person is entombed in an **anechoic chamber** they aren't without sound, not even for a single moment.

Sound without visuals can be processed and spatially located even when not in the field of view. As humans we've learned that birds can be heard chirping during the day and crickets stridulating at night. With outdoor soundscapes audio designers can cycle the time of day with the sound of birds or crickets, which will provide the listener with the expected sonic information.

Psychology of Sound

Understanding the psychology of emotive sound design will help you to create immersive soundscapes. There are plenty of studies that show how happy and upbeat music can have a positive boost on mood while tense music can have the opposite effect on the listener. Luxury brands have been using sound to influence consumers for years. A luxury automobile manufacturer might work with a sound designer to create an engine design that gives the user a sense of power. In games, the audio designer should explore the psychology of sound to implement the perfect emotive influence on the player.

Little Nightmares, a puzzle-platformer horror adventure game developed by Tarsier Studios, has a soundscape that toys with the player's emotions. The story is that of a little girl cloaked in a yellow raincoat who is trapped in a mysterious vessel which is home to wildly creative and somewhat grotesque creatures. The ambient soundscape makes the player feel vulnerable and afraid. If this were a happy tale with fuzzy teddy-bear-like **NPCs** (non-player characters) a different approach to asset choice and implementation would be required.

Audio plays, podcasts, and radio shows are proof that great sound design doesn't always need visuals to evoke emotion and tell a story. A narrative isn't even necessary as we can often hear a singular sound and process it as positive, neutral, or negative sonic feedback. With audio and its implementation as a partner, the visuals and narrative can take on a whole new level of immersion.

Audio designers for *Rockstar's Red Dead Redemption* put massive amounts of thought and work into the detailed soundscape that drops the player into the Old West. It can be difficult to find the perfect balance to transition the listener from an indoor to outdoor setting, or to introduce new musical themes. As creatives we must tame the urge to cram as much of our work into the project as possible. There are plenty of scenarios of games that require minimal sonic elements. The sound of silence can play a big role in provoking emotion.

A soundscape near a campfire in the Old West will require sound to enhance the visuals we see in the scene but also requires sounds off in the distance, beyond what we can see, to fill out the soundscape. The audio designer should examine the scene and determine which sonic assets will make the world feel natural and alive.

Sounds that come from specific sources *within* the scene are called **diegetic**. Diegetic sounds, like **PC (Player Character)** movements are just one part of the soundscape however. The rest is ambience. As well-practiced listeners we expect to hear a variety of aural cues, even if they are not visible. These ambient sounds are **non-diegetic** (not visible in the scene), and they serve the function of filling out background noises. Without light prairie wind or crickets and coyotes in the background, the scene would will feel incomplete. Of course, with a plethora of sound emitting from the scene a well-balanced mix will be necessary to avoid any muddiness.

Fire up *Red Dead Redemption 2* and ride a horse for a few moments in a non-combat game state. The tiny details that include creaking of the saddle's leather and the clinking of the buckles heighten the experience. Imagine how the experience would differ if the personality of the horse was sonically characterized by random neighing instead of distinct hardware details shaping the character.

The total level of immersion in a game is a result of each of the elements, listed above, working together. Audio is sometimes overlooked as an immersive tool, but the reality is that audio is a core element of immersion. The reason for this is that audio intersects with each of the other elements in important ways. It may be "invisible," but without audio the other elements cannot stand up on their own when it comes to immersion.

Let's take a look at how audio connects and supports each of these game elements to create an immersive experience.

Story/Narrative

To create an immersive soundscape, the audio designer should focus on supporting the game design. Sound should provide the necessary information to the player to guide her through the game or support various **gameplay mechanics** and **narratives**.

Let's imagine we are back in a forest scene. We hear birds, wind, and maybe even some animals making sounds off in the distance. This sets the locale and helps us identify with the graphics of the forest. But what is the emotion or *mood* in this *particular* forest? If the birds are happily chirping away and the wind is light and wispy we might feel that there is no danger and think of this as a safe area. This location can be turned into a darker place without swapping the visuals. Changing the sounds so the birds chirping become owls hooting or hawks screeching, and modifying light wind to a darker howl would be evidence of potential danger. This demonstrates sound as a powerful element that can modify the player's perception of the environment, even with the same visuals in place.

Either way, the *character* of the audio causes us to absorb details about the environment, which *immerses* us into the setting. This is a very basic example, but this idea can be taken to levels of extreme detail in more complex games. The more audio information the player has to absorb, the higher the level of immersion is likely to be. By extension, if the level of immersion is high the player has a greater chance at being successful in gameplay tasks. The player will also have a higher likelihood of being impacted emotionally by the narrative of the game.

Game Mechanics

Game mechanics themselves are an important aspect of most games. The mechanics of a game determine *how we interact with it*. Audio creates immersion in this area by offering *expected* aural feedback to the player. By this we mean that in our daily lives when we perform an action, we expect certain sounds to occur. If they don't, it is a strange and attention-grabbing experience. This is also true in games. When we unlock a door in a game, we *expect* to hear some clicks and a door open. When we shoot a gun in a game, we *expect* to hear a click and a firing sound that reflects the material and quality of the weapon. The more detailed the gameplay mechanic, the more detailed the audio must become to reflect this.

This is not only true of realistic games. This is in some ways more important for abstract games. Because most of us have opened and closed countless doors in our lives, we already are familiar with what to expect. The mere suggestion of the sound of a door opening or closing will allow our brains to fill in the gaps, and evaluate the experience as natural. In other words it takes a *really bad* door sound to take us out of an otherwise immersive experience. By contrast, in a game that is completely abstract, the mechanics are completely new to us. Our brains don't exactly know what to expect, they just expect *something*. In these cases, it is even more important for sound designers to use sounds that simultaneously reinforce the abstract visuals of the game while grounding the player in some kind of familiar sound to associate with. An example might be an otherworldly portal opening to another dimension. Players have literally no idea what this should sound like because it doesn't exist, so the design needs to suggest this surreal quality of the portal. At the same time, whatever sound source is being used to design the portal sound must also *feel like a door opening*. By that we mean that the **gesture** of the sound should in some way emulate more realistic sounds, like the opening of a door or a gate. There are many ways to do this (we will explore some of them in later chapters), but to put it simply it usually involves volume and pitch **envelopes** as these are two characteristics of sound that our ears are very attuned to.

Environmental Interaction

The sounds making up the environment in a game can set the mood, offer a sense of locale, and suggest to the player specific emotions at points throughout the game. This is how audio impacts the story of the game, as mentioned above. Sound can also direct attention and actively draw players through an experience. We mentioned earlier that sounds are great ways to attract the attention of a player. This is one method of teaching players where their next objectives are, but sounds like this also drastically increase the level of immersion in a game. If you want to introduce the player to an area of the game or to provide a sense of placement of an object in the world, simply set a **positional sound emitter** on the object with proper attenuation settings (see Chapter 8 for more information on sound emitters).

Reverb reflections or **delay slaps** of triggered sounds can provide a sense of acoustics to let us know if we are outdoors, in a tunnel or a cave, or in any other location. Sonic details such as this spatial and acoustic information add another dimension of immersion in addition to task-related information. This is especially true for first-person or VR (virtual reality) games because the player's perspective is literally the **3D space** of the game scene. Our ears are sensitive to false spatial and acoustic information, so accurate spatial and acoustic information is key to a deeply immersive game environment.

It's also important to remember that the more interactive the environment is, the more immersive a game is. This goes hand in hand without audio. If something in the environment makes a sound, it tricks our brains into thinking there is more depth than there actually is to a game scene. Keep this in mind as you plan your audio design. The more detailed the environment, the more immersion you will likely achieve.

Graphics/Art Style

Sound can provide information about materials and objects as well as setting the mood or providing a sense of space. This is integrated very heavily into the graphics or art style of a game. The sound needs to fit in some way with the *mood* of the art style. Games that are highly realistic and three dimensional usually call for highly realistic and spatialized audio. Artistically rendered two-dimensional games usually have more creative freedom for designers. For example an enemy zombie with a detailed and grotesque visual style may be designed with gruesome levels of audio detail. This would work fine if the game is meant to be a horrifying experience. Another option is for the audio design to be ironic, which would add some fun and lightness to the zombies. The sounds would be less threatening or even funny. This would work well if the game is more of a casual experience. Either way, there must be some palatable logic and coherency to the audio design for the game to be immersive.

Audio also has to provide relevant detail on the objects and environments themselves so that they convey practical information to the player. A large robot walking across a metal catwalk will have movement and footstep sounds attached. These footsteps will communicate the weight of the bot and provide a sense of how powerful the character might be. This is highly relevant to the gameplay because it influences how the player might treat a combat situation. If the sound is menacing and heavy, the player will take that to mean the robot is a serious threat. Situations like this can be made to either supply the player with valuable information to succeed in an upcoming task, or can be exploited to add a sense of surprise and challenge to a scene.

ESSENTIAL SKILLS FOR SOUND DESIGNERS

At this point you may be wondering if you have what it takes to be a Sound Designer for games. The good news is if you have a passion for audio with an interest in recording, collecting, and breaking down sounds then you are on the right path to *learn* what it takes. That said, sound design for games can get pretty complicated. There are a variety of essential skills that apply specifically to sound design that should be on your radar to master if you are serious about working in the industry. We have listed a few such examples below. For more information on general game audio tools and skills refer to "Essential Soft Skills and Tools for Game Audio," Chapter 1, page 16.

Skills for the Game Sound Designer

A Well-Developed Ear

An indispensable attribute for a game sound designer is the ability to listen to sound critically and analytically. An understanding of how sound interacts with the real world is key to translating that into game environments. If you aren't listening and analyzing it can be harder to understand how sound should work in virtual worlds and more difficult to understand how to recreate sounds.

Having a good "ear" for sound doesn't mean having perfect hearing. The ability to listen to a sound and break it down, picking apart frequencies and being aware of noise are more useful abilities for our purposes. Training your ear takes practice, just like learning an instrument or building your composition skills. You can practice listening to how effects change a sound source by applying them yourself and analyzing the results. There are quite a few apps and websites that offer ear training for sound designers that help flex your frequency muscle memory.

SoundGym[2] offers gamified learning and practice with frequency, space and time, quality, and dynamics. It can be really handy when you want to manipulate sound with effects if you have a good understanding for how the effects alter the sound. For example, SoundGym's Filter Expert game plays audio with various EQ settings applied. You can bypass the effect to listen to the original as you decide which frequencies were affected. In a way it is like reverse engineering the effects of equalization. It will help you gain a solid understanding of how EQ affects a sound, which in turn gives you confidence in using EQ. A well-trained ear should be able to listen to a sound and hear whether it needs to have 200 Hz–400 Hz reduced by 6 dB, or perhaps it needs a change in the room size of the reverb, or the **pre-delay** adjusted, and so on.

Reading about building critical listening skills can help get you started but putting them into practice is the best way to develop your ear. Practice by going to a location and sitting still while you listen. Bring a field recorder with you if you have one, although you can certainly use your smartphone with a recording app. Record the sounds you are hearing and make some notes about the sound on a technical and emotional level. How does the sound make you feel about the environment? Cheerful bird chirps in a park create happiness and peace while a wolf howl in the same setting promotes a mood of fear. On a technical side try to guess the frequency ranges of sounds and later check the recordings against a spectrum analyzer to see how close you came.

Consistency and Practice

Although this is not something often discussed, arguably the most important skill for new game audio designers is the ability to *consistently* and *frequently* create sounds within a given timeframe. Great sound design can start from inspiration but in reality that is the first step in a long process which requires intentional and consistent practice to bring great sound design to fruition. The best way to develop this essential skill is to set aside a minimum timeframe, and (without any expectations or judgment about what you are doing) challenge yourself to design a sound *every single day* within that timeframe. Just as musicians must practice their instrument daily, audio designers need to stay in practice and build their skills. The biggest challenge you may find is being tempted to seek perfection in your own work. The important thing is not to micromanage every detail, but to learn how to harness your creative abilities quickly and consistently to meet deadlines and to raise your output capabilities to the highest possible degree.

Thinking Outside the Box

A sound designer should be technically inclined and have a willingness to solve problems. Being creative and thinking outside the box are also necessary attributes. Often in-game audio issues that are seemingly simple need a bit more research and experimentation. For example being tasked with capturing gameplay with computer sound seems pretty straightforward, but some software solutions require additional routing of the systems internal audio channels to capture computer audio. Being able to work with your audio interface to route the computer audio, plus a line-in, will offer you the option to add commentary to the screen captures. This is a very basic example and you will find much more challenging tasks ahead, but it outlines the kind of self-starter problem-solving skills required to work on a team.

Rule Breaker

By this we mean being willing to step outside of standard workflow to experiment with processing and editing sound. Forget the general "rules" for effects chains and try reordering your plugins to hear how a pitch-shifter will influence a compressor and then reverse them and listen to how the affected sound changes.

When trying to place sound in fantasy worlds, experiment with more interesting reverb and delay plugins like Valhalla SpaceModulator or FXpansion Bloom to go beyond what the player might be expecting to hear based on the visuals of the room.

Microphone and Recording Techniques

A sound designer can certainly generate source from synthesizers and/or sound-effect libraries as we discussed above, but often additional source is needed for adding additional details for a unique sound. The best way to generate these details is by recording them yourself. Therefore it is essential for sound designers to understand which microphones best suit the sound they are trying to capture as well as the placement that will get the sound they are after (see "Essential Soft Skills and Tools for Game," Chapter 1, on the companion site and "Microphone Choice and Placement," Chapter 3, page 75, for more details). If your options for microphone selection are limited, you should have a solid understanding of how to get the best sound from the mic(s) you do have access to.

Manipulating Sound

This includes the ability to clean up or restore raw source, trim, layer, and process audio into a high-quality and unique asset for use in game. Good sound design goes beyond pulling a sound from a library. It's all about gathering source that fits into a palette which fits the visuals and narrative. It's great to have a load of plugins but it's important to know what to use and when and how to use it. In Chapter 3 we explore in detail using effects processing as a sound design tool, layering and restoring source.

Music Theory

A sound designer should understand the basics of tuning, rhythm, and harmony. A good sound designer should be able to distinguish when something is in or out of tune and time and how to fix it when necessary. Some sound effects might have tonal elements which would need to be tuned to the games background music or other musical elements. When tones clash the user will pick up on this sound and it can make for an uncomfortable in-game experience.

Looping

The ability to create a seamless loop is an essential skill for both sound designers and composers, resource management being a big factor for using looped assets in games. In order to manage memory and CPU usage on the game platform, loops are used to continue a sound or musical track until the game state changes or an event is stopped. Ambience, diegetic and non-diegetic music and positional sound emitters (see Chapter 8) all make use of loops to manage resources.

Achieving a smooth loop is a task that sounds easier to achieve than it is. In order to provide an immersive experience for the player we want to avoid pops and clicks or gaps in our loops. Editing the start and end points of the file at the zero crossing will help you achieve a smooth loop. In "Essential Soft Skills and Tools for Game," Chapter 1, we asked you to review some basic skills in the Sound Lab (companion site). These resources included *looping music* and *sound effects*. Take some time to review the looping assets tutorials. This will be useful for beginners as well as intermediates or professionals in the industry who may be looking for a different technique to implement into their workflow.

File Formats

Just as a graphic artist understands compression and file formats, an audio designer must have a solid understanding of file formats and their encoding processes to master the delivery of files and to meet the implementation needs of a project.

For example, did you know mp3 files add a space to the start and end of the file during the encoding process? This is true and explains why you may have encountered gaps of silence at the start and end of mp3 files. You will want to look to other formats for delivery of looped assets to avoid gaps in the loop.

Let's quickly explore why the encoding process adds silence to the assets. The encoding process of mp3 files works in a way that it needs to fill blocks of data. If your file doesn't exactly fill the blocks of data (and it usually doesn't) the process adds silence to fill the space. This leaves you with silence at the start and end of the file. When an asset is played on loop you will hear that silence in an awkward gap. You can, however, speak to your programmer and see if they are knowledgeable in compensating for the gap when triggering the asset in game. So all hope isn't lost for mp3 files, but unless you confirm your programmer has a plan you should try to avoid this format for loops.

Implementation

In "Essential Soft Skills and Tools for Game Audio," Chapter 1, on the companion site, we briefly discussed middleware and game engines and in Chapter 8 we dive deeper into the process. For our intended purposes here we want to express how essential it is to be familiar with and understand what it means to implement audio into a game. A game sound designer will at some point land a project which requires working with a game engine. Understanding how to install, run, and add sound to a scene in a game editor is essential to the process. Further understanding of the availability of audio middleware and audio engine plugins and the benefits of each will help you along in the process. Having the ability to integrate your sounds avoids the "asset cannon" approach, and this could have a more positive impact on the final product.

Mixing

Another part of implementation is the final mix of all sound and music instances in game. Mixing a game goes beyond volume control, it also includes ensuring resource management and optimization of sounds per platform, managing the perspective of sound and balancing frequencies (see "Dynamic Mix Systems," Chapter 8, page 286 for detailed information).

Command of Essential Tools

In "Essential Soft Skills and Tools for Game Audio," Chapter 1, on the companion site we reviewed the basic tools necessary for designing and implementing sound effects. Command of these tools is necessary in the competitive world of game audio. Even though it's possible for sound designers to work without much outboard gear, a solid understanding of routing and signal flow is a key skill. Setting up multiple monitoring solutions and controlling them through the computer's audio devices or the audio interface mixer is necessary for listening on different speakers. This is important because players listen on vastly different sound systems, so a game mix needs to sound clear and compelling on all of them.

Those who are looking to get into game audio often question if they have the right tools and setup to get started. The studio or environment sound designers work in can vary in size, shape, acoustics, and equipment. A state-of-the-art studio isn't necessary to design AAA sounds. In a talk at the Game Developers Conference (GDC) in 2012, Darren Korb, audio director and composer from Supergiant Games, explained how he created the sounds for *Bastion* as a one-man team on a shoestring budget by recording in his closet.[3]

Time, practice, and experimentation can result in well-polished sounds regardless of location and equipment. A properly treated room will allow for less time and energy cleaning up noise from sounds, but in the end you can make do with what you have. Don't use your lack of space or equipment as an excuse to delay getting started.

Research and Practice

There are plenty of sound designers who have YouTube channels or blogs with great tips on sound design and implementation. In Chapter 1 we directed you to the Sound Lab (companion site) where we discussed many learning resources. Schedule some time each week to read a blog post, watch a video, or a take an online course to improve, build, and expand your skills in sound design.

* * *

Let's wrap up this section with some tips for aspiring sound designers.

- Start experimenting with different microphones and different sound sources.
- Undertake further research to gain a solid understanding of key fundamentals of sound such as physics, loudness, dB, and frequencies.

- Start recording and building your own SFX library. Invest in a field recorder and library sounds for any source you that don't have the means to capture.
- Build up a database of future source material in your head. Always be listening to and analyzing sounds. Remember the sounds you might want to record later. Record them on the spot if you can, otherwise make a note and go back later to capture them.
- Keep your technical game well oiled. Know your DAW, multiple game engines, and audio middleware as if they are your best friends. Keep up with the latest technological advances and continue to improve your workflow.
- Understand game audio theory and integration systems.

* * *

The Sound Lab

Before moving on to Chapter 3, head over to the Sound Lab to wrap up Chapter 2 with some exercises and further reading suggestions.

NOTES

1 S. Böcking, "Suspension of Disbelief."
2 Sound Gym, "Audio Ear Training and Learning Center for Producers and Engineers."
3 D. Korb, "Build That Wall: Creating the Audio for Bastion."

BIBLIOGRAPHY

Böcking, S. (2008). "Suspension of Disbelief." *The International Encyclopedia of Communication.* Retrieved from https://onlinelibrary.wiley.com/doi/abs/10.1002/9781405186407.wbiecs121

Korb, D. (August 18, 2014). "Build That Wall: Creating the Audio for Bastion." Retrieved from https://youtu.be/jdnMqjLZz84

O'Donnell, M. (August 24, 2014). "Riven Foley." Retrieved from https://youtu.be/-FU_gMFW7Uk

Rayne, C. (November 16, 2013). "Playstation 4 Audio DSP Based on AMD's PC TrueAudio Technology." Retrieved from www.redgamingtech.com/playstation-4-audio-dsp-based-on-amds-trueaudio-technology

Sound Gym. "Audio Ear Training and Learning Center for Producers and Engineers." Retrieved from www.soundgym.co/

Usher, R. (April 18, 2012). "How Does In-Game Audio Affect Players?" Retrieved from www.gamasutra.com/view/feature/168731/how_does_ingame_audio_affect_.php?page=2

3 Designing Sound

In this chapter we provide a more detailed look into sound design for game audio as a process. Here we cover more intermediate topics, which include the purpose of sound in games, sourcing sounds, and designing sounds. We will explore the process of layering, transient staging, frequency slotting, and effects processing within a framework that's easy to understand. We will also break down sound components into categories, which we have found simplifies the process and ensures detailed designs.

DYNAMIC AUDIO

Dynamic range is the ratio between the loudest and softest sound. **Dynamic audio** is a term that we use in regard to games to categorize audio that *changes based on some kind of input.* We have left this definition intentionally broad because it encompasses other kinds of audio systems. **Adaptive** and **interactive audio** also fall under the dynamic audio umbrella. Adaptive audio reacts to changes in the game state. In this way players can influence adaptive audio *indirectly.* Most of the more complex game audio examples such as *Red Dead Redemption 2* are adaptive audio systems. By contrast, players can influence interactive audio *directly* through some mechanism in the game. For example *Guitar Hero*, *Rock Band* and *Parappa the Rapper* are examples of interactive audio.

Sound effects are not simply there to fill space. They serve the purpose of giving feedback to the player and providing a sense of space or setting, which immerses them into the game world. As humans we hear sounds all around us 24/7. These sounds provide information on the space we occupy and warn us of the level of danger. Since we are so spoiled by sound in everyday life, as players we expect to hear the same level of immersive audio to inform us of important factors in gameplay. In a real-world setting we wouldn't just rely on sound and visuals for immersiveness. We would also smell, touch, or even taste the environment! But in a virtual setting we can only see and hear (at least at this point in time). To compensate for this, we exaggerate the sound to enhance the immersiveness of the experience.

In Chapter 2: The Basics of Nonlinear Sound Design we provided an overview of sound design as a process. In this chapter we will aim to provide a solid understanding of the theory and practice of game audio to allow you to apply these ideas to your own workflow and process. We will explore how to create immersive sound design for games by breaking down the tools and techniques used by professionals in the industry. While reading, keep in mind that designing the sound is only half the battle. An asset may sound great out of context, but in the game it might not be a good fit. Having an understanding and appreciation for how sound *functions* in games goes a long way in helping to produce quality and immersive audio for the medium.

SOUND EFFECTS

Sound effects are broken down into a few subcategories:

- Hard/literal sound effects
- Designed/non-literal sound effects
- Foley (performed sounds)
- Ambience

Hard Sound Effects

Hard sound or **literal** sound effects are literal sounds that represent realistic objects or actions in game. These are common sounds that are easily recognizable. Hard effects include car engines, doors opening and closing, weapon fire, explosions, and wood creaks.

Designed Sound Effects

Designed sound effects or **non-literal** sound effects are sounds that typically stem from the imagination. They are abstract sounds that you wouldn't associate with the sound that makes it, like UI sounds. In reality, the action of moving a mouse over a button doesn't have a sound tied to it. Designed or non-literal sounds are often found in sci-fi or fantasy games. Objects like space lasers, fantasy game special abilities, and cartoony bonks and boinks fall into this category. Getting familiar with the work of Ben Burtt, a sound designer noted for his work on *Star Wars*, *Wall-e*, and *E.T.* is a great way to get some inspiration for sci-fi and fantasy sound design.

Foley (Performed Sounds)

The idea behind Foley is that sounds are performed by sound-effect artists or **Foley walkers**. These sounds are recorded in synchronization (with linear media) to complement or emphasize a character, object, or action. In games designers use Foley techniques to create footsteps, armor, cloth movement, and prop handling among other things. Foley sounds are often recorded while the performer is watching a video clip of the gameplay or character animations.

It's important to point out different uses of the term Foley. Film sound designers might insist that Foley only describes sound recording performed in sync to video or animation. In this pure definition, a library of footsteps shouldn't be referred to as a Foley library. In game development you may find the term Foley used to refer to a blend of performing in sync with video for cinematics and recording sound effects without a video reference. Film sound designers would argue that this is simply sound-effect recording but nevertheless you may hear the term Foley used in game audio.

Ambience

Ambience or background sounds are the glue for the rest of the soundscape. They provide an immersive setting which gives players a sense of the space their character is in. Sounds in this category include nature sounds, real-world spaces, or fantasy spaces.

As sound designers are tasked with designing aural experiences that may not always reflect reality, they often take some creative liberties while designing the ambient sounds. For example, outer space is a common setting for games. In reality, the vacuum of space precludes sound waves from traveling. Yet sound designers still design appropriate sonic environments to serve the game and aid the players in their tasks. The idea is to create a believable and immersive atmosphere, even though the result may not be true to life. While sound designers have creative freedom during instances like this, the craft still requires quite a bit of knowledge and mastery of tools and techniques. Sound designers have to know what reality sounds like before they can craft fantasy sounds.

THE FUNCTION OF SOUND IN GAMES

As we have mentioned earlier, the method you choose to design a particular sound effect will be determined by its role or function in the game. Below are a few common functions of game sounds to keep in mind.

- Provide sonic feedback to the player
- Communicate emotion/set the mood

- Provide a sense of space
- Define realism through literal and non-literal sounds
- Establish a sonic identity or brand
- Establish structure, narrative, and pacing

This is not an exhaustive list, but these functions work together (and overlap in many ways) to increase the overall immersion of a game. In order to fully realize the immersion in a soundscape, these functional roles need to be understood. Although games are nonlinear, the techniques adapted from the linear world work well as a starting point for designing assets. Games come with other technical challenges, which we will continue to explore throughout this book. For the moment, let's break down and explore some specific functions of sound in games.

Sonic Feedback

Sonic feedback refers to the communication of information to a player when she takes an *action in game*. The feedback is relayed via a sound effect that is triggered during or after the player's input and typically provides the player with some information on the action itself. For example if a player swings a sword, she will likely hear a *swoosh* sound. If she swings the sword again and hits an enemy she will hear a swoosh *as well as an impact*. This aural feedback tells the player that she hit her target. This is a simple example, but games take this concept to varying levels of complexity as it is an integral part of the player experience.

A **Heads-Up Display** or **HUD**, a part of the game's **user interface**, visually relays information to the player but may also utilize sound to offer feedback. These sounds may provide negative or positive reinforcement as the player interacts within the game. Similarly in-game dialogue can advance the story and provide the player with instruction for the next mission or objective. Other sonic cues can help the player anticipate a change in intensity. For example, as ambience darkens or music adopts a tense tone, the player will become suspicious of what is to come. When the sound adapts prior to seeing the action change on screen this helps the player anticipate and prepare for what is ahead.

Communicate Emotion/Set the Mood

Communicating emotion is something you should be very familiar with from the film world. Sound in games will work to set the mood and induce physiological responses during situational scenarios. This is especially true in games where the player character might be facing a significant risk. For example, a boss battle typically has heightened music and sound effect elements to inform the player of the challenge ahead. After a battle has been completed the sound drops down to a calmer tone to imply safety for the moment.

Provide a Sense of Space

Spatialized audio provides a sense of directionality and environmental setting. Whether the game is set in a realistic or fantasy setting, sound is responsible for immersing the player into the space. Sound can help the player recall settings or characters as well, and it helps them situate themselves in a scene. Spatial feedback is also necessary for the player to determine the size and geometry of each area of the game.

Define Realism through Literal and Non-Literal Soundscapes

Audio can be further broken down into diegetic and non-diegetic sounds. Diegetic sounds are emitted from *objects within the game world*. Non-diegetic sounds come from sources that are external to the game world. Sounds that the player character and non-player characters can hear in game can be considered diegetic: a door creaking as it opens, weapon sounds, footsteps, ambient insects, music playing from an object in game, etc. Any sound whose source you cannot find in a game scene and which the characters cannot hear are non-diegetic: underscore, user interface, narration, etc. An interesting third category is meta diegetic, which is sounds that only specific characters can hear (telepathic communication, voices in your head, earworm music, ringing after a loud noise, etc.).

The use of diegetic and non-diegetic (literal and non-literal) sound stems from the film world. Similar to film audio production, game audio designers will blend a mix of diegetic and non-diegetic sounds in a scene. Diegetic sound will present what the player character hears while non-diegetic sound guides the player's emotions. Defining realism through audio doesn't necessarily mean the game world translates directly to the real world but that the game environment is made believable by the player through immersive visuals, narrative, and sound.

Establish a Sonic Identity or Brand

The strategic positioning of audio assets to reinforce the sonic identity or branding is something companies have been capitalizing on for ages. When you turn on an Xbox console, the logo startup sound is an excellent example of establishing brand identity. Brian Schmidt, sound designer, composer, and creator of *GameSoundCon*, crafted the original Xbox startup sound. This audio experience was not only sonically pleasing but was a technical challenge to implement due to the hardware limitations of the time. As a guest on the Twenty Thousand Hertz podcast (Episode #54: Xbox Startup Sound), Brian describes his experience getting over the technical hurdles.[1]

When a company decides to expand their market often the audio experience needs to be revisited. For the premier of the Xbox 360, Microsoft brought on Audiobrain, whose Michael Sweet,[2] composer and sound designer, helped design a startup sound which set the logo's sonic identity and also offered a detachable audio experience to be played at the end of marketing promos and commercials. With each new version of the console, the sonic logo adapts to the marketing and brand needs but still stands out as a recognizable audio experience.

Another example is the simple four-beep sequence known as the Halo. This start beep has stuck with the series because it easily identifies the "3-2-1 GO!" countdown and has become a sound players expect to hear. If we break down the sound, it seems to be three very quick 500 Hz sine wave beeps followed by a 720 Hz sine wave played slightly longer to emphasize the "GO!"

An audio experience, whether intentionally branded or made iconic over time, doesn't have to be a user interface or logo sound though. Weapons, vehicle sounds, and even ambient soundscapes can become a sonic identity that creates a link between the player and the game.

Establish Structure, Narrative, and Pacing

Sound (or lack of sound) can provide a narrative structure for the player. Games that have a distinct transition between scenes often employ sound effects that emphasize those transitions. Likewise, fading the ambience into a new ambience is very meaningful to the player because it offers a sneak peak of the environment that is about to be explored. It can even alert the player to potential threats. The opposite (a continuous ambience between scene changes) may signal to the player that the environment she is now entering is similar to the current environment. In horror games this schema is often exploited by adding threatening sounds to an otherwise non-threatening ambience, or by adding an unseen enemy to an area with a very "safe"-sounding ambience.

The function of **acousmatic**[3] sound is to provide foresight to the player. In film, when we hear a sound off screen we can't adjust the camera to pan to the direction of the sound. In games the player not only can but usually *will* adjust her viewpoint to investigate an off-screen sound. By using acousmatic sounds we can draw the player to an important gameplay event, or an object which will aid in the completion of a puzzle or task. A strong example of directing attention through sound can be found in Playdead's 2D side-scroller *Limbo*. At one point in the game the player will walk past a trap sitting up in a tree. A spider then attacks, and as its large leg stomps the ground a sound is triggered to signal the trap falling from the tree off screen. A sonic cue then directs the player back to the area where the trap fell. Without this sonic cue the player would have to rely only on the vague memory of the trap she walked past in order to defeat the spider. This particular use of acousmatic sound supports both the structure of the gameplay (the spider puzzle in particular) as well as the narrative (the order of events that transpire).

DESIGNING THE SONIC PALETTE

Just as visual artists choose a color palette before they begin their work, sound designers need to consider the sonic "colors" that will best paint the aural picture. Establishing a sonic direction prior to the actual design sets the path toward a unified soundscape. During the pre-production stage of development the sound designer can select the palette and decide on a direction. This is not usually a one-person job as the development team often has input. In Chapter 2 we covered the pre-production process and discussed questions to ask to define your overall aesthetic. The answers to these questions should inform the choices you make about your palette. Gameplay, story, and narrative all play an important role in these choices as well. To illustrate this let's take a look at a few examples of sonic palettes.

Blizzard Entertainment's strategy card game *Hearthstone* is an excellent example of a well-thought-out sonic palette. Before entering a match, the player is fully immersed into a tavern-like soundscape. Tavern patrons' **walla**, wooden drawers, leather material, paper and books, and metal bits and chains are all elements present in the palette. After pressing the shop button the sound quickly takes the player through a wooden door with a little bell jingle to signal the entry.

A well-designed and implemented soundscape is defined by players *not* noticing the sound. Players should be so immersed in the game world that they don't particularly notice any audio at all. However, with a poorly selected palette, players will pick out the sounds that bother them or don't feel right in the scene. The wrong sound in the wrong place can cause the player to lose their immersive connection. While technical items like noticeable loop points and sloppy transitions can quickly break immersion, overloading or underloading the sonic palette can easily yield the same result.

Tonal Color

Music is often described by tonal color, but sound designers can use this to define UI (user interface), ambiences, and other sound effects as well. A well-crafted sonic palette will have a unified tonal color with similar qualities that tie the sounds together. You can think of these similarities as the *tonal flavor*.

Let's look at a fictional **FPS** (first-person shooter) game with a pixel art style for the sake of this exercise. The menu music is a synth rock loop and the UI sounds were created from various switches clicking on and off. In game the player is greeted by realistic weapon sounds, no background ambience, and an orchestral music track. The change in music feels a bit off (but there must be a reason behind the switch, right?) so the player forges on only half immersed in the game world. Firing away at NPCs (non-player characters) and running around the battlefield feels a bit like a typical shooter, but the player intuitively notices that some sounds have a lot of reverb and others are dry. On top of that, the player's footsteps have noise that fades in and out as each step is triggered. Suddenly, the realistic, yet low-quality soundscape is topped with an 8-bit retro level up stinger that signals the player's progress in the game. At this point the player pulls back from the partial immersion and questions the sound choices. What is the sound palette trying to tell the player? The confusing mix of sounds is not easily enjoyable and does not always fit with the experience. Next, the player visits the in-game shop to upgrade weapons. A pistol upgrade is rewarded with a low-quality "cha-ching" sound and its tone clashes with the music. The entire experience is a mixed bag that doesn't offer the player much immersion.

This scenario, while fictional, is not that far-fetched. This example can serve as a "things to avoid doing" reference during the sonic palette-building stage. For starters, there is a mix here between realism and low-fidelity sounds. The art is pixel-ated, so a lo-fi and retro palette might work well if it were ubiquitous throughout the game, but the inclusion of an orchestral gameplay track and the realism of the sound-scape send mixed signals. The lo-fi sounds are also seemingly unintentional. Realism and lo-fi don't usually mix. If the aim is realism, then the sound palette should be as high quality as possible. If the visual aesthetic is more abstract (as is the case with pixel art) there is more room for intentional synthetic or lo-fi sound palettes.

Less is More

KISS[4] is an acronym for "keep it simple stupid." This principle has been around since the 1960s and it can be applied very appropriately to game audio. Imagine a game where the sounds and music are wall to wall (in other words everything makes a sound, and the mix is always densely packed). In everyday life noisy envir-onments pull our attention in too many directions and cause us to lose focus, and the same is true of game scenarios with too much audio. When this happens, valu-able information meant for the player will get lost in the chaos of the mix. Too much aural feedback means that none of it is breaking through the wall of noise. Important sonic details meant to influence the structure of the experience will be inaudible. By planning your palette ahead of time you can make practical use of the KISS principle to determine exactly how much or how little audio your game needs in each scene. More importantly, you can plan out what priority each of those sounds should take in the mix to ensure that nothing of value gets lost in translation. This applies to music as well.

Since we don't have a soundtrack following us around in real life, it makes sense to strip back the game score and be mindful of the soundscape. The score and sound design can both provide the emotional setting and appropriate cues to the player with-out being too overt. This can be a difficult task for a content creator. Often times we want to show the world what we are made of, which results in grandiose audio elements mashed together in a wall-to-wall masterpiece. As we now know, this is not ideal to support gameplay. Dialogue fighting with sound effects cuts the information pathway to the player and makes the words unintelligible. Music on top of other audio elements interrupts the immersion. Scenarios like this are to be avoided like the plague. Sometimes we need to put our egos aside and create a mix that is less outlandish and more suitable for the game.

The KISS principle can also be useful when it comes to layering sounds, plugin effects, and implementation (all of which we will cover later in the chapter). It feels good to create a complex system of asset implementation, and middleware certainly offers the tools to do so, but it is only worth doing if it supports the game experience. It can be helpful to ask yourself the following questions before designing *any* sounds: "Will this support gameplay? Will this provide the proper aural feedback? Is this essential to the immersive experience?"

Naughty Dog's *The Last of Us* is a shining example of "less is more." The intro starts out with a clock ticking and cloth Foley as the character Sarah lays in bed. Joel unlocks the door and speaks into his cell phone. Sarah and Joel talk about his birthday gift as the clock gently ticks in the background. Delicate Foley accents the character movements on the couch. Joel briefly turns on the TV and a low musical pad makes a short appearance. After putting Sarah to bed the game cuts to a scene where Sarah is awakened by a late night call from her Uncle Tommy. With the music stripped away, the Foley moves to the background while the dialogue and beeping phone take center stage. The player doesn't know what to expect but understands something isn't quite right. An explosion outside interrupts the eerie silence after Sarah turns on the TV. The soundscape slowly builds as Joel and Sarah meet up with Uncle Tommy. The low string pad is reintroduced to score the tension.

This scene is effective because it is dead simple. The soundscape is almost entirely Foley, and the music is as sparse as the soundscape. Because the mix leaves room for the sounds that support the mood and narrative, the audience is able to project its feelings of tension as the scene develops. If the score was more complex, or the soundscape lacked the intricacy of the Foley elements, the scene as a whole would have had far less of an impact.

Visiting Artist: Martin Stig Andersen, Composer, Sound Designer

Sound Design: Less is More

Even though it's sometimes tempting, and your teammates may persistently request it, adding sound to every object in a game often causes nothing but

(Continued)

a cacophony of unfocused sound, obscuring valuable gameplay information. Next time you watch a movie, notice how just many things in the picture don't make sound. This is a conscious decision made by the filmmakers aiming to help direct the focus of the audience. One of my favourite examples is the *Transformers* movies in which sounds of explosion and mass destruction are sometimes omitted entirely in order to allow the audience to appreciate smaller scale actions and sounds, such as lamp-posts being run over or a window smashing. In games we can use this trick in order to highlight gameplay-relevant sounds, and when doing so, not surprisingly, the player starts listening! When you've helped the player solve a couple of challenges by bringing the most relevant sounds to the fore, you've caught her attention! You may receive bug reports that certain sounds are "missing" but remember that you are the sound designer, not the graphics guy who populated the level with potential sound sources (without ever thinking about sound).

CHOOSING THE RIGHT SOURCE FOR YOUR SONIC PALETTE

Effects plugins and other software are a great help when designing sounds, but even the best of such tools need a solid sound source to work their magic. Selecting the right source material is an important part of the process. In Chapter 2 we briefly discussed ways to obtain source material; in the following sections we will break down the process even further by exploring how to source from sound libraries, synths, and recordings. We will also examine ways to be creative about selection, editing, layering, and processing when filling out your sound palette.

Pre-Production Planning

In "Production Cycle and Planning," Chapter 1 (page 18), we covered the basics of pre-production. Here we will discuss mapping out the sonic identity of the game with the development team. Identify the intended genre and create a list of keywords that fit the mood(s) of each level or area. Before creating assets ask some questions to determine how sound will be most effective:

- What is the overall aesthetic?
- How will sound provide feedback to the player in the game?

- How will sound drive the narrative?
- How will sound be used to reward the player?
- If there are weapons in the game how will you ensure the player's weapons feel satisfying and powerful?
- How will sound set the mood or environment?
- What will be the sonic brand?

There are plenty more questions that can be asked before getting started, but the basic idea is to understand the *intent of the developer*. By understanding what the developer is trying to accomplish, and what the game is meant to convey, you will have a better idea of how sound will fit into the game. Keep in mind that this process is very personal – some designers don't need much information to get started while others like to really immerse themselves in the development process. In truth, each approach is valid and it changes based on the needs of the project and the sound designer. Over time you will develop an intuition about what games need and the intent of the developer.

Examining Aesthetics

When examining the aesthetic of the game think about environments, tech, character types, and gameplay. If you are working on a sci-fi game where the player commands spaceships, try to uncover how the ships are built and the mechanics behind them. There may be varying levels of ships where some are designed with the highest tech available while others might be pieced together from scavenged parts. The sound detail will be telling the story of each ship, where it came from, and where it is headed. In other words, the source you choose will serve as the foundation of the game's narrative. If only high-tech source is chosen, then the scavenger ships will not sound believable within the context of the game. Conversely, if only metallic clinks and clanks are sourced for all ships, the high-tech warships might sound like space garbage.

In short, pay attention to the aesthetics of the game during pre-production and log as much information as you can about everything that makes a sound. This will inform your choices for source material, and you will be on your way toward an effective sound palette.

Research

Once you have an understanding of the game it's time to do some research. Gathering references from other games and even movies in a similar genre is a great way to generate some ideas. A lot of great artists and designers find their inspiration from other works. Mix engineers often use a temp track of their favorite mix in the genre as a guide while they work. Taking one or more existing ideas and combining them or experimenting with new ways to expand on them is a great way to generate new ideas. The point is to generate a new and unique idea without copying directly from references.

Organization

During the bidding process the developer may have provided a general list of assets the game will require. Play through the latest build, if you have access, and identify additional assets that may be required (there often are plenty). If you are working as a contractor, it's a good idea to create a spreadsheet to track your assets and share progress with the development team. Solid communication skills are an important part of the process. Always communicate progress and provide detailed implementation notes via spreadsheets and in implementation sessions where applicable. (For more information on the bidding process see "Business and Price Considerations," Chapter 11, page 369.)

In Chapter 1 we directed you to the Sound Lab to check out the example asset list. Take some time to review it before moving on. While each team may have its own preferred method of organization, the asset spreadsheet should contain headers such as:

- Sound asset/game event
- Game location
- Description
- Loop
- 2D/3D events
- File name
- Schedule/priority
- Comments/notes

The spreadsheet could be shared over a cloud-based service like Google Docs or Dropbox Paper to allow for collaboration. The developer (probably) isn't a mind reader, and will need the sheet to keep track of your progress for milestone deliveries, to provide feedback on delivered assets, and even to set priorities for assets. All of this organization will go a long way toward making your palette appropriate and effective. For more information about asset management refer to "Essential Soft Skills and Tools for Game Audio," in Chapter 1.

Transitioning from Pre-Production to Production

As you plan for asset creation it can be helpful to build a list of source material that fits the game's sonic identity. This source can be used as a building block for your sound effects. You might find your SFX library can cover some of your source needs, but you may have to record assets as well. Create a list of specific items that you can realistically record. Projects never have unlimited time or resources to work with, so think logistically. Planning all of this in advance allows you enough time to gather your source without bleeding into your production (asset creation) time.

Once you have a list of source material, you are ready to move on to the production phase.[5] In this phase you will record, synthesize, or pull material from libraries to build your palette. In the following sections we will cover various ways to accomplish this.

Revisions and Reiteration

Creative work is a subjective medium and therefore feedback or criticism can hold a bit of bias. Regardless of how much effort you put into creating or mixing sound for games, feedback is often inevitable so be sure to plan time in your schedule for revisions.

Understanding sound as a subjective medium can help you more willingly leave your ego aside when delivering work. It's something that can be learned over time and you will get better at pouring your heart and soul into a creative work only to have someone pick it apart. Often times, after rounds of feedback and revisions, the audio asset might sound that much better. In the end teams build games and being open to working with feedback will ultimately improve the game's audio.

You won't get very far if you fight tooth and nail on every bit of feedback you receive so learning to digest and implement feedback is a necessary skill.

SOURCING AUDIO FROM LIBRARIES

Library Searches

When searching sound libraries for appropriate source material start by thinking creatively about key words and phrases to search the **metadata**. Designers can accumulate hundreds or thousands of hours of audio from library purchases. Using generic search terms can leave you with far too many audio clips to listen to, and many of them will miss the mark in terms of direction. By using specific and unique search terms you can make things easier on yourself by narrowing the results. A search for "metal" will yield a wide range of generic audio clips. Be specific and try using phrases isolated by quotes to produce more useful results: "metal snap," "trap," "spring action," "squeak," "metal close," or "metal impact."

Adding Custom Content to Libraries

Another point to consider with sound-effect libraries is that audio designers add *their own* sounds to their library. This is useful because it allows them to search everything at once. The downside is that personal recordings can become a nightmare to sort through if they are improperly tagged and labeled. When adding recorded source material to the library, be mindful of how you categorize and name *each and every file*. A buzzing synth sound is best categorized by how it sounds rather than what it might be used for. "Energy_LowPitched_Saw_01" is a more useful title than "Fun_Buzz_01." Avoid using vague file naming or taking the easy road when naming your sounds. It can be difficult to find the time for this in the midst of a busy day, but in the end you will save yourself a lot of time and a bad headache when you have to search for sounds.

It's also good practice to leave your library intact and copy assets to the working project directory. Ideally you are working with multiple hard drives so your sound-effects library and your project sessions are on separate drives. When you pull a sound from a library and add it to your session it should be copied to the session so you aren't overwriting the original library asset. The sound-effect library should be something you build and grow over time. You will certainly find use for sounds in multiple projects so keeping the raw library source intact is good practice.

Backups

When you are first starting out a simple backup solution might work just fine. But as you continue to work on larger and more complex projects, and as you build your library out, your backup solution should also evolve.

> In the past I have used RAID storage solutions but more recently I am using a mix of cloud and external-drive-based backups. I have an SSD locally in my computer to run my current projects. My SFX library is hosted on an external drive. I have two external backups of the SFX library and one of those sits in a firebox. I also have an external backup of my OS drive and a multi-stage backup of my project's drive. Current projects and recently completed projects are backed up nightly to a cloud drive. A monthly backup of all projects are backed up to a set of external drives sorted by project year and stored in a firebox. It seems a bit complicated but it offers peace of mind.
>
> Gina

SOURCING SOUNDS THROUGH SYNTHESIS

Synthesizers are an invaluable source for generating sound design layers. In Chapter 2 we discussed ear training and the programs and tools available to build this skill.

A similar method can be used to train your ears for knowing which **oscillators** or parameters to use when designing synth-based sound source. Syntorial[6] is recommended for tuning your ear to the fundamentals of synthesis and for a solid understanding of subtractive synthesis. Although the topic of synthesis can fill a whole book on its own, in the sections below we have outlined some basic ideas for generating source through synthesis.

Tools for Synthesis

There are many options when selecting a synthesizer for sound design, and each synth has its own strengths, weaknesses, and specialized workflow. Your "go-to" synth will depend on the type of sounds you are looking to populate your palette with, as well as your preferred method of design. Synths like Absynth are great for Match 3 genre game sound effects. We used a combination of Absynth and xylophone and chimes when designing sound for Breatown Game Shop's *Sally's Master Chef Story*. It also works well for magic or special-ability source layers in fantasy-game sound design.

Absynth's semi-modular approach allows for some creative routing and effect chaining. Native Instruments' Massive is a fantastic to tool for ambient, UI, engine, humming, and buzzing sources. Omnisphere can also be a great tool for ambient sound design due to its intuitive user interface. However, as with all source material usage, make sure you are familiar with the license agreement. In a quick review of the Omnisphere agreement, it allows use in music but requires an additional license for sound design. Again, each project will call for a unique sound palette, which will inform your choice of tool or tools to use for synthesis.

In the past synths were limited to hardware models, and there were only a few manufacturers. Today there are many more hardware synths, but also a tremendous amount of software-based synths with a variety of synthesis types. This means that your options are plentiful. Keep in mind that it's not about having all these tools. As a sound designer you will come to understand how to manipulate parameters in just about any synthesizer to create the source you are after. Your understanding of the fundamentals of synthesis, and your ability to quickly learn a synthesis workflow, will have a far greater effect on your sounds than any synth in particular.

Although the theory behind synthesis should be your primary focus, at some point you will have to select one or more synthesizers to work with. This can be overwhelming if you are new to the process. Start by investigating a few synthesizers and their architecture. Become familiar with the available components and eventually this will lead you in the right direction. Selecting a synth comes down to what you feel most comfortable working with and what best fits your needs. Every synth, whether hardware or software, has its own setup, interface, and working parameters. Most synthesizers offer a very broad range of possible sounds, so the one you choose should best reflect how you like to work and what feels intuitive for *you*. Keep in mind there is a range of affordable or free software options to get you started. It's best to give some of those a try before diving into an expensive option.

Hardware vs. Software

Before we move on to synthesis, an important distinction that needs to be made is between hardware and software synths. These categories of synthesizers will change your workflow the most, so here we outline a few key items. First, hardware synth can be expensive, large (usually), and often have a steeper learning curve. Because many of them are analogue, and often lacking the ability to save presets, makes them harder to control. Despite this, some designers choose hardware options because they prefer the tangibility of the knobs and sliders as well as their lack of impact on **CPU**. Hardware synths often have a very "personal" sound as well, which it is argued can be difficult to emulate with software. However, other designers find they don't have the room for hardware synths and are perfectly happy using a mouse or MIDI controller to edit the UI on a virtual synth. Software synthesizers are also great for learning the basics because they are cheap, and usually come with presets that sound great out of the box. Software synthesizers have the added bonus of almost always having multi-voice **polyphony** (more than one note can be played at a time), which many reasonably priced hardware synthesizers *do not* have, as well as flexible or even modular architecture.

What Can We Do with Synthesis?

Synthesis is capable of emulating sounds and well as generating unique sounds. In Gordon Reid's *Sound on Sound* article "What's in a Sound?"[7] he poses the question "What are harmonics, and where do they come from?" In the article he explains how the sound of many real-world instruments could be synthesized with a number of harmonic waveforms. He also discusses *Fourier analysis*, in which a unique waveform can be crafted from a set of harmonics. The key point here is that synthesis is capable of just about anything if you understand the way sound works.

To better illustrate this point for those new to synthesis head over to the Sound Lab (companion site) to review the fundamentals of synthesis. We will cover some theoretical basics, and key terms like oscillators, **waveshapes**, envelopes, and filters as well as provide additional educational resources.

Using Synthesized Sounds Alone

Human ears are great at matching a sound to visuals. When designing sound effects it helps to use source sounds that are captured in the real world so the sound can offer a bit of credibility to the player. This technique works well across many games but there are specific instances or scenarios that require all synthetic source. A good example of this is retro-style games with chiptune soundtracks. Synths can be used to generate retro-styled sound effects to go along with the chiptune background music. There are some retro synth emulators that do a great job of providing presets which recreate classic NES and other early gaming console chip sounds. Plogue chipsounds[8] is a VST that emulates 15 different synths 8-bit with a nice amount of detail.

Having a synth with pretty close presets can be great when you are in a time crunch but these retro sounds are fairly easy to create from scratch using basic waveforms like saw, triangle, square, and a noise generator. With any of these waveforms assigned to an oscillator the attack, sustain, and release can be adjusted to find the right accent for the sound, and standard processing like reverb, delay, chorus, and phasers can give the sound a bit of space and depth since the waveform or noise will be dry. Most virtual synths have an arpeggiator which will give the sound a bit of a melodic pattern to it. Lastly, a resonant low-pass filter will warm up the sound (see "Sourcing Sounds Through Synthesis" on page 58 for more information on generating source with synths).

Using Synthesized Sounds with Recorded Audio

Other designers are probably using the same synths and sample libraries that you are, so you will want to rely on layering and processing to create unique sounds. In working with fantasy and sci-fi sounds, their character and properties are subject to imagination which means you have something of a creative license, but don't rely solely on synth instruments for these otherworldly sounds. As previously mentioned, layering real-world or mechanical sounds with synthesized or synthetic source can add credibility for the player. Even though the sound is processed and blended, the listener can identify and feel comfortable with the sound.

The question is "How can we use synthesized source along with sounds captured from real-world objects?" The answer to this goes beyond the standard laser shot created with a single oscillator and pitch envelope. Synthesized source can be used in a lot of ways like filling out the frequency spectrum of a recorded real-world sound by adding a bit of weight in the low end or packing more punch on the attack. Compressing a weapon sound doesn't always give it the raw edge that is often sought after. Adding tonal elements from a pitched-down synthesized saw wave can ensure the frequency ranges are covered and add a bit more low-end weight to the weapon. Overall, adding a synth source to a mechanical recording can create a thicker and wider sound if that is what you are after.

A "noise sweep" is a common synthesis technique which utilizes a raw white-noise signal and a resonant low-pass filter. Opening and closing the cutoff and resonance creates whooshes, impact effects that can be layered with a mechanical recording to add a bit of an otherworldly or sci-fi effect. For example, layering a recording of kicking a metal object with a noise-sweep-generated swoosh can be used for impact sounds in a robot-centric game.

Mixing Synthesized Material

Critical listening and analyzing reference sounds is always a good way to ensure you are using the proper source in creating the sound you are after. Mix engineers often keep a reference track in the session to inspire the direction for the final sound. Understanding how to use tools like an oscilloscope will offer a view of the waveform and general amplitude envelope and spectrum analyzers will reveal frequency content. Some DAWs and **2-Track editors** offer these tools but they can also be acquired as a VST plugin. With these tools a sound designer can try to construct a similar envelope, wave, and frequency spectrum.

When working with soft synths the source layers created via MIDI could be rendered or bounced out and imported to a new track in your session. Working with a rendered audio file will allow you to edit and process the layer to better combine it with your other source material.

Light compression can "**gel**" these mixed source layers together and make them sound more cohesive. This is a process generally referred to as "bus compression" or even "**glue compression.**" It's a rather simple process but can go a long way in making independent layers processed through a compressor gel together. When two or more sounds are run through the same compressor their transients are shaped similarly and the resulting sound feels more like a unified element.

Heavy compression isn't necessary for this process as a very slow attack, quick release, and subtle gain reduction will provide a transparent process that will get your layers.

A lot of virtual synths have built-in effects such as delay and reverb. Try bypassing those effects and bouncing out the dry sound into a new track. Then process all the layers with the same or very similar reverb and/or delay in your DAW to help them sound cohesive when triggered together.

Layering a very dry synth source with a heavily delayed real-world recording will be difficult to gel using the compression method described above alone. Always think about your sound palette and the spatial properties your sound requires. Take time to listen to the mix of layers to ensure nothing sticks out or feels like it is from a different space.

In summary, synthesis is a great way to create source layers for all kinds of sound palettes. Engine sounds, explosions, UI, sci-fi, fantasy, energy, power sounds, and many more types of audio can be generated very believably using synthesizers. Going out and recording an actual explosion sound is dangerous if you try to do it on your own, and expensive when you hire someone to do it in a controlled environment. If the budget allows for the recording session then by all means go for it. However, most of us don't have that luxury. By mixing layers of library explosions, some custom synthesized energy sounds, and original debris recordings you can create some truly amazing explosion sounds.

SOURCING AUDIO THROUGH STUDIO AND FIELD RECORDING

In Chapter 2 we discussed microphones and field recording equipment. Here we are going to discuss some specific uses, ideas, and techniques for recording original source for your palette.

Take a moment to note the sounds that currently surround you. Experiment with objects in your current space to see how they might tell you a unique story. While making your bed, the flap of sheets could bring a winged creature to life. After baking, slamming an oven door shut could be a layer in a mechanical robot's footsteps. When recording source, pretty much anything can be useful. Keep your ears open and think creatively!

> The objects around you are waiting to tell you a story through sound … start listening, analyzing, experimenting, and recording.
>
> Gina

Foley and Sound Design Props

In Chapter 1 we defined Foley as a process. Here we will explore recording Foley and sound effects to generate source layers for sound design. Effective Foley will sound like it was part-recorded during the production, and not after the fact. Finding the right source to record and using the proper techniques are core ingredients to a recording session. A bit of creative editing can also go a long way toward making the Foley sync believable with a game.

There are an unlimited number of objects that could be used as sound design source or to recreate a sound for use in game. Here we are going to explore Foley and sound design props that we have used over the years to record source. This should get your gears turning by opening up an abundance of ideas for recording and creating sounds.

One fundamental tenet about Foley sound design is that an audio designer will often exaggerate a sound in order to make the scene feel believable. This means the object or action you are seeing in the visuals is *not always used literally to record source*. The sound of a sword plunging into an enemy was certainly not recorded by acting out the exact action. More likely it was recorded by stabbing vegetables or fruit. This is true even with subtler sounds. The sound of footsteps on snow may have instead been recorded with kitty litter to emphasize the "crunch" of each footstep. The takeaway here is that the best way to record source for a sound is to use your imagination and find *sonic similarities* rather than visual similarities. As sound creators we are interested in *soundalikes*, not lookalikes. It is also important to keep in mind that Foley and sound-effect recording is not magic by any means. It is really about being creative and building a mental database of interesting sonic characters by listening to and analyzing the sounds around you.

Fruits, Vegetables, and Foods

Produce makes for an excellent sound design prop.[9] Vegetables and fruit can be used for fight scenes, gore, weapon impacts, and more. As you are preparing food or watching food being prepared, take note of the types of sounds you are hearing and experiment with those sounds. They can be used to recreate or simulate many real-life and fantasy sounds. In the Sound Lab (companion site) we provide additional ideas for Foley and sound-effect props and source. For now let's explore some additional categories of props and source.

Household Items

There are many other miscellaneous items that can be used for Foley. If the fruit smashing wasn't gooey enough for you, wet sponges or water-soaked dish towels can add more of a watery splash effect. There are plenty of items sitting around your house or studio that work very well for a Foley session. Don't overlook anything! Remember that we are after *soundalikes* and not lookalikes, and it can be difficult to predict what an object will sound like when a microphone is placed in front of it. Keep an open mind and a critical ear.

The key is to experiment, try new things, and always be listening to the sounds around you. There really is no limit to the kinds of household props that will prove useful for recording source. A simple toy like a Slinky can be used for laser source material (see Figure 3.1).

Tools and Machinery

Garage sales, thrift shops, and flea markets are great places to source used tools and unique items. Vintage machinery such as adding, sewing, or dowel machines can provide interesting mechanical clicks and ticks to use as source. UI sounds and other mechanics are great places to apply this kind of source material. For instance, you can recreate the sound of opening a safe or lock box by processing these same clicks and

FIGURE 3.1 *(Left)* Recording session with a Slinky using various household items as "amplifiers" and recording with a contact and shotgun mic. *(Right)* Recording session with Deep Cuts Gear (@deepcutsgear) "SOV" trucker hat trimmed with coins for jangly source. Photo: Armina Figueras.

clanks. Puzzle games that have interesting locks and treasure mechanisms could benefit from adding this level of detail.

Power tools like drills or electric screwdrivers produce servo motor layers for machines, robotics, or other motorized devices. Elevators and dumbwaiter motors can be designed with pitched-down power tool layers. There are numerous powered tools that can produce source for these kinds of layers. Compressors, industrial fans, chainsaws, power saws, and welders to name a few are all great starting points. Experiment with a mix of contact microphones and placements to give yourself a few perspectives to work with.

Clothing and Materials

Recording Foley in a home studio can produce very usable source material. In Chapter 1 in the Sound Lab (companion site) we touched on building a Foley pit.

Here we will talk about how to use it to record effective source. Professional footstep Foley artists are known as "Foley walkers." They are well practiced in matching the pace and feeling of the characters' movement in games and film. Footsteps are recorded with the artist walking in place and positioned inside a Foley pit, which is filled with various terrains. When recording footsteps for games, walking away from the mic and then back to it *won't produce usable source*. Walking in place in front of the mic *will* capture the proper perspective but it can be difficult to pull off because it is awkward walking in place without naturally stomping your feet. Take it slow and work on the **heel-to-toe roll**.

The type of shoe you wear for the Foley session should match the character's footwear in the game. The terrain types throughout the game environment will also need to be matched. Hardware stores are a good source for obtaining tiles, turf, wood chips, stone, gravel, and more. A little patch of grass or a piece of broken concrete from around your home could also work. As a side note, professional Foley studios often *dig out* their ground terrain pits so that the footfalls don't sound hollow from lack of depth.

> Start building up your Foley props by collecting items that maybe useful in the future. In a call with my brother, he mentioned a recent storm blowing over and shattering the glass table in his backyard. Living only a block away I said, "I'll be right over." I grabbed a metal tin and a small broom and headed over to collect some glass shards. I didn't have a project that required glass debris but I knew it would come in handy in the future.
>
> Gina

The player character in the game *Journey* by thatgamecompany traverses miles of desert sand. The sound of the footsteps were created by poking fingers into a pile of sand.[10] Using fingers instead of a foot allowed for a softer step to fit the soundscape. This is an important concept to understand: *The Foley must match the aesthetic of the game*. The point of Foley is to produce perfectly synced sound material. It cannot be perfectly synced if you are recording generic footstep sounds for a uniquely portrayed character. At that point you might as well use a sound library. Go the extra mile and pay attention to the details of the visuals so that your Foley can be unique and well synchronized.

Continuing from the above line of thought, to recreate non-human character footsteps try slapping a hand or slamming a fist onto different surfaces. By processing these sounds in interesting ways the result can be surprisingly effective. A large robot footstep can be created by slamming a metal door or stepping on a sturdy baking pan. To avoid the hollow metallic sounds of the pan, turn it over so the bottom of the pan is facing up and place some books underneath. This will ensure the sound has thickness and density to it, which is an important characteristic of the physics of any character. With the pan set on the floor, step with a hard-soled boot or punch down onto it with a fist. Be sure to protect your knuckles if you choose the latter.

Characters are usually outfitted with clothing and equipment. These kinds of accessory sounds should be layered with the footsteps to add more realism and detail. Leather and denim make great cloth movement source when rubbed against itself in front of a microphone. Additionally, few feet of chain can be purchased at any hardware store and used for armor movement. The chains can be recorded by manipulating them in your hands or by wearing them around your waist or arms and moving around in front of the mic. Be mindful of the clothing you wear so you don't capture unwanted cloth movement along with the chains. We highly recommend reading Ric Viers' *Sound Effects Bible: How to Create and Record Hollywood Style Sound Effects*[11] as an additional resource.

> One project required some light metallic clanking to accent a fantasy character's movement in game. I taped some pennies, nickels, and quarters to some sewing thread and taped them to the brim of a "Deep Cuts Gear" trucker hat (see Figure 3.1). I carefully spread them out so they didn't become tangled or overly jingly. Wearing the hat, I shook my head side to side and captured the coins lightly connecting with each other.
>
> Gina

Unique Instruments

If you travel internationally you can find some really interesting handmade instruments from various locales. We have a collection of instruments from travels of our own as well as gifts from friends and family. Just because the intended purpose of the item is musical doesn't mean we can't use it a bit more creatively!

> I once used a small steel drum to create Foley for a knight-in-armor movement. The steel drum was about 10 inches (25 cm) in diameter and had a hollow back. With the drum turned over, face down, I put in a sturdy piece of cardboard which was shaped like a circle. The diameter of the cardboard was slightly smaller than the drum so I placed it inside the hollow body of the steel drum. With the cardboard piece in place I grabbed the drum with my hand and moved my wrist back and forth causing the cardboard to clink against the sides of the metal. With a small amount of processing I was able to recreate this armor movement for the knight character.
>
> Gina

Idiophones (instruments that produce sound by vibration in the body of the instrument) are examples of musical instruments that can be used as fun and interesting sound sources. We once obtained a unique idiophone called a "dance rattle." A dance rattle is a bamboo stick with dried coconut shells attached by plant fiber strings. When the rattle is shaken it produces a clack sound as the coconut shells bang together. This was used in a project which required skeleton bones breaking apart and dropping into a pile on the ground.

In general percussion instruments and accessories are great for sound design source. Wood blocks and wooden mallets can generate tonal wood hit sounds. They can be used to accent movement in cartoon-style games or as UI sounds in learning apps or kids' games.

The waterphone is a highly unique instrument to be used as sound source, and it has been used in horror films to create eerie and haunting background ambiences. It is an inharmonic percussion instrument which has a bowl at the bottom acting as a metallic resonator. A metal cylinder stems from the middle of the bowl and there are metal rods of various lengths and diameters attached to the rim of the bowl. The resonator (the bowl) contains a small amount of water, which makes the sound dynamic and ethereal. Waterphones can be played with a bow or mallet and used to create complex source layers.

Bowing

Once you have recorded most of the objects or instruments around you, try recording them again, this time armed with a cello bow. You will find the resulting sound to be very different when objects are bowed. Instruments that are commonly bowed and recorded for source are electric basses and other string instruments, cymbals, and various types of glass. Be sure to use rosin on the bow to create friction with the object you are bowing.

Field recordist Thomas Rex Beverly has captured the sound of bowing cacti, which is a very distinctive and organic source (see Figure 3.2).

Visiting Artist: Thomas Rex Beverly, Field Recordist

Which Cacti Make Cool Sounds?

Generally, I look for cacti with strong, thick needles that aren't too dense. This is more difficult than you would imagine. Most easily accessible cacti have short spines that aren't strong enough to bow. However, after testing many cacti and after many failures, you'll find the perfect cactus. You'll take a violin bow to the cactus spine and hear guttural screeches with intense, physical energy and hear thick bowed spines growling like supernatural animals. Cactus sounds are incredibly soft and intimate in real life, but when recorded from two inches they morph into otherworldly creatures brimming with ultrasonic energy.

(Continued)

Recording Techniques

For Bowed Cactus 1 I used a Sennheiser MKH50 and 30 in Mid/Side. In Bowed Cactus 2 I used a Sennheiser MKH8040 and MKH30 in Mid/Side. The MKH8040 is rated to 50 kHz, so it picked up more ultrasonic content than the MKH50 did in the first bowed cactus library.

I positioned the microphones as close as possible to the spines and was careful to not stab the sharp spines into the diaphragm of my microphones while I was frantically bowing. With close proximity, I was able to capture as much of the short-lived ultrasonic energy as possible. Then, with endless experimentation I gradually found the sounds of supernatural cactus creatures!

FIGURE 3.2 Bowed cactus. Photo: Thomas Rex Beverly.

Procedural

We won't go into detail here because there is already a great book that covers procedural audio. We recommend reading *Designing Sound* by Andy Farnell[12] as it's a great book to reference for working with Pure Data but also worth a look for those interested in MAX/MSP. The idea behind the book is interactive sound design in Pure Data, which can be useful for games, live performances, and interactive experiences.

Electromagnetic Fields

All electrically charged objects around us emit electromagnetic fields, which can lead to otherworldly or ghostly source recordings. Below we will describe some tools with which you can begin experimenting by capturing the sounds of televisions, computer monitors, cell phones, electric toothbrushes, blenders, transformers, shredders, game consoles, and any other device you have access to. Not every device will produce a pleasing or useful sound but part of the fun is uncovering the unexpected sources. When you are recording your mobile phone, be sure to capture the audio during standard processes like switching between apps and sending or receiving data. You will be pleasantly surprised by the sounds emitted from the device. Spectral repair plugins like iZotope RX can help clean up any unwanted noise or frequencies to make the recordings more useable.

Slovak company LOM has developed boutique microphones with a collection that captures electromagnetic waves. The Elektrosluch by LOM will take you into a secret world of inaudible source material. The instrument is sold as a DIY kit or as a fully manufactured device. All of the equipment is made available through LOM's website in small batches. It also sells Uši microphones and electret condensers that set out to capture delicate sounds the ear can identify, which also go above or below the frequency range of human hearing.

If you can't spare the budget for the Elektrosluch, you have the alternate option to make your own device with inductive coils that will convert changes in the electromagnetic domain into electric signals. A single coil guitar pickup plugged into a DI or amplifier will also work well. Any old FM/AM radio will also pick up electromagnetic signal. In fact, that's exactly what they are designed to do! They will pick up actual stations of course, and plenty of white noise, but you can tune *between* stations and poke the antennae to listen to smaller devices emitting EM.

This type of source is widely applicable for sci-fi ambiences, alien ships, energy bursts, articulations, or the humming of more realistic objects in a game. A little bit of processing can go a long way toward making this source a more usable and polished asset since the raw recordings will already possess the otherworldly aesthetics.

I first read about picking up EM sounds in Nicolas Collins' book *Handmade Electronic Music*. I was so obsessed with the technique that I immediately grabbed an old AM radio and recorded about an hour of static from light bulbs, computers, and even my Wii! You'd be surprised at the diversity in sound that can be captured, and once it's all processed and sampled it works amazingly well for horror games.

Spencer

Circuit Bending

Taking electronics or electronic kids' toys and altering the circuits to generate new sounds is called "**circuit bending**." These sounds are strange, but wonderful. By nature, most sounds generated by circuit bending are glitchy and noisy, but sometimes you strike gold and find a sound that is usable in one of your projects.

Before forging ahead into the world of circuit bending, you should have a basic understanding of electronics, circuits, and soldering. Start your experimenting with devices that don't mean much to you, as you can easily fry the circuits with a misstep. Be cautious as you work with live electronics and take precautions to ensure your safety. Most importantly, *never EVER circuit bend electronics with a wall outlet*. Use battery operated electronics *ONLY*.

Circuit bending is unpredictable. The point isn't to control sound; it's all about finding new and unique sonic characters. The variations on sonic possibilities are endless. Pitch, speed shifting, automating patterns, and modulations are all performable bends. The voltage drop crash is a common bend that simulates the sound a device might make as the batteries are dying.

At its most basic, bending is just another way to find interesting sounds. You can read deeply about circuit bending in Nicolas Collins' book *Handmade Electronic Music: The Art of Hardware Hacking*.[13] If you aren't into taking devices apart and bending, you can always try to distort or warp sound by sending signals through vintage outboard gear or software plugins.

Vocal Mimicry

It's in our nature to make sounds with our mouth, so don't underestimate the power of your own voice and mouth when sourcing sounds. **Vocal mimicry** can be a great way to quickly generate source for UI, creatures, wind, vehicles, and weapons. The pops, clicks, and clucks that can be made by positioning the tongue and lips a certain way are highly useful for these situations.

Let's take a close look at our body's process for creating sound. Air flows from the chest through the larynx and into the vocal tract, and then exits the body via the mouth or nostrils. Speech and various sounds are created as muscle contraction changes the shape of the vocal tract and pushes the airflow out into the atmosphere. The larynx contains our vocal cords, which vibrate. Muscles in the larynx control the tension on the vocal cords, determining the pitch of the sound produced. The nostrils, mouth, and throat are resonators like the mouthpiece on a horn. As the resonators change shape, sound is created.

We apologize for the seemingly tangential biology lesson, but it's necessary to understand the physics of how we create sound. Professional voice artists often study the muscle contractions and air-flow process to understand how to control the sounds they make. It's just like understanding a musical instrument, effects plugins, or synthesizer architectures. If you really know the tool you can control the end result instead of shooting in the dark in the hope of generating a useful sound.

When tasked with creating subtle UI sounds, using your mouth and voice to generate sound source can be a lot quicker and easier than trying to tweak a synth. A good number of casual games use mouth or tongue pops for notification sounds and button tap sounds. Another simple noise that works well for button taps or clicks is the mouth pop. Suck your lips into your mouth with your jaw closed. Then open your jaw quickly to generate the sound. The harder you suck while opening the jaw, the louder and more defined the sound will be.

Vehicle and weapon sounds can also be designed using mouth and voice source. A resonating buzz produced in the vocal tract can be a solid source layer for a vehicle engine or RPG pass-by projectile. Although these vocal sounds might seem to stick out of an AAA quality sound effect, they usually don't. Remember that we are creating *source*, and then manipulating the sounds and adding layers later in this chapter in the section on "Sound Design Techniques." Applying plugin effects like pitch, chorus, distortion, and reverb will also allow you to manipulate your vocal sounds so they match the visuals, and don't stick out of the mix.

To gain some vocal mimicry practice, position your tongue so it is touching the roof of your mouth with your jaw in an open position. Once your tongue is in place, pull it down quickly to generate the pop or cluck. Next, make an "ah" sound and try shaping your tongue in different ways. Alternate between a "u" shape and and "eeh" shape. This will divorce your association between your jaw, lips, and tongue to allow for sounds that range beyond the vowels. Also try playing with varying degrees of lower jaw position and you should be able to sculpt the sound to your liking.

This may come as a surprise, but animals can make useful vocalizations as well. Animal sounds are not only usable for creature sounds; they are also great source material for vehicle and weapon sound design. The roar of a tiger or lion can be pitch shifted, time stretched, and blended in with engine sounds to add a ferocity to the vehicle. Growls work well in vehicle pass-bys or engine revs to give them a bit more bite. In general, animal sounds add a very intuitive (and usually threatening) mood to well-crafted sound effects. We'll discuss layering animal sounds in more depth later in this chapter in the section on "Sound Design Techniques."

FIELD RECORDING

In Chapter 2 we discussed microphone types and in this section we will explore field recording for new and unique source.

Recordings captured outside of the studio are referred to as **field recordings**. These recordings are a crucial element of your sound palette, and often spell out the difference between a generic soundscape and an immersive one. Common targets for field recording are ambiences, sounds found in nature, human-produced sounds that are difficult to capture indoors, large sounds like vehicles and technology, and electromagnetic sounds.

Field recording is an art form just like photography. When you are sitting out in nature with an array of microphones and field recorders armed to record, your headphones become your eyes as you monitor the lush soundscape that surrounds you. There are many challenges when recording out in the field that you won't encounter in a studio setting. We will discuss some of them in the "Location Scouting and Preparation" section below. For now, let's focus on the variety of source that can be captured and used in your sound palette.

Finding Sounds in the Field

Sound designers are always looking for unique and fresh sounds for their games. In the field there is a whole world of interesting sounds waiting to be captured. Some sound recordists capture audio out in the field to create libraries of SFX collections to be used by other designers. Field recordings can also be used with minimal processing to create realism in a game, or manipulated into an immersive and otherworldly soundscape.

FPS games like EA's *Battlefield* require realistic sonic elements from battles off in the distance. Field recordings can truly capture the realistic background ambience required of games like this. There are a good number of re-enactment groups that might be willing to allow a recording session during one of their performances. If you know someone in the military you may alternatively request access to a training base to garner more authentic recordings. If you are lucky enough to have an opportunity like this, you will certainly want to have the right equipment. At a minimum you should bring several microphones to record at different distances. The audio team at DICE used field recordings of military exercises as ambience in *Battlefield*.[14] More focused recordings of specific objects at the base were then used as details to fill out the soundscape.

Animals are a very common target for field recordings, but they can be unpredictable. Setting out to capture seasonal sound like bird call can be tricky depending on your location and the time of year. Game production doesn't always adhere to nature's schedule! There are also other difficult factors that may prove to be an obstacle in producing usable field material. You may scare off the birds if you come marching in with all of your equipment for example, so it's a good idea to research your subject for a more successful session. Perhaps visit the location once or twice before the session with minimal gear to see how your subject behaves and to check for noise or other obstacles.

Visiting Artist: Ann Kroeber, Sound Recordist, Owner of Sound Mountain SFX

On Being Present in the Field

To me the most important thing is to be present and all ears when recording. Have good sensitive microphones and recorder then turn gear on and be completely in the moment. Forget thinking about technique and expectations. Simply listen and point microphones where it sounds best. Find a way to get away from background noise and the wind.

When recording animals really be there with them, tell them what you're doing and how the microphones are picking up sounds … you'd be surprised how they get that and what they can say to you when they know.

Ambience and Walla

Ambiences and walla are two more commonly targeted sounds for field recording sessions. Games generally need quite a bit of both, so the more distinguished and original ambience and walla that you have in your library, the better prepared you will be. Try starting out with local field trips and practice capturing some environment sounds. To continue improving your skills as a sound recordist the trick is to always be recording. Just get out there and explore new areas to record. If you like to travel internationally, be sure to pack up your gear so you can grab some local sounds for your library. Recording urban ambiences in different cities and in different countries will provide a wider library of ambiences for your collection. An understanding of your game's sonic needs will define the recording format in which you will capture the sound; mono, stereo, **quad ambience**, surround, and ambisonic are all options for ambience recording.

Creativity in the Field

Regardless of the source you are looking to record, there are many sounds just waiting to be captured in the field that you could never recreate in a studio. Let's say you have found a fence near a train yard. Connecting a contact mic to the metal fence or placing it inside one of the posts as a train rushes past can be striking. This shift in recording perspective makes a huge difference in the resulting sound and how you can apply it to your palette. If you have a few mics to work with it's often a good idea to capture multiple perspectives at once.

Recording water and other fluid sounds may present a challenge. The fluid will sound different depending on the container you use. If you set out to recreate the sounds of a character sloshing through water you may run into some trouble. Filling a metal bucket with water outdoors and setting up a mic to record your hands splashing around sounds like a good idea, but the recorded audio will have a thin and resonant sound as it reflects off the metal material. This only works if the sound was intended for a character confined to a metal space. A better option would be to use a larger plastic container or a kiddie pool. You might also try lining the metal with towels to reduce the metallic resonance in the sound. Experimenting with the amount of fluid can help as well. Too little water will sound like puddles rather than shin-deep water. Too much water will sound like the character is swimming in an ocean. With field recording it's best to listen critically, be flexible enough to try different approaches, and be prepared to think on your feet.

Getting Started

When selecting source to record in the field be mindful of your equipment and what it can handle. Wind can be a big factor whether or not you plan on capturing it in the recording. We will further discuss ways to deal with wind in the "Consider Conditions" section later in this chapter. If you have a noisy preamp in your field recorder, you may want to avoid trying to capture very quiet settings. Similarly, you will need microphones and preamps that can handle extremely loud sounds if you want to record rifle sound effects in the field.[15] Practice makes perfect, so record as much as you can and analyze the results. You will eventually gain an intuition about which recording setups work best for the environments you record.

MICROPHONE CHOICE AND PLACEMENT

There is a whole host of things to think about to ensure you capture a usable sound. The quality of the recording as we stated above is dependent on the environment, technique, and equipment. Choosing the right mic, polar pattern, sensitivity, and proper gain settings along the input path and monitoring those levels are key components. Mic placement could be viewed as more important than the microphone choice. Let's explore some options.

Proper placement really depends on the sound you are trying to capture and the end result you are looking for. Placing the mic too far from the source may introduce too much room or environment sound into the recording. Close-miking can cause **proximity effect,** a bump in the 200 to 600 Hz range in the recording, though the proximity effect can be useful if you are looking to thicken up the sound at the recording stage. A bump in the low mids can be handled with EQ more easily than working to remove the sonic character of the room.

Since there are a large number of choices when it comes to microphones, it can be difficult to know which one to arm yourself with. Knowing your source will help you determine the best mic, pick up pattern, placement, preamp, and use of pads or limiting. We suggest testing your recordings with several of these combinations to determine the right fit, and after some extended practice you will have some go-to techniques to use in the field. Experienced sound recordists typically have their go-to mic picked out after years of experimenting. Some mics will color the sound in a way you may or may not prefer so let your experience and practice in the art of recording guide you on your journey.

Consider Conditions

When choosing a mic be sure you consider its limitations in regard to very cold weather, high humidity, wind, dust, and even really loud sounds. **High SPL**[16] mics are best suited for capturing loud sounds like weapon fire or explosions. In cases of high humidity, higher quality mics might fair best as they are crafted with better **RF technology,** which helps quickly acclimate to changes in climate conditions. In this case, the mic should be wiped down and placed in its wood box with a desiccant bag to draw out any moisture. Protection against rain can also be tricky as putting a plastic covering over the mic can degrade the recording quality.

Wind moving across the microphone capsule can ruin the recording. There is a variety of manufacturers like Rode and Rycote that manufacture wind protection devices and there are also plenty of **DIY** options, which we discuss on the companion site in Chapter 1.

Research your mic before moving it into different conditions so you can be prepared to capture the best quality sound but also protect your investment.

The project's budget will determine the equipment and location of the recording sessions. When the budget permits, equipment can be rented or purchased to handle all the needs of the recording. If you are just starting out and working on projects with smaller budgets you can make it work with what you have. As you work on more projects, your equipment list will expand. We suggest looking into renting mics before you buy to see if they are the right fit for your workflow.

Microphone Choice Considerations

In Chapter 1 on the companion site, we discussed different types of microphones and the pickup patterns along with ways to avoid wind and handling noise during a recording session. Here we will explore some practical uses.

There are typical mic choices, patterns, and common practices you will read about or hear about in the community, but it's always good to experiment and uncover new ways of capturing sounds. Microphone selection is a subjective process defined by each individual's tastes and sensibilities. We have favorite placement and positions and they make great go-tos under tight deadlines. When a project has some time in pre-production it's good to play with different polar patterns, frequency responses, and mic positions to experiment with levels, phase, and delays between microphones. There are many factors that affect the state of the object being recorded and the environment, which can change the end results. Here we will explore some choice considerations.

You are probably familiar with or have used dynamic mics at some point in your audio journey. While they are used in a lot of live recording situations, dynamic mics are good for capturing loud source. Condenser mics will always offer the ability to capture smaller details but can be too sensitive and even damaged when capturing loud sounds. A combination of the two could be useful for capturing specific source material. For example, the recording of a door closing could benefit from a dynamic mic for capturing the low-end impact of the door meeting the frame while a condenser mic could capture the higher frequency content like latch movement. The two sources can be blended during editing to produce a full-bodied yet highly detailed sound.

While a narrower pattern offers direct capture of the source, omni patterns are a better choice when wind is an issue. Omni microphones also tend to have low self noise which is useful for recording very quiet sounds. Polar patterns are basic knowledge that all recordists should be familiar with, knowing the pros and cons of each. Once you have an understanding of your source and the expected outcome of the recording session you can explore different microphone types.

A sensitive large diaphragm condenser mic with a narrow cardioid polar pattern is great choice for picking up subtle nuances from certain sources due to its lower noise floor. Microphones that lack good frequency response and noise floor will require more restoration of the source during editing and may not be as processing friendly as higher quality source.

A shotgun mic is very versatile for sound-effects recording. We use the Sennheiser 416 and Rode NTG2 for animal recording, gears and gadgets, doors, impacts and more. A shotgun is great for recording source, with rejection on the side and rear working best for sounds that need to be captured close and without room ambience in the recording. It's also a great choice for Foley source and SFX recording where small details need to be captured. Keep in mind the quality of the mic pre is important when recording quiet Foley or sound effects, for example cloth movement. The noise level of the mic pre may mask the quiet sound of softer material cloth movement like cotton for example. Of course, it is equally important to have an acoustically controlled room when recording quieter sounds.

The different formats of shotgun mics from short, medium, to long, and mono or stereo, make for more confusing choices. A longer shotgun is often the best option for capturing distant sounds with the highest off-axis rejection. A shorter shotgun will have a slightly wider pickup range but will still be great at focusing on the source. The longer shotgun mic isn't always physically the best for maneuvering around a session.

Pencil or small diaphragm condenser mics or SDCs are great for capturing more high-end detail. They also offer a more natural sound whereas large diaphragm condensers often color the sound. The SDCs also have a fast transient response, which is great for producing brilliantly detailed source. We have used the Oktava MK12 with a hypercardioid capsule for capturing delicate cloth movement which requires more detail in the sound.

Lavalier mics are commonly used for capturing voice while placed on the actor but these small mics can also work well for stealth recording. They can also be used for boundary-style recording by securing the lav mic to a reflective non-porous surface.

Hydrophones are microphones made for the specific purpose of recording in water. They can be used for picking up animal sounds like dolphin and whale calls in large bodies of water. Hydrophones may not always capture the sound underwater that we might expect to hear as a listener and it might sound a bit too thin to our ears. Mixing the hydrophone recording with a recording from a condenser mic above the water can work well if you apply a band pass filter to the mids. There are plenty of DIY options on the internet or you can invest in a manufactured option like the JrF Hydrophone,[17] which can be purchased for under $100 USD to get you started.

Contact microphones are piezo transducers designed to pick up sound from an object it is physically touching. Signal is derived from mechanical vibrations instead of airborne soundwaves. You can DIY (do it yourself) a contact mic and there are also other budget-friendly options such as the Zeppelin Labs Cortado MkII. When using this type of mic you might be surprised to find not all objects you think will produce beautiful resonances will create an interesting sound. These microphones are best for creating abstract sounds but can also be layered with airborne sound-wave recordings to add an extra dimension to the sound.

We were contracted to design sounds for a mobile app in the kids' learning space that required the user to tap on various virtual objects on a tablet. Our recording of tapping on various surfaces wasn't giving us the tactile sound we were looking for. Using a contact mic we recorded tapping the same surfaces and layered them with the original source. It provided the sound we were looking for.

Spencer

Binaural and other spatial microphones are becoming more widely used with VR (Virtual Reality)/AR (Augmented Reality) audio and other immersive experiences because they capture sound similar to the way humans hear. Our ears and brains can perceive accurate spatial positioning in just about a 360-degree sphere. These mics offer multiple diaphragms in one mic capsule. The audio team at Ninja Theory used binaural microphones to record abstract voices to simulate a "voices in your head" effect which creates an **ASMR** experience.[18] A binaural mic can also be used to capture reference recordings which can be useful for building accurate sonic spaces. Soundfield microphones are useful for capturing ambisonic audio, which offers mimicking of how sound is perceived from all directions. This can be very useful in building immersive sonic experiences in virtual reality. The price of binaural microphones can vary dramatically, starting at $100 USD and reaching to thousands of dollars.

Another interesting device for capturing sound is a parabolic mount. An omni mic is placed into the center of the polycarbonate dish to assist in capturing long-distance sounds, birds, or sporting events.

While it's a good idea to understand the pros and cons of microphone types and polar patterns, hands-on experience will allow you to use your ears to define your go-to tools for capturing different subjects.

Placement Considerations

Now that you have some considerations for different microphones to work with, you can begin experimenting with placement. In a single-mic scenario, start off with getting the mic very close to the source, about 3 inches (2.5 cm), then back it up as you monitor with headphones, using your ears to find the best location. If you feel the sound is too narrow, switch to a different pickup pattern with a wider range or back the mic up a bit. As an experiment, listen to how the sound changes at 6 inches (15 cm) and 12 inches (30 cm) from the source. Then try different mics and pickup patterns at each of those positions on the same source.

If you are looking to capture a stereo sound, start by setting up a pair of small diaphragm condenser microphones, preferably a **matched pair**, configured in an XY or right-angle (90 degree) position. This is a great technique for recording ambiences but can also be good for a more direct sound when you are after more depth in the sound. The 90-degree angle doesn't offer a super-wide stereo image but does offer minimal **phase issues**. Experiment with widening the mics beyond the 90-degree angle to increase the stereo width, but be mindful of the loss in the middle as as you adjust the position. An **ORTF** position is a take on XY and allows for a wider stereo image without sacrificing the center. This technique uses a stereo pair of mics placed at a 110-degree angle from each other to mimic the way human ears hear. A stereo bar is used to position the mics and attach them to one stand or boom pole. Marking the bar can help easily identify the XY or ORTF positions. Since the mics are small condensers you can fit them into a **blimp** or will have to use separate wind protection when you record outdoors.

Ambience recordings can benefit from **Mid-Side technique**. This consists of a cardioid mic as the mid and a figure eight mic as the side. The mid mic acts as the center channel and the side mic pics up the ambient and reverberant sounds coming from the side. This technique will produce a recording that needs to be decoded before the proper stereo image can be heard. The decoder is often referred to as an **MS decoder**. The recording will produce two channels but the side channel will need to be split into two and panned hard left and right respectively. Changing the blend between mid and side in the mix will offer more control over the stereo width. This can come in useful when you want to adjust the room in the mix. Another bonus of MS recordings is that they play back in mono without any phase cancellation (unlike stereo recordings). MS shotguns are a great choice when you want to focus on one specific sound but you want to capture the ambience around it without committing to a true stereo recording.

Binaural mics are often attached to a dummy head to properly position the mics at ear level to capture sound as humans hear. There are also wearable mics that can attach around the sound recordist's ears. One needs to be mindful of breathing and head movement when doing it this way. Ambisonic mics captures four-channel, 360-degree sound information that not only positions on the same plane as our ears, but also above and below our head. These techniques offer greater immersion as they give a more realistic sound based on the way we hear sound naturally.

When using a lavalier as a boundary mic it should be placed as close as possible to the surface without it touching the surface. This will allow you to capture a full sound, as the reflection off the surface doesn't have enough time to go out of phase since the gap between the surface and the mic is so tiny. It's not something you might use all the time but a useful thing to know. For example, recording ice-skate blades scraping across ice can be done by attaching a lav mic to the boot, but you can also attach it to the plexiglass wall surrounding the rink to try out the reflective "boundary" technique.

A contact mic can be placed directly on the resonant body. Attach the mic with double-sided sticky tape, sticky tack, or gaffer tape. You should do some test recordings as you secure the mic to the object as too much tape can deaden the resonance you are trying to capture.

Regardless of the placement you choose, having a backup mic in position in case you bump the main mic or clip the input signal is useful. A backup mic can also offer a slightly different color to the sound so you can decide in **post-production** which to use.

Input Gain Considerations

The microphone is connected to a recorder in order to capture the soundwaves. Poor preamps, input levels, and sampling/bit rate can also make or break the recording.

Remember, you want to choose a high-sensitivity mic with low **self-noise**. The mic's circuitry will introduce noise into the audio path. Self-noise is the noise introduced to the audio which will result in an audible hiss when capturing quiet sounds.

Ensure you have phantom power set up if your mic requires it. Do a test recording to check levels ahead of time. In digital equipment, 0 dB[19] is too loud and could lead to clipping of the signal. It's best to leave headroom of around -6 to -12 dB. A lot of field recorders make checking levels visually easier by including a marker in the middle of the meter.

When recording quiet sounds the tendency is to boost the gain which, in turn, introduces system noise. This is where gain staging comes in to play. The output level of the mic should be considered and switching to a higher output mic might be necessary if you can't make the source any louder. Additionally, you can try moving closer to the source or remember you can achieve a bit of gain with compression, limiting, and EQ in post-production.

In-line mic pres like the Triton Audio Fethead or Se Electronics Dynamite have ultra-clean, high-gain preamps. When used on a low-output mic they can save the day by providing that extra boost of clean level to quiet source.

PERFORMING SOUND WITH PROPS

The idea behind performing Foley or sound effect with props is to tell a story with sound by capturing the movement so it best fits the visuals or animations. For example, a character walking through a haunted mansion on a wooden floor could benefit from wood creaks randomly triggered along with footsteps on wood. To really achieve a believable result, the wood creaks should be performed and captured. Let's say you have a wooden floor that creaks with the slightest movement. Standing with your feet firmly planted about 12 inches (30 cm apart), start to shift your weight from the left to the right. Adjusting the speed at which you shift the weight will provide a drastically different creaking sound.

Later in this chapter we will discuss recording footstep source in "Sound Design Techniques." There we will examine how Foley artists or Foley walkers produce varied heel-toe footfalls with some scuffs and minor weight shifts. A sound-effect recordist new to performing Foley may tend to walk with a flat footfall which generates a stomping or plopping sound that isn't very convincing. Others new to the field may overcompensate and the result will be overdone heel to toe.

The same can be said for other actions in game which require synced sound. If a character tosses papers off a desk, capturing the sound of simply dropping paper on the floor won't satisfy the player. Capturing the hand sliding the papers across the desk should be the first action you start with. Having animations or video capture to work with will allow you time to re-enact movement to produce believable sonic movement. The idea behind it is to actually perform the sounds. Control the sonic output of the object you are working with to get the best fit for the animation.

We have discussed sound being exaggerated to create a believable experience for the player or listener. An example of this is picking up an object. In real life we often don't hear a sound when picking up an object but in game audio we may enhance the action to add drama to it.

Let's look at this in another scenario where we have the player character in a game outfitted with a backpack. What does it sound like when you wear a backpack? It may not sound like much at all but in game we want to exaggerate the sound a bit to accent the character's movements. To achieve this you could fill a backpack with some books and jingly items like a set of keys and shake it in front of a mic. Try moving and shaking the backpack in various ways and at various speeds to generate source that will best tie in to the visual action.

If you are performing with the props be sure to tie back your clothing or wear something that has less swish or swoosh when you move around to avoid introducing unwanted sound into the recording. Also be mindful of your breath as you get close to the mic to perform with your props.

Summary

The idea is to capture the sounds or performance with enough duration to ensure you have enough to work with in editing. Variation in the frequency of events if being performed will offer more to work with in the editing process. For example, when recording footsteps performing slow, medium, and faster paced steps along with soft, medium, and hard impacts will offer enough variety to match various needs in game. When recording door opening and closing sounds, you can apply this same principle. Open the door gently and then more aggressively so the handle and lock mechanism will sound different in each approach. A door creaking open can also benefit from this process as some creaks may need to be quick and high pitched while others need to be long and low pitched. Always record additional takes so you don't have to worry about a lack of source in the editing process. Finally, it is good practice to "**slate**" your takes by recording your voice calling out some descriptors to later identify the recording.

LOCATION SCOUTING AND PREPARATION

Location, location, location ... choosing an adequate location to record is a vital part of the process. There is a huge list of obstacles that can get in the way and spoil a recording session. Preparation to reduce or eliminate unwanted noise and smooth out logistics will ensure that quality source is produced.

Noise Sources

If you are recording in your home studio you will have to think about outdoor and indoor noise sources. Here we've listed a few sources starting with some of the more obvious ones and moving onto others that are less obvious.

Outdoor

- Traffic and vehicles
- Landscapers (lawn mowers, leaf blowers, snow blowers)
- Humans (conversation, coughing, kids playing)
- Air traffic (proximity to an airport)
- Animals (crickets, dogs, cats, squirrels scratching, birds, etc.)
- Weather (wind, rain, leaves rustling, branches falling, humidity, thunder)
- Time of day/season

Keep in mind that a lot of these noises can be used as source as well.

Indoor

- Other humans or pets
- TV, radio or phones (these can also be listed as outdoor if you have neighbors who enjoy watching or listening at inconsiderate volumes)
- Faucet leaks
- Tiled or empty rooms
- Pipes, heating and cooling systems or units

- Creaking floors, sticky floors
- Sump pumps
- Cloth movement from your own outfit
- Breathing
- Clocks
- Fans on computer and other equipment
- Refrigerator humming
- Reflective surfaces
- Frequency build-ups

Scouting the Location

The ideal recording location is a room that has a very low **noise floor**, a balanced frequency response, and that is extremely quiet. Unless you are in a dedicated and acoustically fitted space it will be difficult to find a dead space to record in. To avoid these issues, prepare by scouting out the location first. To scout the space, sit quietly in the space you've chosen with recording equipment and a pair of closed back headphones and monitor the feed. Listen carefully for any unwanted sounds. Be sure to listen long enough to give less frequent sounds a chance to crop up. Make sure to record the clap of your hands in the area and listen for reflections, **flutter echo**, and any frequencies that tend to accumulate.

Scouting external locations also includes finding interesting places to record *legally*. Nobody needs to take on jail time for grabbing some ambience. Train stations, automobile repair shops, junk yards, shipping container lots, construction sites, and many other public locations can yield amazing source material. Seek out small businesses that can provide unique source and make your case with the owners. You'd be surprised how generous people can be with their time and resources when approached with a professional and kind attitude.

Locations to Avoid

Avoid recording in empty rooms, especially if they are symmetrical. The lack of objects to absorb sound reflections will leave you with too much **room sound** in your recording. This can give recordings a tinny flavor, or cause the build-up or drop-out of certain frequencies. Filling the room with absorption and diffusion materials can help reduce these issues. A full bookshelf is an effective natural diffuser and can provide some degree of absorption. The books won't completely fill the space however, so you may end up with absorption at one frequency and none at others. Researching and testing your indoor space will take some time, but it will save you time cleaning up recordings later on.

Bathrooms are even worse for recording than kitchens because the entire room is usually one giant reflective surface. The idea is to find a place to record clean source that offers numerous processing options afterward. By recording in a heavily reflective room you are limiting those options. Of course, if you are after the reverberant sound of a bathroom then by all means give it a go. Just be aware that reverb is easy to add after the fact, but when it is "baked" into the recording it takes a lot more effort and patience to remove it.

Location Preparation

It may be necessary at times to record various materials in a location that is not ideal. In situations like this, consider the reflective surfaces and added noise around you before pressing the record button. In a kitchen for example, you'll need to unplug the fridge to reduce hum, but try to expedite the recording as much as possible to avoid spoiling your groceries. Use towels to line reflective surfaces to increase sound absorption and decrease reflections. A **reflection filter** or **baffle** situated around your mic and sound source can also be useful. They can be purchased online at relatively low cost. If you are crafty, you may be able to whip up your own in a DIY fashion. Hanging a blanket over two mic stands or over a door can provide some further sound isolation. Pretty much any thick and malleable material (blankets, comforters, pillows, jackets, etc.) can be set up or hung around the room to act as a makeshift acoustic treatment. Pillow forts were fun to make as kids, and now you can build them for work!

Outdoor Locations

Recording outdoors requires a lot more navigation because you have far less control over noisy sound sources. Sometimes it's possible to find a quieter space to record in if you do a bit of exploration. For outdoor ambiences you may find a quieter spot in the woods rather than out on the grass by a soccer field. When the source of noise is intermittent (in the case of cars driving by, or a crowd of pedestrians) you may also be able to simply wait it out, or chop out the extraneous sounds back at the studio.

Recording quieter soundscapes outdoors can be even more troublesome. If the noise floor is louder than the sound source, the recording will be unusable. Environment, microphone technique, and equipment are all a factor that especially impact outdoor recordings. When capturing quiet soundscapes any background noise at all will interfere with the usability of the audio in game. To bypass this issue the sound recordist must have patience and willingness to stake out a location to find the most quiet of times. This can take hours or days, or even relocating and trying a different location. Signal to noise ratio can be a problem when trying to capture quieter soundscapes since quieter sounds will require you to increase the input gain quite a bit. A large diaphragm condenser mic will offer a lower noise floor and adding in a low-noise, in-line microphone preamplifier to the mix could offer about 20 dB of clean gain. You can also try a different microphone if the mic you chose has a low output. A recorder with cleaner preamps can also help you capture a higher quality recording of quieter sounds.

Surprise Sounds

Although preparation is extremely important most of the time, there are situations where you will be pleasantly surprised by a completely unpredictable sound. If you are ever caught by a captivating sound and you don't have the time to prep in advance, capture it anyway. Having a noisy sound is always better than having no sound at all. However that doesn't mean it's okay to be a sloppy recordist and "fix it in post." Do your best to produce the best clean and dry recording as possible, and fix only what you need to afterwards.

Miscellaneous Factors

There are a number of miscellaneous and unpredictable factors that can get in the way of a quality field recording. Equipment can malfunction and add unwanted noise or hum into the recording. Stomach gurgles can strike if the wrong food (or no food at all) is eaten prior to the session. Humidity can dampen the high frequencies or even damage the microphones. Similarly, a very windy day can make it tricky to capture quality source. Luckily, most of these are predictable and preventable. Bring backup equipment, eat a nutritious meal beforehand (not a gas station burrito), and check every weather app you can to find a location where the conditions will be right for recording. Even if wind is unavoidable, using a windscreen can reduce its effect.

Here is a list of things to consider when recording.

- Pre- and post-roll a few seconds for each new track.
- Slate your take with enough detail to work with in post.
- Capture more source than you think you need and include variety.
- Monitor and spot check levels (it's good to take monitoring breaks by taking off the headphones and listening to the environment in real life).
- Experiment with mic placement and pickup pattern.
- Consider the environment and work to protect your equipment.
- Think outside the box.

Processing Field Recordings

There are a vast number of processing techniques you can use to clean up audio after the session is through. If you don't own a windscreen, wind can often be reduced or removed completely from a recording by using a dynamic EQ (low shelf) and spectral repair instead. A de-plosive can also help.

There are also many ways that you can creatively process field recordings to create otherworldly sounds. Martin Stig Andersen, composer and sound designer of the brilliant soundscapes in Playdead's *Inside* and *Limbo*, uses low rumbles and pulsing sounds to envelope the player. These sounds are manipulated in a way that creates such an ominous soundscape the players feel tethered directly to the game world.[20]

You can achieve similar results by experimenting with some field recordings from earlier in this chapter. For example, taking a recording of crickets or cicadas and running it through various effects like delay, tremolo, or a granular **pitch shifter** can produce an ambiguity in the sound that will make the non-literal atmosphere almost "come alive." Players will consequently feel immersed in the detail that these techniques add.

Another way to really mangle your field recordings is with a sampler. With the recording as an oscillator you can transform the sound through filters, modulation and envelope shapes to create surreal pads and granular tones.

Don't forget about pitch shifting up and down to get a sense for how your recording will respond. While extreme time stretching can introduce artifacts into the sound it might just be what you are looking for if the game's soundscape calls for a non-literal steampunk atmosphere.

Later in this chapter we discuss various ways to use effects processing as a sound design tool. There you should experiment and practice with your field recordings by running them through various effects and chains of effects. Be sure to play with the order of effects in the chain and take note on how running one plugin into another changes the final sound.

Field Recording Summary

To summarize, selecting and preparing a location is an important factor in how your source recordings will turn out. To produce an effective sonic palette, you will need quality source. To get quality source, all of the above preparations and considerations should be taken seriously. Whether you are recording indoors or outdoors there will be various noise sources to contend with. By researching the location ahead of time, listening with recording gear, and applying some of the sound absorption techniques we have mentioned, you should be able to grab some high-quality, low-noise audio source for your palette.

Discover and record sound source to use as source layers and recreate sounds. Practice with various location scouting, preparation, and mic placement.

> I had the opportunity to record at a shipping container site. There were numerous sound sources to record and I managed to capture a good deal of them. Large

compressors that provided power to the tools, forklifts, dropping one large metal container on another while inside, metal latches, welding, saws, metal impacts, fork-lift backup alarm, and hydraulics were some of the sounds I was able to capture. The session wasn't a great success but it did yield some really useful source. I didn't have the ability to bring anyone else along with me, due to my limited permission, and I could only bring limited equipment due to safety regulations at the site. I had to work with a Rode NTG3 shotgun mic, Triton audio FetHead inline mic pre, Tascam DR100 MKIII recorder, and Ultrasone Proline headphones. The mic was held by a pistol grip or on a boom pole and the recorder hung around my neck using a Sachtler recording pouch. I was asked to wear a safety helmet and a large safety vest which made a bit of noise when I moved. In the end I would have liked to have multiple mics set in different positions, time to adjust input levels and to check a few recordings for quality control. I was happy to have done the session and captured what I was able to get under the restrictions.

Gina

DESIGNING SOUND EFFECTS

At this point we have covered building a sonic palette as well as sourcing material via libraries, synthesis, and recordings. In this section we will explore ways to take source material and use it to produce specific sound assets for games.

Getting in the Mindset

Have you ever asked yourself when "creative sound design" is appropriate? Do realistic graphics call for outside-the-box thinking when it comes to designing sounds? Or do they require perfect realism all the time? In truth, fantasy and sci-fi aren't the only genres that rely on experimentation to deliver a quality audio experience. Even the most realistic AAA graphics require unique and innovative approaches to designing sounds in order to satisfy the needs of the game, and create an immersive experience.

When starting a project you will almost always have some visual assets or design documents to work with. Use these assets to begin thinking creatively about how you want to design the sounds. When you have access to a build of the game take video capture and import it into your DAW. This will allow you to quickly see and hear whether your designs fit with the visuals or not. In other words, it will streamline the creative process so that you can easily evaluate your work. You will find more often than not, that the more inventive your sound design is, the better it will suit the visuals *regardless of graphical style*. Games almost always require an original sonic personality for the player to feel immersed, so don't undervalue the merit of creative design.

The video capture is purely for design inspiration and timing, so keep the interactive nature of games in mind. Some animations in game may require multiple parts exported from your DAW later on during implementation, while others may require all the parts baked into a single sound file. Understand that the creative aspects of design are separate from implementation, so make it a point to work out with your programmer how you will deliver files after your design direction is approved.

Deconstructing Sounds

In order to design sounds from scratch, we must first be able to deconstruct them. This process is vital to a sound designer's technical ability because it trains the ear to listen for details in the sound that most people will overlook. It also triggers our brains creatively to think of how we can begin to build new sounds from scratch using our source audio.

Every sound can be broken down into three categories: physics, material, and space. Because all sound is fundamentally oscillating air molecules, the **physics** of a sound is incredibly important. The physics of a sound will tell us how heavy the sound source is, how fast it is moving, how large it is, and anything else related to how it interacts with the environment. A ball flying through the air has a definite mass, speed, and size. In turn, this will impact the sound it makes. The larger and heavier an object is, the more likely that lower frequencies will be present when it is tossed through the air. Without those low frequencies our brains will tell us the object is tiny and light.

The object will also be comprised of specific *material* which affects the sound as well. Let's use a ball as an example. If the ball is made of metal it will fly through the air with very little wind resistance compared to a ball of yarn. The effect becomes more pronounced if you imagine the ball actually landing somewhere. If it lands in the palm of someone's hand the metal ball will have a sharp attack and a metallic resonance. The ball of yarn will be much gentler, with the attack being soft and fluffy. If you were to compare waveforms of each of these examples you would notice a clear difference in the moment of impact. To add another level of detail you might even take into consideration the material of the ball *as well as* the material of the object the ball hits. In this case the ball is hitting the skin of the "catcher's" hand. In a game, it might land anywhere, so all possible materials must be accounted for in the resulting sound.

Finally, our hypothetical ball also occupies a particular position in *space*. There are two ways we can look at the defined space of our ball. First, we can see it as a standalone sound effect having width, height, and depth in the stereo field. In sound we can create the illusion of three dimensions in a two-dimensional space. Humans subconsciously equate higher pitch to a higher vertical placement in a space. So we can assume that defining height can be done through pitch. Width can be achieved with panning of the sound or layers within the sound. This is particularly useful for adding movement within the sound. Finally, depth will define through sound how close or far away the object is. Louder and brighter sounds will sound closer while sounds lacking higher frequency content will appear to be further away. More reverb on a sound will also push it back in the mix. Second, since we are going to implement our sound effects into a game engine we can define the process of positioning a sound effect in a game as **spatialization**. Spatialization is largely dealt with in the implementation phase of development (see Chapter 8), but for now think of it in terms of the audio engine applying panning, volume, and reverb on an object. The farther away our ball is, the lower the volume will be on the sound it emits. If there are reflective surfaces in the environment, we will also hear more of the reflections and less of the original dry sound source. Depending on the nature of the game the sound may require having these properties baked into the asset or will rely on the audio engine to handle it. For example, sound effects attached to a first-person character will require the sense of space to be baked into the sound while a third-person perspective may require mono sound assets, which will allow the audio engine to control the spatialization.

Reconstructing Sounds

Now that we have deconstructed the ball sound, let's focus on reconstructing it to be used as an audio asset. To do this we must decide on the relevant physical, material, and spatial information to include in the sound. The ball is moving fast, so we make sure the sound is quick and zippy. The ball's surface is smooth, so we don't need to include too much air friction. Finally, we make sure the sound pans from left to right as that is the (hypothetical) direction it is being thrown in, and attenuates as it gets further from the listener. Great! Are we finished with our asset?

If this was an asset to be used in linear media, then the answer is likely yes. However, since we are working on a game, there is one more element we need to take into account – gameplay. This sound needs to serve a *function* for the player. Perhaps the ball is actually a grenade, meaning the player has to feel a sense of excitement and urgency when throwing it toward an enemy. This can easily be accomplished by adding a satisfying "click" or pin-pulling sound just before the swoosh of the throw. We call this **narrative design**. Narrative design can take many forms, but in games it is often used to describe a sonic element that is added to spice up the sound, or add a bit of interest to the player's auditory experience. Here we increased the satisfaction by throwing in a click to show that the grenade is active. The key point is that we add the click only to add interest to the game event, regardless of what occurs in the animation. We are adding sound, not only to provide credibility, but to add depth to the gameplay and narrative. This is the fourth element of sound that really only comes into play when designing sounds for games.

Getting Started

Earlier in this chapter in the section "Designing the Sonic Palette" we discussed creating a playlist of sounds as source material for use in your design. Here we will explore the four elements of sound (physics, material, spatialization, and narrative design) as helpful tools for designing sounds. If you are completely new to sound design it can be overwhelming to begin from scratch, and you might not know where to start. These four elements can be a framework to get you started. Ask yourself first, "how heavy is this object, and how fast is it moving when it emits a sound?" These physical qualities will likely tell you to start looking for low-frequency or higher frequency elements. Then ask "what is the object that's making sound made of?" This will help you hone in on various sound sources in your palette to choose from. These two answers alone will give you plenty to work with. When you've found your groove you can then start adding spatial details and planning relevant functional aspects of the sound.

SOUND DESIGN TECHNIQUES

Now that we have an effective framework for designing sound, let's dive into the important techniques and methods for designing audio assets for video games.

Layering

Music composition is an art form in which composers combine and layer groups of instruments to create a fully developed arrangement. This technique has been employed for centuries to tell the sonic story. This is similar to sound design in a lot of ways. Both disciplines allow the artist to inform the audience's emotions and influence their mood. Just as a composer combines layers of instruments to create an engaging score, sound effects require various layers to produce a well-crafted effect.

Remember that sounds in the real world are complex and include physical, material, and spatial elements. Beyond these elements, we must also include **temporal** (time-based) development of sound. In other words, every sound contains information about its **attack, decay, sustain, release**. If you were to break a pint glass your ears would pick up a huge number of clues. The attack of the sound would consist of a quick impact, followed by a sustained shattering sound as shards of broken glass fly about. The release, in this case, would be bits of broken glass settling on the floor and table after the initial sounds have decayed. In order to include all four elements of sound, as well as the relevant temporal information, sound designers use a technique called **layering** for important or complex sound effects. In short, designers stack audio source tracks on top of each other, and process them in ways that result in a complex, cohesive sound. You would be shocked at how many layers are used for your favorite sounds in a game. Some more complicated sounds like weapons or special abilities use up to and above ten layers of sound, all of which need to be present in the mix!

When choosing source to be used as layers you will want to pick sounds that pair well together. As we mentioned above, there may be dozens of layers that all need to add something to the mix, so selecting the right source to use is crucial. Consider the frequency information of the layer so you don't stack sounds that are heavily weighted in the same range. Adequately including highs, mids, and lows in a sound can make it feel more complete and satisfying. If you are missing a range of frequencies in an asset it might sound unfinished or thin. With that said, some sounds call for one particular frequency range over others. Use your ears to determine the best direction. As a rule of thumb, always think of the function of the sound in game when choosing appropriate layers.

> When designing even the simplest sounds like a button click or press I always use layers to help sell the sound. By adding those little details in the sound, it creates a unique effect that feels believable to the player. Pitch shifting and stretching or editing the sound also helps match it to the object's movement.
>
> Gina

Often times sound designers new to the field don't experiment with layers, and often pluck one sound straight from a library and use it as an asset. Chances are that a single sound won't tell the full story. By employing multiple layers we can mold the sound we expect to hear. For example, the velociraptor roar in the film *Jurassic Park* was a combination of animal growls. A recording of a dolphin's high-pitched underwater squeaks blended with a deeply resonant walrus roar serve as the DNA of the roar. Mating tortoises along with some geese calls and chimpanzee noises were used to create the raptor's barks, while its growls were a combination of tiger sounds and Spielberg's dog.[21]

Complimentary vs. Competing Layers

Being mindful of "less is more" is again useful when stacking layers. Getting a bigger sound doesn't always mean more layers. In fact, when you stack too many layers with the transients all triggering at once, the sound might still feel small. This is because the layers aren't *complementing* each other. By stacking similar waveforms right on top of one another the sound gets louder, but less detail will be heard (see "Transient Staging" below).

Another way to avoid competing layers is to avoid utilizing too many sub layers around 50 Hz. It seems logical to go to low-frequency layers to add punch to a sound, but this can be achieved in other ways, which we discussed above. Sub layers (or too many of them) can cause the mix to sound muddy. In other words, the "mids" might sound unclear and overloaded because there is too much going on in the low end. Reducing the mids might hollow out your sound, so it is better to carve out the low end in this kind of scenario.

Important Frequency Ranges for Layering

In general one of the best ways to ensure layers are complementing each other rather than competing is to balance the sound across the entire frequency spectrum (see "Frequency Slotting" later in this chapter). Using layers with a low-end emphasis (55 Hz–250 Hz) can help add density and power to your sound. When working with sounds in this range be sure to reduce the tail and preserve the attack. This will allow the sound to come in fast and add its punch without lingering and taking up valuable headroom. As mentioned above, even if you're looking to add punch, it's good practice to avoid layering too many sounds in this range.

Layering sounds in the 255 Hz–800 Hz range can help glue the low and high end together. Lack of representation in this range can result in a hollow, small sound. By contrast, too much high end 7 kHz–10 kHz can add an undesired brittle element to your sound. This will make the sound effect feel harsh to the listener.

Other Considerations

Always keep the development platform in mind when layering. Too much low or high end won't translate well on mobile speakers. We recommend using an EQ such as Fab-Filter Pro-Q, which has a spectrum analyzer. This will help you visualize areas in your mix that may have too much or too little happening in a given range.

Understanding the target platform and purpose of the sound in game will help define if you need to create a stereo or mono asset. Creating a stereo sound effect with a lot of panning will lose the effect when folded down to mono.

It is equally important to plan how your sound will work in terms of the game genre and design. A battle arena game, for example, will require careful thought when designing weapon or ability sounds. Each of these sounds must be able to stand out among all the others happening simultaneously in game. In some cases these sounds affect gameplay dramatically, so players need to hear the aural cues assigned to things like reloads, or impact sounds. By using EQ (as well as processing techniques mentioned in the following sections) you can carve out an effective sonic niche for each weapon, thus supporting the player and facilitating exciting gameplay.

Lastly, be sure to clean up and process your sounds to give them a unique flavor. This will lower the likelihood of someone picking out a library sound effect in your game. This is exactly how the **Wilhelm Scream** has become a sort of inside joke for film sound editors.

Transient Staging

With multiple layers making up a sound, **transient stacking** needs to be considered. This workflow can be particularly useful with weapon, ability, and impact sounds. If all layers were to be stacked up with the transients in line, there might be numerous sonic artifacts in the resulting sound. Sometimes phase issues can be heard, or unwanted spikes in loudness. In general, the asset will be lacking in detail if not clarity altogether. When layers are directly stacked the sound has no unique identifiers as it's just a wall of sound. Staging the layers strategically into a pre-transient, transient, body, and tail will allow for all the extra details in the sound to stand out therefore creating a unique sound. This is called **transient staging** (see Figure 3.3) and it is a more specific way to convey the temporal pacing we mentioned earlier using attack, sustain, and release. Sounds all have a natural flow to them and this should be maintained.

The order of stacking can be done in a way that best fits your workflow, but be sure to map out the lows, mids, and highs of the sound. The idea is to allow each element to have their time to shine in the mix. Use your ears to listen to how the sound moves from one element to another. It's also important to decide how much of the tail in each layer should ring out. You may end up with single or multiple layers to make up the spectrum ranges (see "Frequency Slotting" below) so in the end it's really whatever fits the sound.

A pre-transient or impulse sound can be added just before the transient. Adjusting the timing of this pre-transient in 5–10 ms increments can really make a huge difference in the sound. Experiment with the space between all layers by adjusting them in millisecond increments until you find the cadence that generates the sound you desire.

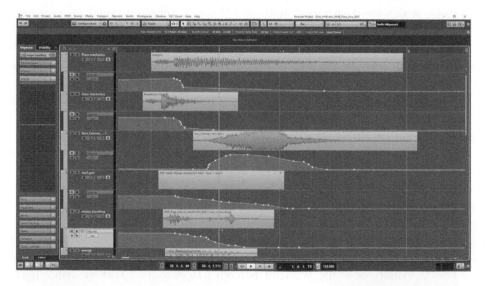

FIGURE 3.3 A screenshot demonstrating an example of transient staging in Steinberg's Nuendo.

After finding the best cadence for your layers, the next step is to manually draw volume automation on each track. The aim is to provide punch at the head of the sound, and then dip the volume afterward to allow room for other layers to peak through the mix. This will help set the cadence of the sound and allow other details space in the mix. How aggressive you get with the volume automation depends on the sound you are designing. If it is a weapon you may automate heavily. If it is a magical ability you may decide to make the automation subtle. This kind of transient control can be done with **side-chaining** as well, but manual automation offers more precision.

Frequency Slotting

Frequency slotting is a critical technique that sound designers use to add definition to the numerous layers used in sound effects. As previously discussed in the "Layering" section, all of the source you use needs to fit tightly together to sound like one cohesive sound. Layers add detail and transient staging allows sound effects to develop over time. Frequency slotting will then ensure that your layers appropriately map across the frequency spectrum so that all elements are heard. It will also keep layers from **masking** one another.

Masking is when one sound clashes with or covers up another. This isn't to say that masking is a bad problem to have as some layers need to blend and overlap to create a denser arrangement. When masking is an issue, it can be resolved with re-arranging layers (transient staging), subtractive and additive EQ, and adjusting dynamics.

A specific type of masking is **frequency masking**, which happens when two or more sounds have similar frequency content causing them to fight for their spot in the frequency spectrum. As we discussed in the "Transient Staging" section, when mapping out transient arrangement, placement choice and lowering the sustain of certain layers can allow other layers to shine through. This isn't the only way to manage frequency masking though. Panning elements so they are not sitting in the same exact space can also be a useful method for allowing each sound its own room in the mix. However, when rendering mono assets for use as three-dimensional events in game, panning will be of little use. This is where frequency slotting using EQ comes in to ensure your layers are filling the parts of the spectrum you might be missing. If all of your sounds have high-frequency content, the sound will feel unbalanced and incomplete.

Sound Design Techniques Summary

 In the Sound Lab we offer a video tutorial on designing sound effects using layering, transient staging, and frequency slotting. This information will help your understanding relate to the information in the rest of the chapter. After viewing the tutorial come back here to explore effects processing as a sound design tool.

EFFECTS PROCESSING AS A SOUND DESIGN TOOL

In this section we will explore the use of plugins as it applies to creative sound design. Working with plugins to polish or shape your sound sometimes means using the plugins in ways beyond what they were intended for. Here we will provide some ideas on how to use plugins creatively and effectively, but it is always up to you to experiment on your own to find the best practices for your particular workflow. There are simply too many plugins to explore each one in depth, but this section will give you a headstart. As you read, keep in mind that having a handful of plugins that you know in and out is much more useful than having hundreds of plugins that you are unfamiliar with. It's difficult to use a plugin correctly and creatively if you don't understand the basics first. It's also important to understand the effective use of processing order or chaining effects. Inserting EQ before compression can have a different outcome over compression before EQ.

There are numerous sources available that offer a general understanding of various plugin effects and their typical uses. In most cases third-party plugin manufacturers will provide detailed manuals of their products. These are usually freely downloadable, and offer in-depth information on the plugins themselves as well as the basic function of plugin types (EQ, compression, etc.). We highly recommend downloading these manuals and watching video tutorials from the manufacturers themselves to familiarize yourself with these tools before you buy them. Make it a point to research and experiment with a variety of plugin manufacturers per effect type because each has its own sound and workflow. For example if we look at limiters as an effects processor, Fabfilter L2 offers more transparency than iZotope Vintage Limiter, which adds thicker tonality and more weight to low-end frequencies.

Here is a brief list of some of the topics we will be covering in the Sound Lab (companion site).

- Creative use of reverb: Instead of using an IR of a concert hall or stage, you can record the IR of a glass vase or a plastic bucket. This will yield distinctive spatial information to use with your source.
- Frequency slotting: A critical technique that sound designers use to add definition to layers and ensure they appropriately map across the frequency spectrum.
- Advanced synthesis: Learning to take advantage of intricate routing and features in synths like Native Instruments Absynth. This will allow for control over multiple parameters to generate far more sonic possibilities.
- Advanced Plugin Effects: Adding more punch to your sounds with transient designers by controlling the contrast between silences and the loudest moment of the sound.

Effects Processing Summary

 Before you move on, head over to the Sound Lab for a discussion and tutorial on outside-the-box effects processing for creative sound design. This information will help you understand and relate to the information in the rest of the chapter. After viewing the tutorial come back here to continue reading about the sound design process.

In summary, so much can be done with the tools that come with a DAW. Those who are starting out in their careers need to practice with the tools on hand before moving on to bigger and better things. Develop a deep understanding for how each plugin works and be in command of your tools. When deciding on a new plugin, do your research to find what works best for you DAW and workflow.

Before purchasing that shiny new thing, think about how this tool will work for you and what problem it will solve. Stick to your budget as more expensive doesn't always equate to better. Lots of practice and experimenting will help you make the best of the tools you have on hand.

PUTTING IT ALL TOGETHER

More often than not sound designers have to work from animations to ensure that the audio is synchronized to the visuals in a game. Now that we have a framework to break sounds down into their four components (physics, materials, space, and narrative) and the tools (layering, transient staging, frequency slotting, and effects processing) to design sounds from scratch, let's take a quick look at breaking down and synchronizing animations.

Breaking Down Animations

Breaking down an animation requires patience and attention to detail. Realistic AAA-quality animations are often full of nuance, which means that it takes focus to absorb every subtlety of the graphics. Start by focusing on an important object within the animation and list every part of it. Then think through how sound can bring each of those parts to life. Start with the more literal, realistic elements (physics, materials, and spatialization) of the object and then move onto the narrative design elements.

It can help to do a few iterations of design over an animation. Watch it enough times that you have a good idea of all of the mechanical details and movement the animation possesses. These are all great points to add character and intricacy to the sound design. In general, the more detail in the animation, the more detail needs to be present in the sound design.

Step-by-Step Design

Now that we have a set of tools for creative sound design, let's look at some common game sound effects and how to design them from the ground up. We will break each sound down into its basic sonic elements from all four of our categories: physics, materials, spatialization, and narrative design. We will use these categories as a starting point to choose source layers. Then we will use layering, transient staging, frequency slotting, and effects processing to transform the source into an asset that is ready for implementation.

Gun Sounds

Before we break down the parts of our animation, let's start by exploring what gunfire really sounds like. If you've ever been to a shooting range you may have a good idea about this. If not, do a YouTube search to familiarize yourself. In reality gunshots sound like a sharp crack, similar to a firework exploding. This will *not* work well in a game. It lacks detail and character, so using a realistic gunshot sound will add virtually nothing to the player experience. Instead we need to design this pistol so it sounds powerful, adding a sense of perceived value to the fire power. Effective weapon sound design will make the player feel rewarded, thus improving gameplay. We will focus on power, punch, and spectrum control.

Let's begin by viewing the gameplay capture of our pistol animation on the Sound Lab, our companion site (feel free to generate your own gameplay capture and work with that instead). With the visual fresh in your mind let's start choosing source layers by breaking down parts of the animation that will potentially make a sound. After creating our list of source material, we will return to the Sound Lab for a practical exercise in layering, transient staging, frequency slotting, and effects processing to design the sound.

Physics

- Sharp transient fire
- Main body of the shot is short, almost non-existent
- Tail (release as the bullet flies away)
- Mechanical kickback of the weapon
- Mechanics of the barrel
- Mechanics of the reload

Materials

- Metal
- Plastic
- Possible sci-fi energy source (the yellow/orange glow)
- Laser shot

Spatialization

- Gun – front and center
- Bullet impacts – spatialized (via implementation, see Chapter 8)
- Use mono source or adjust the stereo width to keep the detail focused

Narrative Design

The first three categories are pretty comprehensive, but there are a few things missing. This is a small weapon, but it needs to sound sleek and dangerous. The animation seems to suggest some kind of assassin character, using high-tech weaponry. This means the shot needs to sound sharp, silenced, and futuristic. The sound of a .357 Magnum would be totally inappropriate here. In short, this pistol should sound powerful but controlled in the low end, and the high end should be sharp, piercing, and include a elements of futurism.

Gun Sound Design Practice

 Now that we have our list of source sounds let's head over to the Sound Lab where we will outline how to design your own pistol sound using the sound design framework we discussed earlier.

Explosion Sounds

Let's begin by viewing the gameplay capture of our explosion animation on the Sound Lab, our companion site (you are free to generate your own gameplay capture to work with instead). With the visual fresh in your mind let's start choosing source layers by breaking down parts of the animation that will potentially make a sound. After creating our list of source material, we will return to the Sound Lab for a practical exercise in layering, transient staging, frequency slotting, and effects processing to design the sound.

In this particular explosion animation a few things are happening. For starters, there are multiple explosions, so we know that these sounds need to be capable of chaining together without phasing or doubling. This means there should be adequate variation (usually pitch and source materials) so that our ears can differentiate between them. Second, there are specific *objects* that are exploding. Our sounds need to include information about the object *as well as* details in the explosion itself.

Let's break it down as we did with the pistol animation to help direct us toward the best source material to use.

Physics

- Pre-transient "anticipation" layer (added a few milliseconds before the transient)
- Transient attack of the explosion
- Body of the explosion
- Tail (in this case, reverberation and release of the sound)
- Various impacts

Materials

- Fire
- Metallic creaks, groans, and movement

Spatialization

- Wide range of space, possible stereo sounds for explosions
- Mono sounds for specific objects, spatialized via implementation

Recording explosions can be dangerous if not controlled. Dedicated sound record-ists like Watson Wu books time with professionals to record various explosions with a variety of microphones and positions. He can be hired by a studio to record unique source for a game or he might put the source into a library and make it avail-able for licensing. Raw recordings will still require editing and processing to ensure they fit well into the game.

Spencer

No two explosions are the same, which means you can get very creative with source when designing these sounds. Designers sometimes call stories of stumbling upon a unique sound or source through the process of experimentation as a "happy accident." Thinking outside the box is a great way to approach explosion sounds. Sometimes this means using source layers that are unexpected just to hear the end result. Other times it means using creative effects processing. Either way, make it a point to experiment here when recreating your own explosions.

Gina

Narrative Design

In an actual explosion it would be hard to hear much detail. To make these sounds more suitable for the gameplay we have to balance the "hugeness" of the explosions with valuable detail that the player should hear. The materials need to be present so the player knows what exactly is exploding, and there should be a "pre-transient" as mentioned in the physics category, so that the player can *anticipate* the explosion, adding drama to the event. Lastly, we will need to ensure that there some seriously punchy impact layers in the sound to add power to the explosions.

Explosion Sound Design Practice

Before moving on, head back to the Sound Lab where we will explore designing an explosion sound (on a limited budget) by sourcing household objects for recording and constructing layers.

Spells and Special Ability Sounds

"What should this sound like?" That is the big question one must answer when tasked with designing sound for something that does not – and cannot – exist in reality. The producer or game designer may not even know what they want a spell or special ability to sound like. But they may be able to offer a reference from a real-world sound or another fictional example. This direction can be quite nebulous however, but the plus side is it leaves a lot more creative freedom in the hands of sound designers.

The style is an important factor in how you choose source material for a spell like this. A fantasy game with special abilities may not work with entirely realistic source. Synthesized source can be really useful, but be sure to keep the synthetic elements in line with the sonic character of the sound effect overall. Processing synthesized source through an organic convolution reverb can help bring a bit of reality into the sound by softening harsh overtones. Make sure to blend real-world source as well to help the listener "relate" to the sound of the otherworldly visual.

Processing organic source through spectral synthesizers like iZotope's Iris 2 can be helpful for creating more magical layers. Iris 2's spectral selection tools in combination with modulation sources (like LFOs or envelopes) will change the sound over time and provide a sense of movement. With magic or fantasy sonic elements this type of movement can be the difference between a sound that is special and compelling, and one that is completely bland.

If the game requires a darker or more action-packed quality, relying on heavier and harsher layers will add some density to the asset. Whooshes, impacts, animal growls, hisses, explosions, and fireworks all provide detailed layers that can add size and depth to a sound effect. Also try using a transient designer or even a distortion unity (like iZotope Trash 2) to add some "umph" to the spell as it meets its target. Trash 2 has an impulse named "Creep," which can be used as a convolution setting. Try it with a mix around 10–20 percent wet and it will likely increase the power behind of your source layer.

For lighter, less intense spell casts try sparklers, sizzles, broken glass, chimes, wind gusts, and lighter whooshes. These elements can add a mystical character, which is often useful for buffs and less damage-oriented abilities. A convolution reverb can also help soften the sound along with a 10 kHz + rolloff.

Let's get started by viewing the gameplay capture of our spell casting animation on the Sound Lab, our **companion site** (as always, you are free to capture your own gameplay video to work that instead). With the visual fresh in your mind let's start choosing source layers by breaking down parts of the animation that will potentially make a sound. After creating our list of source material, we will return to the Sound Lab for a practical exercise in layering, transient staging, frequency slotting, and effects processing to design the sound.

Physics

- Pre-transient layer (as spell is being cast or charging)
- Transient (the initial sound as the spell is cast)
- Body of the spell (sustain)
- Release (pay close attention to the VFX as the spell dissipates as this can change with the animation)

Materials

- Blue energy

Spatialization

- Begins right in front (possible mono)
- Quickly spreads out and around the listener

Narrative Design

This category is absolutely crucial for spells and magical abilities because it is often the only real point of reference we have as designers. The important thing to consider is *what the spell does*. In other words, what is the function of the spell? Is it a heal and attack, or a buff? Attacks will almost always sound more threatening and harsher, and buffs need to sound pleasing in order to provide accurate aural feedback for the player. In this case, this blue energy looks to be a shield, so we will build our sound to be a powerful and ethereal energy that protects our player character from harm.

Spells and Special Ability Sound Design Practice

 Now that we have our list of source sounds let's head over to the Sound Lab for some practice designing fantasy spells and special ability sound effects using the framework we outlined earlier.

Creature Sounds

As with all other categories of sound design, the designer will need to answer some preliminary questions before jumping in. An important example is whether or not the creature you are designing speaks or has some language elements. If the answer is yes, then the dialogue will need to be intelligible, and processing cannot hinder this. In these cases, less is more. You may pitch dialogue up or down, but take care when adding heavy modulatory effects.

When going through your library to select source for creature sounds be sure to listen for anything that might stand out. Don't overlook samples of the human voice. Voice artists can be brilliant with manipulating their voices to sound inhuman and beastly. Generally speaking, animal sounds from lions, pigs, bears, and tigers have become standard source material for many of the creature designs out there. They make for great base layers but you will need other elements to give your creature a unique sound.

Don't forget to add in small details to your creature sounds. These are the elements that make your creature sound unique. Try including lip smacks, snarling, and snorting breaths. A wet mouth filled with gooey substances like pudding, jello, or oatmeal can be a valuable source for vocalizations. Tasers and gears also make great high–mid frequency layers that blend well with inhales, and they can also be placed at the tail end of roars. It's also important to cover the non-vocal elements of creatures. How do they move? How big are they? Details like this will go a long way toward creating a memorable creature sound as well.

Another useful technique is to chop up the "expressive" parts of an animal sound and use it to humanize the creature. If you are working with a voice artist you can also direct them to perform some emotive expressions without using language. This can give the player an affinity toward the creature, helping to cultivate an emotional bond. If you aren't working with a voice artist, try automating pitch to manufacture this sense of emotionality.

Even if you don't plan on using your own voice in the final product you may want to use it to lay out a guide track. This can help you hit the mark on the emotive side of the design, or to convey ideas to other performers. Software like Krotos Dehumanizer works well with human voice and is especially helpful when you are on a tight deadline. Dehumanizer is also perfect for breaths and hisses as well as roars and growls. Keep in mind that even when using Dehumanizer to perform layers, there is always other source blended into the final sound.

> Breath is an important (and often overlooked) part of creature design because it conveys a clear sense of size, even if the creature is off-screen. These vocalization techniques are much easier when you can perform the grunts or growls yourself, as opposed to searching for and chopping up source to make it work. Even if you don't have Dehumanizer, try recording some yourself and processing them to see what you can come up with. You'd be surprised how well it works!
>
> Spencer

Start by viewing the gameplay capture of our creature animation on the Sound Lab, our companion site (you are free to capture your own gameplay video to work with instead). With the visual fresh in your mind let's start choosing source layers by breaking down parts of the animation that will potentially make a sound. After creating our list of source material, we will return to the Sound Lab for a practical exercise in layering, transient staging, frequency slotting, and effects processing to design the sound.

Physics

- Vocalizations
- Giant body
- Multiple long sharp legs scraping
- Heavy impacts
- Body breaking apart

Materials

- Crab-shell-like flesh (bulky)
- Tree-bark type material for the legs
- Ice and snow

Spatialization

- Huge creature, so these sounds pretty much cover the stereo space

Narrative Design

This creature animation is somewhat open to interpretation as it does not have any place in the real world. You can try (as we did) comparing its material to animals in the real world, but the bottom line here is that this creature needs to sound threatening and huge. All of your layers should be chosen to fully realize that direction. The size can be handled nicely with some LFE impacts and sub-frequency hyper-realism. Also try creating some chilling shrieks!

Creature Sound Design Practice

Now that we have our list of source sounds let's head over to the Sound Lab to get some practice designing creature sounds using the framework we outlined earlier.

Vehicle Sounds

As we mentioned earlier in the chapter, when working with realistic vehicles it's a good idea to do some research first. Car engines in particular can be divisive because of their memorable sonic character. Car enthusiasts will easily be able to recognize a Corvette engine sound from a Mustang, so you should too. Start the process by getting into the specifics of vehicle type with your producer or designer and familiarize yourself with the sound of the engines.

It helps if you have contacts with the cars you are looking to record. If you don't know anyone with the specific car you are after you can try auto forums on the internet or auto shops to see if you can make any solid connections in this regard. Keep in mind that games have deadlines, so you may need to work quickly. The project you committed to won't allow for all the time in the word to locate and record vehicle sounds. If you run short on time you can reach for some library SFX. When the budget allows you can look into hiring field recordists like Watson Wu or Pole Position Production for custom sessions.

For this particular gameplay capture we will focus on functionality rather than direct realism to engine samples. Head over to the Sound Lab (companion site) to view the racing gameplay capture video. After you have watched the video come back here and we will break the vehicle down to help us find some useful source material.

Physics

- Engine idle
- Engine RPM (revolutions per minute)
- Engine load (pressure on the engine)
- Gear shifts
- Exhaust
- Skids
- Collisions
- Suspension rattle

Materials

- Vehicles (metal, rubber tires, glass)
- Obstacles and barriers (metal, wood, water)
- Surfaces (asphalt, dirt, etc.)

Spatialization

- Perspective (interior and exterior views)
- Doppler
- Mono sounds for specific objects, spatialized via implementation

Narrative Design

Racing games can be broken down into arcade and simulation categories. In simulation games the player may have specific expectations of the vehicle sound while arcade games may offer the sound designer a little more creative liberty. The scene suggests a Formula One open cockpit vehicle with asphalt and grass terrains. The barrier around the racetrack looks to be concrete. A bit of internet research will show that the engine has a low-end growl on startup and revving. The engine of a standard sedan will not provide a credible sonic experience for players.

Visiting Artist: Watson Wu, Composer, Sound Designer, Field Recordist

On Recording Vehicles

There are various areas on a car that produce unique sounds. My role has been to secure microphones to those areas in order to capture onboard sounds (sounds heard from driver and passenger views, versus external pass-bys). These sounds are always thought of from the driver's point of view: engine left, engine right, inside the air intake box, cab left, cab right, and exhausts. All of the mics are covered by windjammer etc. wind-suppressing items and firmly secured by cloth-based gaffer tape. The mic cables are then taped on and routed to my where I sit, next to the driver. On my lap is an audio bag housing a multitrack field recorder that allows me to selectively listen to each of the mics while recording. I go back and forth listening to each of the inputs and adjust the recording levels throughout the session for the optimal sounds.

During the initial tests I usually have the driver perform high revs then we go on to recording ramps. A ramp can be done by putting the car into first gear and drive from idle smoothly up to the red line of the RPM. This ramping up to red line can be done between 10 and 15 seconds long. After reaching the red line, the driver will engine brake (letting go of the gas pedal while in first gear) back to a complete stop which is also between 10 and 15 seconds long. Some cars can easily perform this maneuver while others cannot. I stress to the driver that recording ramps is the most important part of the recording, which makes them more eager to do their best.

While recording a Ferrari 488 supercar, the ramping downs weren't smooth at all. The owner, being an engineer, thought about performing the ramps using third gear. This worked out so we were able to capture smooth ramp ups as well as smooth ramp downs. After ramps we do high-speed or aggressive drivings. The driver is to delay shifting so that we can capture high RPMs (usually where the loudness of the vehicle shines). Once I have the correct overall adjustments (staying below the loud recording peaks for each of the inputs), I make best use of time by exiting the vehicle to grab another set of gear to capture external sounds. The onboard recorder in the audio bag is seat-belted to the passenger seat so that I can do both, record onboard and external sounds at the same time. Other items on the shot list will include driving casual, driving

(Continued)

reverse, pass-bys in various speeds, approach-stop-away, startup, shutdowns revs. Other options are burnouts, skids, cornering, drifting, etc. stunts.

What to be mindful of:

- Avoid external recordings when wind exceeds 15–20 mph depending on your microphone and windshield kit.
- Always watch the oil temperature. If a vehicle overheats, use regular gear and do light drivings at 45 mph for a few minutes. Then, shut off the vehicle and raise the engine hood for better cool down. Often times, this method cools the engine better than shutting off the vehicle when the temperature is way up.
- Always record Foley sounds first (heated cars make bing/ding etc. unwanted sounds), onboard sounds second, then external sounds.
- Always record on smooth roads unless you need off-road sounds. Avoid driving over road reflectors, don't use turn signals.
- Have long enough roads for safe braking.
- I typically start a recording with a startup and end with a shutdown. This way there are plenty of variations of those sounds.

Watch my "Watson Wu Jaguar" YouTube video for some examples of how I record vehicle sounds.

Vehicle Sound Design Practice

 Let's take a look at vehicle sound design. Head back to the Sound Lab for some practice designing vehicle sounds for the racing gameplay capture using the framework we outlined earlier.

As a side note, real-world vehicle sound design is growing into another audio career option. Vehicle manufacturers have been tweaking the sounds of exhausts for years but with the recent rise of electric vehicles, manufacturers are hiring audio designers to create the engine hum of these rather quiet mechanics. It comes down to sonic branding and marketing to find the best sound to match the demographics the make and model would target. Currently, it seems that consumers are still very attached to the engine sounds we are all so very used to but as time goes on and we become less attached the sound possibilities can be cracked wide open. I can see a vehicle that might allow several options of sound choices. Having a bad day? Set your car to purr like a cat. Or feeling empowered? Have it blast off like a jet fighter. I could even imagine an adaptive system that changes the sound of the vehicle based on game data like season, time of day, or even age.

Gina

UI and HUD Sounds

The term "auditory icon"[22] was coined by Bill Gaver in the early 1980s when he researched the use of sound for Apple's file management application "Finder." As we march into the future the various devices we choose to embrace require support for the visual information as well as a tangible conformation or feedback to the user. This is what it means to design sounds for UI, or user interfaces.

Designing sound for UI and heads up displays is often thought of as being the easiest part of sound design. How difficult can it be to produce and implement a few simple blips, bloops, and bleeps? In reality, effective UI sound design is extremely difficult because it is a critical element of the sonic branding. This means that every sound that a UI or HUD makes needs to give the player a sense of mood and satisfaction. In other words, as sound designers we have to squeeze every ounce of emotion out of each and every blip, bloop, and bleep in a game. How simple do they sound now?

The purpose of the user interface is to provide data feedback (health or option settings) and to facilitate interaction between the player and the game. UI/UX designers take the time to carefully create user interface visuals that won't get in the way of the player's visual field. Too little or too much feedback to the player can ruin an immersive experience. Since so much thought and effort is spent developing the interface's visuals, comparable amounts should be spent on the audio portion.

The game style and genre will play a large role in deciding how the UI sounds will be crafted. Generally speaking, UI sounds need to be on-theme. A sci-fi game may require electronic, or glitchy, granularized textures to accompany the hologram screens and futuristic visuals. Using organic layers like wood and grass will sound out of place. The same can be said about a casual Match 3 game that has a bright and cheery forest look. Implementing synthesized UI sounds might feel weird, thus breaking the immersion.

Earlier in the chapter in the section "Designing the Sonic Palette" we discussed how the Blizzard sound team handled the UI sounds in *Hearthstone*. The sounds were crafted so the player would feel totally immersed in the tavern scene. The UI visuals in game provided a sense of value and craftsmanship which is relatable to players. In a 2015 talk at GDC,[23] *Hearthstone* senior UI designer Derek Sakamoto discussed the approach to the visual design of the UI. He talks about making objects look valuable using materials like gold and gems that can make the player feel really good about the time they spend playing. The card pack opening sequences are crafted after the actual experience of collecting physical cards and opening them in real life. The sound needed to amplify that experience.

Gina

Often, UI sound is thought of as being strictly button presses and slider movement. In fact, UI elements can be found in game as well as in menus. These in-game UI sounds can provide feedback to the player without them needing to actually look at the visuals. This is an extremely important concept, and it is why sound plays such a large role in the UI experience overall. For this reason UI sounds should be something the player can identify with without being distracting or annoying if triggered frequently. In an endless runner the player must focus on what is directly in front of her. As she collects coins, the sound effects that trigger will give her an accurate sense of success or failure without even seeing the actual number on the display. Sounds like this often have a pitched element to them. This produces a satisfying, almost musical effect that easily cuts through the mix.

UI sounds in **MOBA** or team-based multiplayer games like Blizzard's *Overwatch* or Hi-Rez's *Paladins* are designed with the players' focus in mind. These team-based games can get very overwhelming visually. Effective UI sounds allow the player to keep their visual focus on the gameplay while relying on aural cues to know when important events are happening. There is so much information relayed sonically to the player in *Overwatch* that technically it could be played without visuals (to a point anyway!). This is actually the basis of *accessibility in sound* (see "Inclusivity and Accessibility in Game Audio," Chapter 12, page 382).

Now that we are familiar with UI sounds and the role they play in games, start by viewing the gameplay capture of our UI animation on the Sound Lab, our **companion site** (as always, you are free to capture your own gameplay video to work that instead). With the visual fresh in your mind let's start choosing source layers by breaking down parts of the animation that will potentially make a sound. After creating our list of source material, we will return to the Sound Lab for a practical exercise in layering, transient staging, frequency slotting, and effects processing to design the sound.

It's important to note that while UI sound design may be thought of as almost complete narrative design we can still use physics and materials that relate to real-world objects to gather information for our sounds. When graphical user interfaces mimic the physical world users can more easily understand how they function.[24] In game, animations show how the object moves in the two- or three-dimensional space.

Physics

- User interaction (slider, button press, highlight)
- Animation

Materials

- These can stem from the visual of the button or visual themes from the game
- Metallic creaks, groans, and movement

Spatialization

- Most UI sounds both in the menu and in game will be in 2D space

Narrative Design

The gameplay in this video seems to be casual and the UI is somewhat abstract and open to interpretation as it does not stem from specific objects in the real world. Since the overall theme of the game involves cooking you can introduce recordings of kitchen items as source material for use as layers in the design. The UI sound in this game should be subtle yet satisfying. An instant upgrade should sound more rewarding than a timed upgrade.

UI and HUD Sound Design Practice

 Let's take a look at UI sound effects design. Head to the Sound Lab for some practice designing UI sounds for the gameplay capture using the framework we outlined earlier.

Footstep Sounds

Have you ever taken the time to focus solely on the footsteps in a game? You might be surprised by the level of detail that goes into the sound design of footsteps. Some people would consider footstep sounds to be non-essential to gameplay, but in reality footstep sounds play an important role in games. Footsteps provide information on terrain type and NPC location, which can make the difference between success or failure in a given task. In multiplayer matches players rely on this information to understand their surroundings and assess danger.

Studying the footsteps of humans will reveal a heel-to-toe movement as a step forward is taken. Although mobile games require fewer assets for implementation, console and PC games with any level of detail will typically have eight to ten randomized footsteps per terrain. Volume and pitch randomization will also be applied adding further variety to the sound. More detailed integration might utilize separate heel and toe asset groups that trigger in synchronization to the footstep animation. This, of course, offers even more variety for the player's ear.

To add an additional level of detail to the character's movement, clothing (fabric, armor, etc.) sounds can be added as a separate trigger. The layers in these sounds should reflect the materials that the character is wearing in as much detail as possible. However, these accents shouldn't get in the way of the clarity of the footsteps. There should also be enough variety to these accents that the player isn't hearing the movement sound over and over.

For one final example, head over to the Sound Lab (companion site) to view a footstep animation (just as before, you are free to capture your own gameplay video to work with instead). With the visual fresh in your mind let's start choosing source layers by breaking down parts of the animation that will potentially make a sound. After creating our list of source material, we will return to the Sound Lab for a practical exercise in layering, transient staging, frequency slotting, and effects processing to design the sound.

Physics

- Heel-to-toe movement
- Footfall velocity
- Walk/run
- Terrain (creaking bridge)

Materials

- Shoe type
- Terrain type
- Outfit type

Spatialization

- First-person perspective – stereo sounds but avoiding a wide stereo field
- Third-person perspective – mono sounds spatialized via implementation

Narrative Design

The character in the video appears to be tall and muscular. From this we can add a bit more weight to his footsteps. We see that he is wearing leather boots as well as leather material around his waist that moves each time he steps his foot outward. Hanging from a few strands of hair is a large circular metal accent. It should be determined if this will make any sound as the character moves. The terrain starts out as stone steps and goes into a wooden bridge. The physical shape of the bridge should be taken into account.

Footstep Sound Design Practice

Let's take a look at footstep sound design. Head to the Sound Lab for some practice designing footstep sounds for the gameplay capture using the framework we outlined earlier.

The Sound Lab

Before moving onto Chapter 4, head over to the Sound Lab for additional reading, practical exercises, and tutorials on the topics discussed in Chapter 3.

SUMMARY

Here we presented a few examples for sound design practice. Using the framework we laid out in this chapter you should continue to practice creating sound design for a variety of game assets.

NOTES

1 Twenty Thousand Hertz, "The Xbox Startup Sound."
2 Michael Sweet, composer, sound designer and artistic director of Video Game Scoring at Berklee College of Music; author of *Writing Interactive Music for Video Games*.
3 B. Kane, *Sound Unseen: Acousmatic Sound in Theory and Practice*.
4 Wikipedia, "KISS principle."
5 When time allows, gathering and generating source could be started in the pre-production phase while editing and mastering will be completed during the production phase.
6 www.syntorial.com/
7 G. Reid, "The Physics of Percussion."
8 www.plogue.com/products/chipsounds.html
9 Note: Wasting food isn't good for the environment or the people living in it. It is a good idea to take some measures to reuse food (or as much of it as you can). Grocery stores will often have

a huge amount of expired produce that you can claim by simply asking a manager. Of course, you also want to be careful not to contaminate the food during the session if you plan on eating it or serving it at a family dinner later on. Use good judgment in either case.

10 www.gamasutra.com/view/feature/179039/the_sound_design_of_journey.php
11 R. Viers, *Sound Effects Bible.*
12 A. Farnell, *Designing Sound.*
13 N. Collins, *Handmade Electronic Music.*
14 www.gameinformer.com/b/features/archive/2011/02/28/war-tapes-the-sounds-of-battlefield-3.aspx
15 Always take proper precautions and go through the proper channels when recording weapons and explosives. There are many professional field recordists who can handle proper setup of these types of sessions.
16 Wikipedia, "Loudness."
17 http://hydrophones.blogspot.com/2009/04/hydrophones-by-jrf.html
18 Turtle Beach Blog, "The Audio of Hellblade: Senua's Sacrifice."
19 How Stuff Works, "What is a Decibel, and How Is it Measured?"
20 D. Solberg, "The Mad Science behind *Inside*'s Soundtrack."
21 D. Shay, *The Making of Jurassic Park.*
22 R. Gould, "Auditory Icons."
23 Sakamoto, D., GDC, "Hearthstone: How to Create an Immersive User Interface."
24 D. Mortensen, "What Science Can Teach You about Designing Great Graphical User Interface Animations."

BIBLIOGRAPHY

Collins, N. (2006). *Handmade Electronic Music: The Art of Hardware Hacking*. New York: Routledge.

Farnell, A. (2010). *Designing Sound*. Cambridge, MA: MIT Press.

Gould, R. (November 24, 2016). "Auditory Icons." Retrieved from http://designingsound.org/2016/11/24/auditory-icons/

Johnson, S. (October 10, 2012). "The Sound Design of Journey." Retrieved from www.gamasutra.com/view/feature/179039/the_sound_design_of_journey.php

Jurassic Park. "Jurassic Park Raptor Effects." Retrieved from http://jurassicpark.wikia.com/wiki/Jurassic_Park_Raptor_Effects

Hanson, B. (February 28, 2011). "The Xbox Startup Sound." Retrieved from www.gameinformer.com/b/features/archive/2011/02/28/war-tapes-the-sounds-of-battlefield-3.aspx

How Stuff Works. "What is a Decibel, and How Is it Measured?" Retrieved from https://science.howstuffworks.com/question124.htm

Kane, B. (2014). *Sound Unseen: Acousmatic Sound in Theory and Practice*. New York: Oxford University Press

Mastering the Mix. (August 16, 2016). "Mixing and Mastering Using LUFS." Retrieved from www.masteringthemix.com/blogs/learn/mixing-and-mastering-using-lufs

Mortensen, D. (2019). "Auditory Icons." Retrieved from www.interaction-design.org/literature/article/what-science-can-teach-you-about-designing-great-graphical-user-interface-animations

Mortensen, D. (n.d.). "What Science Can Teach You about Designing Great Graphical User Interface Animations." Retrieved from www.interaction-design.org/literature/article/what-science-can-teach-you-about-designing-great-graphical-user-interface-animations

Phon2 (August 16, 2016). "The Production of Speech Sounds." Retrieved from www.personal. rdg.ac.uk/~llsroach/phon2/artic-basics.htm

Reid, G. (June 1999). "The Physics of Percussion." Retrieved from www.soundonsound.com/ techniques/physics-percussion

Rodrigues Singer, P. "The Art of Jack Foley." Retrieved from www.marblehead.net/foley/jack.html

Sakamoto, D., GDC. (June 15, 2015). "Hearthstone: How to Create an Immersive User Interface." Retrieved from www.youtube.com/watch?v=axkPXCNjOh8

Shay, D. (1993). *The Making of Jurassic Park*. New York: Ballantine Books.

Solberg, D. (August 23, 2016). "The Mad Science behind *Inside*'s Soundtrack." Retrieved from https://killscreen.com/articles/mad-science-behind-insides-soundtrack/

Stripek, J. (October 14, 2012). "Sound Sweetening in Backdraft."Retrieved from https://cinema shock.org/2012/10/14/sound-sweetening-in-backdraft/

Sweet, M. (2015). *Writing Interactive Music for Video Games: A Composer's Guide*. Upper Saddle River, NJ: Pearson Education.

Turtle Beach Blog. (August 21, 2017). "The Audio of Hellblade: Senua's Sacrifice." Retrieved from https://blog.turtlebeach.com/the-audio-of-hellblade-senuas-sacrifice/

Twenty Thousand Hertz. "The Xbox Startup Sound." Retrieved from www.20k.org/episodes/ xboxstartupsound

Viers, R. (2011). *Sound Effects Bible: How to Create and Record Hollywood Style Sound Effects*. Studio City, CA: Michael Wiese Productions.

Wikipedia. "KISS principle." Retrieved from https://en.wikipedia.org/wiki/KISS_principle

Wikipedia. "Loudness." Retrieved from https://en.wikipedia.org/wiki/Loudness

Voice Production

In this chapter we will cover voice production for games by exploring the process of casting, recording, editing, and preparing assets for implementation.

DIALOGUE VS. VOICE-OVER

In the film world dialogue is typically recorded on location and sometimes re-recorded in a sound booth or on a soundstage in a process called ADR (Automated Dialogue Replacement). ADR is the process of syncing a voice-over performance to a pre-rendered video or animation (much like Foley syncs to picture). For games, however, the dialogue often does not need visible sync to a character. In the cases where sync is necessary, the dialogue is usually recorded first, leaving animators to animate mouth movements to sync with recorded speech.

Sometimes the term voice-over is used in film to describe narration or speech from offscreen characters. These voice-overs are recorded in a sound booth, motion capture studio, or soundstage. In the world of animation since there isn't typically a physical set, the process involves putting a voice-over to the animated characters.

Recorded speech in games tend to fall in line with the voice-over process that stems from the animation world, but dialogue captured by the voice talent on a mocap stage or sound stage could be referred to as dialogue. The phrase dialogue is also used to refer to a gameplay mechanic, often seen in role-playing games (RPG) or adventure games, which uses conversation trees to present the player with a text-based interaction between characters. In summary, you may hear recorded speech in games referred to as both dialogue and voice-over. Here we will use the term dialogue on a macro level and voice-over on a more micro level as it pertains to recorded speech.

Voice-over Director

Voice-over is just as important to games as sound design and music. While pre-composed music and sound effects can be licensed for use in game, voice-overs are almost never "off the shelf" assets. All speech in game require custom assets and a unique voice to provide information to the player, and to further the narrative. As game platforms have evolved and storage has increased, voice-over production has become an essential element of design. It has progressed to become more natural, reactive, and involved. On top of being a major tool to drive the story, dialogue is often one of the most entertaining aspects of the gaming experience.

All this said, budding audio designers should be ready to handle the voice-over needs of a project. Depending on the size of the development team, an audio designer might be tasked with handling all or part of voice production. On smaller teams the sound designer might be responsible for everything from casting to mastering. Larger teams might break the tasks up between sound designers and the audio director, or the dialogue could be outsourced. For our purposes in explaining the process in this section let's assume one person is handling the full voice production process as an acting voice director.

Voice-Over Examples in Games

While some games don't require any spoken word, there are plenty of games such as RPGs and visual novels that rely heavily on dialogue to carry conversations between characters and provide narrative. With the adoption of voice-enabled devices like Amazon's Alexa we have seen the rise of interactive audio adventures like *Jurassic World: Revealed*, *Baker Street Experience*, and *Wayne Investigation*. Quite a bit of dialogue goes into each of these audio experiences.

Visiting Artist: Bonnie Bogovich, Audio Designer, Vocal Artist – BlackCatBonifide

Thoughts on Voice-Enabled Devices

The Amazon Echo helped open up a new world of possibilities for audio designers and voice actors who had previous experience in the world of radio plays and audio drama. When working on traditional video games, there are so many elements to balance in the neverending tug-o-war for space, graphics, processing … with devices like the Echo and Google Home, finally we had a venue where there was no choice but to put audio first. Now, even though we are talking about

(Continued)

a voice-driven device that functions with audio only, the first "skills," a term for Amazon-created applications for Alexa, did not utilize many audio files. You ran everything through the robotic voice of "Alexa" herself, the voice that guides you through the system. It did not take long to realize that listening to the same robotic voice forever had its limitations.

Choose-Your-Own-Adventure™-style games work well with the Alexa Skills system, but without a voice that is emotive, the ears get bored rather quickly. Once designers started taking advantage of the ability to upload audio files to their Skills, which could include any audio files up to 90 seconds long, the possibilities of recording human narrators, and to mix in sound design and music, the quality and entertainment value of upcoming releases shot upwards!

In open world games like Avalanche Studios' *Just Cause*, voice-overs can be heard throughout the game. They often trigger during cinematic scenes or in gameplay during conversation between the player characters and non-player characters (NPCs). Many games use voice-overs in the form of **barks**, short one-lined bits of dialogue spoken by NPCs, to bring the setting to life. An example of their use is in Ubisoft's *Assassin's Creed III* where the player's character (Ratonhnhaké:ton) can hear NPCs conversing as he walks through a marketplace. This adds another level of realism for the player and can can be used to offer feedback or information. At certain points an NPC may offer the direction "He's down there" in the form of a bark.

The audio team and programmers at Insomniac Games also went the extra mile with dialogue in Marvel's *Spider-Man*. Each line of dialogue for the player character was recorded and implemented with two voice-over takes. The engine was programmed to detect the state of the player character and play a voice clip accordingly. An idle state triggers a more relaxed and calm set of voice-overs, while a combat or "swinging" state will trigger a tenser version of the voice-over to give the effect of physical exertion.

While there are many innovative ways voice-overs can be used in games, it is not always possible to use them on smaller projects. In truth, it depends on a few factors in the development process. The choice to include voice-overs adds the need for scriptwriters, voice talent, recording sessions, dialogue editors, a voice director, mastering, and extra implementation. Indie budgets don't always allow for these expenses. The voice production budget should be decided early on in the development cycle, even though casting and recording will happen closer to the end of the production.

Scripts

The process of getting recorded voice into a game starts with a completed script. Writing for games can be likened to solving a complex mathematical puzzle. If the game has any extended dialogue systems (or even just a few barks) a script will be necessary. The script must be crafted, revised, approved, and organized before the recording can commence.

An audio designer typically isn't involved in the script-writing process, but if the game development team is smaller the game designer or narrative designer might ask for external opinions on the script.

Visiting Artist: Michael Csurics, Director – The Brightskull Entertainment Group

Thoughts on Scripts

There is only one universal truth when discussing scripts in voice and performance production for games – there is no one right way, yet there are many, many wrong ones.

Our company works with development teams from every facet of the industry and as such has worked with a wildly diverse gamut of script formats. In *Just Cause 4*, as in most AAA games, there was a mixture of linear narrative storytelling, diegetic audio, and systemic barks. *Bioshock 2* was the same, but all conversations were one-sided (i.e. silent player character). *Masquerada: Songs and Shadows* is entirely linear, recorded ensemble, and unfolds more like a play. *Tacoma* was a vignette-driven diverging/converging/overlapping linear narrative recorded ensemble.

Every game puts unique demands on its script, but there are some commonalities and grave mistakes to be made across all projects.

Here are my cardinal sins of game scripts:

1. **Not having a script**. You laugh, but this happens, and it happens often enough. Before entering the recording phase, you need to have all of the material that is to be recorded written and accessible in a single location: the master script.
2. **Not using a spreadsheet**. There is no excuse for this. Every single line of dialogue is an asset and that asset needs to be tracked, sortable, and able

(Continued)

....

to have flexible fields for metadata. Even if your game has only one line of spoken dialogue, put it in a spreadsheet.

3. **Not preparing theatrically formatted scripts**. You wouldn't tell a carpenter to use a hammer on a screw. Don't give an actor a spreadsheet as a script. It's easy enough to generate theatrically formatted scripts from a spreadsheet using Word's built in mail merge and styles.

4. **Not including action, setting, or leading lines**. If your script is just a collection of words for actors to say out loud, that's exactly what it will sound like. Like walking into a restaurant and ordering "food" or telling your tattoo artist to just do "whatever," without providing context you are relying on the director and the actor to just "figure it out" as they go.

5. **Not identifying projection levels**. Without a doubt the #1 question asked during a session will be "How far away?" If you include projection levels in your master script, you can (and should) be able to sort systemic sessions in projection order and flag any stressful voice work during narrative records to be done at the end of the session.

6. **Not including page numbers**. Put page numbers in every theatrically formatted script you make. Just do it.

7. **Not getting the scripts to the talent ahead of the session**. More of a production issue than a script issue, but yes, do this.

Casting Calls

When a rough script and character descriptions are available from the developer, it's a good time to start the casting process. The casting process depends heavily on the budget for voice production and the amount of time available.

The first step is organizing the casting documents. These should include character art, voice references, and a sample of lines from the script to provide for auditions. Check with the game developer as **non-disclosure agreements** may be necessary before sharing any confidential content with prospective voice artists (see "Navigating Contracts," Chapter 11, page 378). Be mindful of what information is deemed confidential when sending out the casting calls. Some games go under a code name or working title as developers often like to remain anonymous at the beginning of development.

Visiting Artist: D. B. Cooper, Voice Artist

Thoughts on Casting

The first rule of casting is DO IT. Do some casting. This gives you a chance to discover talent you might otherwise not have known who can give you insights into your game script you might otherwise not have considered.

When you create your casting script be sure to include all the aspects of vocalization you are going to need. You might find a great dramatic or comedic actor but that doesn't always translate into someone who can fight, take bullets, howl in agony, and die. It breaks the player's immersion in your game when a strong character is wielding a battle axe but making a sound like he's swinging a tennis racket.

Finding Voice Talent

Voice artists can be be found in a variety of ways. Let's start with low budget indie projects first. There are quite a few forums and social media groups that allow voice talent to post portfolios as well as job ads from prospective clients. Voiceacting. boards.net is an example of this. This site doesn't charge to post auditions, so you may find a lot of beginner talent. But there are also a few seasoned professionals who use the site. Great talent can be found at all stages of a career, so it's a good idea to be open-minded. Remember that you may find that novice voice artists have a lower range of equipment and technical recording knowledge due to lack of experience.

There are quite a few pay-to-play websites like voices123.com that charge the voice artists a fee to post their portfolios and audition for work. As a job poster there is usually no charge but there is a minimum dollar amount for hired talent. The minimums can range from $100 to $250 USD depending on the site. Since the voice talent on these sites are required to pay to audition, you may find more talent that are further along in their career. Again, it's good to keep an open mind as great talent can be found almost anywhere.

If your budget allows for a rate of $250–$1,200 per session, there are voice talent agencies that can offer quality talent and even handle portions of the process for you. Independent voice artists can also be found via their websites with a well-rounded keyword internet search. Many of these artists might refer you to their talent agency, but a good number of them are willing to negotiate their own deals. The Voice Casting Hub[1] is another site where agents and talent can be found by listening to demo reels or posting job ads and receiving auditions. The options we've outlined allow for a far reach of talent with a choice of various skills, accents, and languages.

Union vs. Non-Union

It's good to understand the categories of labor that voice talent falls under. In the United States, talent can be categorized as non-union and union, with the former being protected by a union such as SAG-AFTRA. Union jobs have a predetermined rate, but it's common for professionals to charge more based on variables such as demand, session duration, script length, and number of voices. Alternatively, voice talent can choose to be a part of Fi-Core, which allows them to work both union and non-union as they choose.

Visiting Artist: Rachel Strum, Executive Producer – The Brightskull Entertainment Group

Thoughts on Union vs. Non-Union

The question of whether to use union or non-union actors is an important one. The answer is typically very specific to the project's needs and goals. There is often an incorrect assumption that using union actors will greatly increase the cost of talent for your project. This is not always the case and it's important to do a side-by-side comparison to truly understand the financial implications of your decision. It is also important to consider the types of performances your project requires. As a rule, union actors tend to have more experience than non-union actors. This experience can save you time in the booth. Going union will also typically provide you with a more experienced diverse talent pool to choose from. This can be particularly useful when casting children and characters who are over 60 years of age, as well as when you require specific accurate accents and dialects. There are exceptions, of course, and Fi-Core actors (who can perform in both union and non-union productions) are helping to provide more depth to the non-union field.

With that said, we have had much success both with union and non-union casting and encourage developers to consider the benefits that both options offer to make their decision.

Thoughts on Signatory

If you are choosing to use SAG-AFTRA union actors for your game, you need to either become a union signatory or partner with one. A union signatory has signed an agreement with the union that states they will make sure all guidelines and rules detailed in the current project are being followed for that production. The signatory is also responsible for ensuring that all applicable documents such as contracts and timesheets get completed and signed by

(Continued)

actors along the way, and submits the appropriate materials to the union. They are also responsible for the payroll process. This includes payments to actors and agents as well as making sure the appropriate amounts get contributed into the union health and pension plan. While it may sound like a bunch to manage, it really is a fluid process.

Budget Considerations

With all of this information in mind, the developer will need to decide how much budget is available for the hiring of voice talent. The recording session may accrue additional fees for the recording engineer and studio space. As the budget is discussed, resources such as the Global Voice Acting Academy[2] and the SAG_AFTRA[3] can be used to ensure you are paying industry rates.

Drafting the Ad

Now that the budget has been determined, the ad for the audition must be drafted. The voice artist will need detailed information to assess if the role is the right fit. By adding adequate detail to your post, the better fitting your prospective voice artists will be. Here are the details that should be considered.

- Union or non-union, and contract terms
- Rate (if sending auditions out internationally be sure to include the proper currency)
- How many characters and lines per character are required (this should coincide with the rate)
- Recording location and requirements (include the audition location as well)
- Audition dates and final recording dates
- Details about the game genre (as much as you are legally able to share)
- Character details, including an image that helps describe the voice. This could be mood, size, accent, age, race, pitch, breaths, and any performance or general direction notes that could help the person auditioning
- Character voice reference in the form of a video or audio clip to showcase a specific tone you are going for
- A portion of the script to be read in the audition (it's best to choose a range of emotions from the script to be sure to capture the full character)
- Some details on the scene setting and other characters' lines to be read off of

As the sole person in charge of all voice production you may not have the time or budget to be available for every audition. By opening auditions to remote performances you will increase the likelihood of finding the right talent for the part.

Choosing the Right Audition

After receiving auditions they should be reviewed to filter the most qualified submissions. If callbacks are required it might be necessary to be physically present for the audition to provide direction. The shortlisted auditions can then be shared with the producer or creative director (or the rest of the development team) for final selections. It usually isn't necessary to edit the auditions before sharing them, but in an instance where a specific effect will be applied (e.g. tuning or vocoding for a robot voice) it can be nice to offer a sample to share with the decision makers

Localization

It's important to point out that video game sales have a far reach. A game developed in one country can be played all over the word thanks to expanded distribution systems and digital downloads. If you played a game with recorded dialogue which was not in your native language the information being delivered by the dialogue may not mean too much to you. Sub-titles can help but being able to hear recorded voice-overs in your native language offers a much more immersive and not so confusing gameplay experience.

Dialogue isn't the only part of the game that may go through the localization process. Other game assets may be removed or changed to conform to cultural sensitivities and/or restrictions held by local laws. The process as a whole is an economic decision that is often made by the publisher or games creator.

Getting back to dialogue, there are a few things to think about when working to have all the games voice-overs recorded by a native speaker in the language you wish to include in the localization process. There are quite a few companies like Lionbridge[4] that can provide the translation of scripts and character descriptions, casting and recording/editing process as a package. Smaller teams may rely on contractors to handle translations and talent.

A focus on quality is just as important with localization. Poorly recorded or translated dialogue may end up as an internet meme or a YouTube video. The whole idea is to provide an enjoyable experience to players regardless of locale.

The Session

With the final selections out of the way, contracts can be settled with the voice talent or their agency. Additional art, direction, and the full script can be shared at this point. Refer to Chapter 11 for more information on contracts and rights.

Time and schedules don't always flow as planned but try to budget enough time for the voice artist to do several takes of each line. Dialog editors will want the additional takes to use in case of an issue in the recording or performance. It also leaves room for direction after the first pass of each line. As you monitor, mark up the script. Note which takes you prefer. This will later help the dialogue editor choose the proper takes for editing and processing.

Recording voice-over can happen in a variety of ways depending on the budget. A smaller budget might only allow for the voice artist to record in their home or project studio. In this case you can dial in via audio or video calls over the internet to direct the session. Zoom, Skype, Source Connect, and Discord are all viable options for providing feedback. The plus side of remote recordings is that the artists will usually know which microphone setups work best for their particular voice. They will also have the flexibility to schedule the session at the most appropriate time of day (avoiding large meals and early morning sessions, etc.). Since it is common to work remotely with talent all over the world, you should use a world clock and time zone converter app to coordinate schedules and be considerate of time zones.

When the budget allows for in-person sessions, a studio near the voice artist can be contracted. Typically, a recording engineer and a two- to four-hour session will need to be budgeted for and booked. You and/or another team member who may want to direct the voice artist can then choose to fly out and be present at the session if the budget allows. Make sure you instruct the voice artist to avoid wearing jangly jewelry, a starchy or nylon shirt, or drinking milk right beforehand. Similarly, a big meal before a session (or not eating anything at all) can both cause gurgling in the stomach, which may be picked up by the microphone. Try to have them eat around two hours prior to the session to avoid this.

Check the area they are standing on for floorboard creaks and groans. In the case of a creaky floor with no options to relocate, you or the engineer will have to monitor closely and direct the voice talent to avoid moving. If the location has a hardwood floor or tiles, a towel or blanket should be placed below the artist's feet. If the script is being read from a mic stand, try covering the stand with a towel as well to avoid reflections off the metal. Be mindful of paper movement, microphone stand bumps, and any other little noises that can make their way into the recording.

Mouth noises can be dealt with through careful editing, but it's a good idea to ask the voice talent to remove any gum before you record. An old recording engineer trick is to have the voice talent chew a green apple to clean up the inside of their mouth. We can attest to the fact that this does work!

The ideal place to record voice-over is in a vocal isolation booth to ensure the best signal-to-noise ratio. This will also reduce the amount of reverb present and make the recording as dry as possible. Life is not always ideal however, so if you are preparing a space in your home or project studio consider noises and reflections that might make their way into the recording. Turning pages, clothing movement, and mouth noises are common things that can render a good take useless during a voice-over session.

Preamps and Effects

The microphone preamplifier (preamp) and audio interface are important factors in the voice-over recording process. A preamp amplifies low-level signals into **line level**. Most audio interfaces offer some type of built-in preamp function, which is good enough to get you started, but the quality can vary widely between entry-level and more expensive devices. Low-quality preamps can introduce noise into the recording. The microphone's output and impedance, if low, can cause the user to push the gain too high and in turn introduce noise. Adding a quality preamp to your signal before routing into the audio interface can make a difference, but you also need to consider the rest of your signal chain. Perhaps a different microphone, mic placement, or room can improve your signal.

Preamps can offer additional clean gain to your recording, but they can also color the sound. When you are looking for a preamp to add warmth to your voice recording you might consider adding a tube preamp to your signal chain. Condenser microphones require phantom power, so make sure your audio interface has an option for "48v" as this will affect the recording level drastically.

Some engineers like to set up a compressor on the recording track, but with digital recording it isn't necessary. In the days of tape, engineers compressed the input signal to work within the dynamic range limitations. With digital recording it is best to leave around 10 dB of headroom and hold off on compression until the editing phase. This will avoid being stuck with the compression setting that was baked in during recording. Of course, if you are going for a specific sound it's perfectly fine to use compression on the input signal. In these cases we suggest sticking between 2:1 and 4:1 ratios with a slow ~1–2 ms attack and a ~200 ms release (or use the auto release function when available) to avoid an overly compressed recording. For a refresher on compression, refer to "Effects Processing as a Sound Design Tool," Chapter 3 (page 96).

In general it's good practice to avoid adding any effects when recording because you will have much more flexibility by adding them later on in post-production. For specialized effects like "distant vocals" have the voice artist back off from the microphone and record it dry. Then you can decide on reverb and delay specifics later on.

Microphone Position and Acoustic Treatment

A pop filter positioned a few inches from the mic will help reduce plosives (the popping that can be caused by fast-moving air hitting the microphone). A pop filter is not to be confused with a windscreen, which we discussed in Chapter 2 on the companion site. When using a dynamic mic such as a Shure SM58, a pop filter isn't always necessary as the device has a minimal filter built in. Alternatively, reflection filters can be purchased online at a reasonable price (or crafted with a bit of DIY gumption). Reflection filters minimize reflections from the rear and sides of the voice talent. To minimize reflections coming from behind the voice talent, hang some blankets over a door or on a wall. Be aware that reflection filters can be a bit cumbersome. The larger ones usually work better for absorbing sound, but they can be heavy and awkward on the mic stand.

With your reflection filter and pop filter in place, position the voice talent with their back to a wall with acoustic treatment (professional or DIY).

The microphone type, polar pattern, and placement matter as much with voice-over as they do with recording source sound effects. A large diaphragm condenser microphone with a cardioid pattern is a good place to start. If you have a particularly noisy environment, try recording with a dynamic mic like a Shure SM58. Each microphone model will color the voice in a different way, so try to test a few out to familiarize yourself with their particular coloration. Microphone placement can greatly affect the sound of recorded voice. Finding the perfect source distance from the mic can make the recorded voice sound clearer, warmer, more robust, and less sibilant. Mic placement includes distance from the voice source and axis. There isn't a specific rule for placement, but as we discussed in Chapter 3 you will want to be mindful of having the source too close to the mic, which can introduce proximity effect or exaggerated bass that can cause the recorded voice to sound muddy and less clear. Sometimes this effect is exactly what you are looking for so by all means use it as a technique when necessary. When the voice source is too far from the microphone, room sound will be introduced and the voice can become less intelligible due to reverb and echo from the room being captured. Experimenting is important when you are using a new mic, recording in a new room, or recording a different voice source. With that said a good starting point is about 6 inches (15.24 cm) from the microphone. From there you can try to pull the source back or move the source closer until the desired effect is achieved.

If you don't have access to multiple microphones, be sure to experiment with placement to find the "sweet spot." If you plan on recording more voice-overs as you progress in your career, purchasing a high-quality microphone may be a smart investment. Top studios often use a Neumann U87 as a go-to vocal mic for example.

Monitoring

Monitor with headphones and also have a pair of headphones for the voice talent to listen to themselves. Headphones for the talent should be "closed back" to avoid bleed from the monitor into the recording. Anyone else in the room with the voice artist should also use closed-back headphones. Some voice artists prefer leaving the head-phones off on one ear so they can hear their voice in the room, and this is fine. If this is the case, try panning their monitor signal to the single side that is *on their ear* to avoid bleed.

> The saying "we'll just fix it in post" is a line that lives in infamy in the audio industry. It leads to sloppy recording technique. Poor recording practices will cause more than just a headache for the editor; they can often ruin the final product. Make it a point to ensure you are capturing quality audio.
>
> Spencer

Recording

Time and schedules don't always flow as planned, but try to budget enough time for the voice artist to do several takes of each line. Dialog editors will want the additional takes to use in case of an issue in the recording or performance. It also leaves room for direction after the first pass of each line.

There are a few ways of organizing the recording session in the DAW (Digital Audio Workstation). In the studio, recording each line on one track is effective as long as you are setting up markers along the way, which can later be exported. Each line and the variety of takes can then be captured on separate tracks to make things a bit easier when bouncing everything out. Alternatively, most DAWs offer a system to handle capturing various takes. Pro Tools gives the user playlists to record new takes. Nuendo has a feature named "enable lanes," and in Reaper and Logic the user can record over the top of an existing file and the takes are automatically arranged for you. Find the workflow that suits your DAW before the session begins.

Ideally, you will have a workflow that organizes track names or markers to allow for seamless export after the session. You don't want to have to listen to each take to remember what line in the script it references.

Recording mono versus stereo will depend on each project's needs. Be sure to understand the requirements prior to starting the session. Input level settings may vary depending on the voice source, mic position, and selection. Start out by asking the voice talent to use their loudest voice for the character so you can adjust your input level accordingly. During the session you can adjust the input level as necessary. You can also ask the voice talent to pull back from the mic a bit.

It's a good idea to go through at least a few lines to warm up the talent's voice. You can certainly record the warmup but when you are ready to start the session, the lines read during the warmup should be re-recorded. As you record, don't stop the artist mid-phrase if they have made a mistake. Give them some direction as they re-read the line instead. This avoids adding pressure and stress, which can negatively affect the performance.

Directing the Session

The voice-over session should involve a recording engineer, voice director, and the voice talent. Sometimes you will play the role of more than one (or all three!) of these. In any case, it's difficult for even a seasoned voice artist to direct themselves so it's important to have separate individuals on the performance and recording/directing side. When taking on the role of the director, take the time to offer guidance and direction during the session. The idea of the session is to get the best performance out of the voice artist. Ensuring they feel comfortable both mentally and physically will add to the quality of the performance.

Talking about the character with the voice artist before the session can provide important information to the voice talent. There are specific characterizations that can be used to relay information regarding the performance. For example, a game show announcer has a specific style of voice that a director might be looking for. Additional descriptive words like "warm" or "quirky" help to further define the characterization. Be specific and try to avoid choosing adjectives that are too general or too broad. Relaying the events of the game can also provide context. It is important to make it abundantly clear for the artist what the character's *motivations* are. A voiced line must sound different in stealth mode than it would during battle, for example.

Visiting Artist: Tamara Ryan, Voice Artist

Thoughts on Casting

My favorite bit of info for casting is the personality. Even if the directors aren't sure what type of voice they want, the personality can give me a great jumping-off point while also giving room to play. And when it comes to confidential info, it's always appreciated if the general mood and genre of the game can be shared. Knowing if it's a kids' adventure games vs. a gritty MOBA can help dictate a lot of performance choices.

As for live recording sessions, keep in mind that each actor is different. One type of feedback might not work the same for everyone. While "make that line more blue" might click for one person, it could give a totally different result for another. Try to be patient while both you and the actor figure out how to work with each other.

Avoid the Line Read

When working with kids try mimicking a particular line. This can help offer the young voice artist a direct reference to what you are looking for. However, when working with more seasoned voice artists giving line reads can easily derail the session. The session will have a far better result if you encourage the artist, and allow them to engage their own frame of reference. A line read is a tool of last resort. If the voice artist is truly stumped, that's when a line read is appropriate.

Effective directors will often have the talent act out scenes in the booth as much as they can. If the game character is lifting a heavy object, try giving your artist an object of similar weight to lift. The DICE team had voice actors tossing around cinder blocks as they read lines in *Battlefield 1* to deliver an adrenaline-fueled performance.[5]

Focus on the Positives

It might feel natural to direct by telling the talent what you don't like about the performance, but it's always better to focus on positives and be clear on what you *want* to hear instead. For example, saying "I don't like your low energy" is a criticism, and not a helpful direction. Saying "I want more energy from you, this is an action-packed scene" will provide a solid direction with purpose. When providing criticism is necessary, it's always nice to precede it with a bit of positive feedback.

Variation

Each line should be read at least two times, and then again with a few varying directions. You don't always know how a take will fit into the game until the implementation stage so it's a good idea to have options and avoid having to rebook a session. For efforts and barks, record several takes. Effort sounds are typically written out in onomatopoeia, so allow the voice talent a bit of room to improvise. For example, jumps are typically written out as "hup!" or "hyup!" When directing the voice talent, ask them not to be so literal when reading those lines. Instead, have them mimic jumping and vocalize what feels natural. This will more effectively relay the effort of a character jumping to the player.

Mocap

We discussed in depth the DIY way of recording, but when the budget allows voice-over recording can be done on the motion capture stage. Game developers like Naughty Dog have been utilizing motion capture (mocap) studios to record dynamic dialogue for years on the *Uncharted* and *The Last of Us* series. Mocap is the process of capturing actors and actresses' movements digitally to be translated into data used to animate a character in a video game or movie. The talent gets suited up into a tracking suit, which is covered in dots that look like ping-pong balls. Cameras and special software pick up character movement by tracking the dots. For capturing audio during action scenes, a lavalier mic is attached to the headband of the actress or actor. The lav works well because it captures the authenticity and physical exertion of a performance, albeit at a slightly lower quality than a professional grade studio condenser mic would. In most cases, the uptick in performance is well worth it.

As you direct the session keep in mind that voice actors and actresses are creative individuals hired to do a job. Treat them with respect and understanding and you will find they will put in their best effort.

File Organization

After the session is finished, it's time to step out of the voice director (or engineer) role and into the role of dialogue editor. The dialogue editor's job is to take the audio from the session engineer and produce high-quality game-ready assets. Typically, once you have the session audio (or you have prepared it yourself as the engineer), the next step is to create a "selects" and "edits" folder for editing and mastering. Some recording engineers will be kind enough to break the files into the selects folder for you. If not, grab the best takes from the session and drop them in the selects folder. Include only the takes that were marked on the script and any alts you may have chosen. If the select folder was delivered to you by the engineer, you will still need to use the marked-up script to double check that the correct files were delivered. Also make sure that the raw files were delivered along with the selects as they may be needed later on. Next create an "edits" folder. The audio in the selects folder should be copied into this "edits" folder to be edited. If you make a mistake in the editing process, this will make it very easy to go back into the "selects" and copy the file over again.

Be prepared to receive notes from the voice director or producer in regard to editing and final delivery. These notes will provide requirements on sample rate and bit depth as well as comments on how to handle breaths. Many games will require mono voice-over assets even if they have been recorded in stereo, so a common production note is to convert the files. This will save space in the game and it will also allow voice clips to be spatialized. Misreading or ignoring notes entirely can cause a bottleneck during the implementation phase, so read notes thoroughly.

A naming convention may already be planned out by the programmer and embedded in the script. If the naming convention on the raw files does not adhere to to the programming conventions, use the file names from the script to rename the files. A batch-renaming tool will speed up the process, but you will have to take the time to input the original file name and the new file name into a spreadsheet as you do. This is an important step to keep track of files throughout the complicated process of game development. In the Sound Lab (companion site) we will present an image of a marked-up script.

Editing and Preparing Assets for Implementation

At this point it's time to begin cleaning up unwanted noise from the recordings; if the session was done in a properly calibrated studio setting this shouldn't be much of an issue. Keep in mind that restoration tools can ruin the sound if used incorrectly. Begin by listening carefully for lip smacks, stomach grumbles, cloth movement, microphone stand bumps, and any other mouth or nasal noises.

Sibilance is a tonal harshness or hissing sound produced from words that begin with or end with the letter *s* and in some cases *ch* and *sh*. This can often be handled with a de-esser plugin during editing or by moving the microphone slightly off-axis (or to the side of the artist's mouth) during the recording session. Plosives are a thump-like effect in the recording produced by a blast of air hitting the microphone and creating larger pressure changes. The letters *b*, *d*, *p*, and *t* typically are the cause, but it can be reduced by putting a pop filter in place during the recording or with EQ in editing. These are the artifacts you will need to remove before delivering the audio for implementation.

One way to edit voice-over is via destructive editing in a two-track editor such as Adobe Audition. This method allows for editing and saving the edits without having to take the additional time to bounce the edited file. Another option is editing in your DAW. Either way you decide to do it, you should feel comfortable with the process. Some projects might require you to use a specific DAW so the sessions can be shared with the audio director, but if this isn't a requirement use whichever method you prefer.

"Quick keys" can be set up for shortcuts to allow the editor to blaze through each line in under a minute. If the game has thousands of lines of dialogue, you will have to work quickly to produce 500–800 lines per day. Aim to get the editing time down to a minute or less per line. Not all lines are created equally however, so you may come across a longer line from a cinematic that might take slightly more time to clean up. Speed is a crucial factor for seasoned veterans, but keep in mind it's something to work up to. Start slow and internalize the workflow, and then practice increasing your speed.

iZotope has a full range of tools like de-click, de-plosive, and de-noise specifically for voice. The big question is whether to do a full restoration with plugins, or to edit manually. Some audio directors might not want the artifacts that can be imposed on a sound when using restoration plugins, so the end result must be considered. If you do move forward with the plugins, it's best to use your ears to determine the most effective settings. As a general rule apply light restoration in several passes rather than going heavy in one pass. A lot of the tools offer a "learn" mode which can assist in suggesting settings. If you use this as a guide pull back on the threshold and reduction and apply noise reduction in several passes. It's good practice to think about your processing chain on a macro level so you can run any de-click or de-crackle passes before broadband noise reduction. In Chapter 3 on the companion site we discussed restoration on source, and you might find some of those tips helpful for voice-over as well.

If plugins aren't your preferred method, it is entirely possible to manually remove pops and clicks, and reduce breaths, harsh plosives, and sibilance. Refer to the companion site for a detailed walk-through of cleaning up voice-over. The material will walk you through trimming the heads and tails, as well as cleaning up the issues previously mentioned.

Keep in mind that voice performance and intelligibility should be preserved during editing. Over editing/cleaning a voice-over can make the asset sound lesser quality than if some of the noise were left in the recording. Use your ears to determine how much editing and clean-up is necessary. In the Sound Lab (companion site) we offer a video tutorial on dialogue editing.

Mastering

Once the files have been edited, the assets can be copied into the mastering folder. You can now take off your editing hat and jump into the mastering role. At this stage you should hopefully have edited the select takes to sound clean and free of unwanted noise. The final step is to prepare the files for implementation. This part of the game development process is called mastering.

The game engine needs high-quality audio files, so start by applying EQ. Use an EQ to control the low end and remove any unnecessary or overpowering frequencies. This can be done in your DAW on a mix bus so the effects can be applied to all the edited voice assets. This will allow the files to sit better in the full game mix and sound like they stem from the same source. Starting at 120 Hz, use the "bell curve method" (discussed in Chapter 3 on the companion site) and boost bands of frequencies in the low end to listen for noise that can be reduced. Next, boost from 1,500 Hz to 3,000 Hz to add some clarity to the speech, making it more intelligible to players.

Next you'll have to decide how you want to handle the dynamics and overall level of the assets. Here we will discuss two processes that can be used to achieve even levels using different processes.

A compressor/limiter combination can be used control very drastic differences in volume across a set of voice-over assets. For a more natural sound start with a 2:1 ratio and avoid going above 4:1. The threshold will be used to set the level at which the compressor kicks in. Typically the threshold should be set at -24 dB to start and it can be adjusted from there. The attack and release time will determine how quickly the compressor kicks in and reduces the audio level. A slow attack time can work well on instruments but make voice sound unnatural by only affecting the end of the word or phrase. If the attack is too fast it can squash the entire sound. Aim for 2–5 ms as a starting point and use your ears to adjust. The same can be said about release time. When set too slow it can create a pumping effect. Aim for 10–12 ms and adjust to taste. If the audio is too low in volume after compression the gain control can be used to *make up gain*. Start with a 4–6 dB gain and adjust from there. Compression with a little gain boost in a few passes with a limiter at the end of the chain will help boost the overall volume of the voice-overs without squashing all of the dynamics.

We often use the dynamics processing module in Adobe Audition which offers an expander, compressor, and somewhat of a noise gate in one plugin. In the Sound Lab (companion site) we present a video tutorial explaining our process with the Audition dynamics processing module.

DAWs and audio engines all offer some combination of peak, RMS (root mean square), and LUFS (loudness units) for metering, but often a novice may not fully understand what each does and how one might be more useful than another. Before we jump into measuring loudness as part of the mastering process, let's go back to the basics for a quick refresher. Peak records the loudest part of the signal. It can tell us when we are close to clipping. RMS on the other hand is average signal level over time. It factors the average of the loudest, quietest, sustain and some decay. It is a function of level and time. A shorter window of time or a quicker decay can give you a higher RMS level, a barometer for how loud sound might be perceived. Signals at different frequencies are perceived at different loudness levels. With bright sound vs. dark sound at the same signal level the bright be perceived as sounding louder. LUFS is similar to RMS in that it measures perceived loudness but LUFS integrated loudness is proven to be more accurate and telling in terms of human hearing. Broadcast standards are defined in loudness units. There are plenty of great resources[6] on the internet that explain how to use LUFS when mixing and mastering so we recommend spending some time getting a grasp on it.

Now we have had a refresher we can discuss loudness normalization, which is the process of raising the volume of the entire asset to a user-defined threshold. Most DAWs and editors have a built-in normalizer. Some offer more control over the process. A general pass of normalization can really take the dynamics out of a well-recorded audio file, so think carefully and understand the plugin's features before applying normalization. Our workflow typically includes normalization at -24, -22, and -18 RMS for the respective groups previously mentioned. Audio engines like Wwise include tools such as Loudness Normalization, which will apply normalization at run-time based on the estimated loudness of the source. This process, which is non-destructive, will boost the gain of softer sounds and attenuate louder sounds. It can also be tweaked to preference making it extremely useful. We don't recommend normalizing to LUFS as it isn't great for shorter files.

Both techniques are great for **batch processing** large amounts of assets. When applying leveling over a batch process, it's good practice to organize groups of sound into softer, normal, and louder categories. By grouping whispers, yells, and natural-speaking voice-over assets into different folders they can be batch processed by category. This will offer greater dynamics as a whisper and yell won't have similar perceived volume.

The final assets will always be delivered as individual files. When using batch processing on a large number of assets, apply random quality checks on the batched assets to ensure there were no issues in processing.

With a smaller number of assets you may decide to go a manual route and level peaks by hand. Drawing in the volume curves for each asset can be time consuming though. As with a lot of other manual processes there is a plugin for that. iZotope RX Advanced offers a Leveler Module that can control peaks and contain breaths and sililance. Prior to having a voice-over project record some of your own voice and try out these different techniques.

Other plugin effects like filters and reverb may need to be baked into the audio file so you will want to understand the project's needs before finalizing the assets. In Chapter 8 we discuss the pros and cons of real-time effects versus baking them into the asset.

Visiting Artist: Ayako Yamauchi, Sound Designer, Sound Editor

Editing Dialogue

My role for editorial tasks is not limited to editing but also mastering. Sometimes I am only asked to edit and sometimes I am only asked to master. For foreign-language release in AAA games such as *Uncharted*: *The Lost Legacy* and *God of War* (2018), I have worked on dialogue mastering for the game's foreign-language release. For those titles, there are more than 10,000 lines in each language and we often need to complete the job on a tight deadline.

Whenever I work for different studios, I always ask my supervisors how they listen to audio when checking our work. The majority of engineers use a compressor or a limiter while they are editing to simulate how the final mastering files sound. This allows editors a better chance to catch problems if the files are going to be much louder in the finalized version. Other engineers simply use the volume knob while editing. To reach the final loudness level, it is generally better to listen at a level louder than comfortable. If the audio's dynamic range in the recording is acceptable, I prefer to listen without a compressor/limiter because that enables me to think through the process realistically. However, the recorded file's loudness level may not be considered final as the editor may be asked to adjust for a specific dynamic range. I ask the supervisors how they set up the compressor/limiter and I follow their work along so I am able to listen to the audio as they shape the file I am editing.

I use iZotope RX Advanced to clean up any noise found in a dialogue line's recording. Removing mouth clicks is how I spend much of my time. If consonants such as *p*, *b*, *t*, *d*, *k*, *g* are too obvious and unpleasant, I use clip gain in Pro Tools to reduce the gain level a tiny bit. Some people might use "de-plosive" in iZotope RX. If sibilant consonants such as *s*, *z*, *ch*, *j*, and *sh* sound too harsh, I use a de-esser. When making an audio file louder in mastering, there are parts

(Continued)

of the sound's frequency range, especially around 4–5 kHz, that become harsh. This is what is called "psychoacoustics." When the mastered file needs to be made louder, this appears more frequently. I always structure any restoration process to make the most sense in each stage, for both unmastered and mastered audio. If a problem appears in mastering, I simply fix the mastered audio, and I also check back into the unmastered file of the same line and see if the unmastered file is clean enough, because these kinds of problems are sometimes not obvious or easily discernible in editing.

Before iZotope RX became an industry-standard plugin, there was little else that editors could fix with sound problems. Noise suppressors are great tools but if expectations are too high, the final audio quality never sounds optimal. The major part of the sound quality is still dependent on recording and performance. Editors charge by the hour but an editing budget is determined by lines, so it is very important to discuss what is realistically able to be fixed in editing.

The Sound Lab

Before moving onto to the next part, head over to the Sound Lab to wrap up Chapter 4. We offer some audio examples of sibilance and plosives, an image of a marked-up script, a video tutorial on dialog editing and processing.

* * *

PART I REVIEW

Don't forget to head over to the Sound Lab if you haven't done so already for each of the chapters in Part I.

In *Chapter 2* we provided an overview of sound design for game audio as a process. We discussed the building blocks for those new to the industry and intermediate sound designers looking for a refresher. We also broadly covered concepts such as nonlinearity, tools and skills for the game sound designer, and other foundational components of game sound. We will continue to explore these topics in more detail in later chapters.

In *Chapter 3* we explored the creative side of sound design for games and took a more detailed look into sound design for game audio as a process. We covered intermediate topics which include the role sound plays in games, sourcing sound through libraries, studio and field recording, and designing sounds. We discussed ways to create a unified sonic palette by exploring layering, transient staging, frequency slotting, and effects processing as a sound design tool. We also explained our framework for breaking down sound into its component categories. Visiting artist Martin Stig Andersen discussed his thoughts on "less is more" in relation to sound design for games while sound recordist Ann Kroeber, Thomas Rex Beverly, and Watson Wu discussed capturing sound in the field.

In *Chapter 4* we discussed our approach to dialogue production for games by offering a breakdown of the various stages of production from casting calls to mastering assets. Visiting artists D. B. Cooper, Tamara Ryan, and Ayako Yamauchi imparted their voice production wisdom throughout the chapter.

Chapter 5 begins Part II, the music composition portion of our textbook. We encourage you to read through the entire book, but as we mentioned in Chapter 1, if you are only interested in sound design you can skip Part II and move directly onto Part III, Chapter 8: Audio Implementation. From there we recommend reading Part IV, Chapters 10 through 12, for the business side of game audio.

* * *

NOTES

1 https://voicecastinghub.com/
2 www.globalvoiceacademy.com/gvaa-rate-guide-2/
3 www.sagaftra.org
4 www.lionbridge.com
5 E. Lahti, "Battlefield 1's Voice Actors Played Catch with Cinderblocks while Recording Their Lines."
6 Masteringthemix.com, "Mixing and Mastering Using LUFS."

BIBLIOGRAPHY

Henein, M. (November 17, 2017). "Answering the Call of Duty." Retrieved from www.mixonline.com/sfp/answering-call-duty-369344

Lahti, E. (March 3, 2017). "Battlefield 1's voice actors played catch with cinderblocks while recording their lines." Retrieved from www.pcgamer.com/battlefield-1s-voice-actors-played-catch-with-cinderblocks-while-recording-their-lines/

Masteringthemix.com. "Mixing and Mastering Using LUFS." Retrieved from www.masteringthemix.com/blogs/learn/mixing-and-mastering-using-lufs

Part II
MUSIC

The Basics of Nonlinear Music

In this chapter we will cover a range of basic and intermediate topics. Our goal is to offer a broad perspective on concepts such as nonlinearity (as it relates to game music), requisite skills helpful to game composers, as well as foundational components of game music such as loops, stingers, and layers. We will explore these topics in depth in later chapters.

Many of the following concepts are entry level and can be understood by novice game composers. However, we will quickly move on to more intermediate topics. If at any point you have trouble with material later on, we recommend returning to this chapter to refamiliarize yourself with the basics. You can also use the index and companion site for reference as needed.

WHAT IS NONLINEAR MUSIC?

In order to define **nonlinear music** we must first understand what **linear music** is. We have already encountered both linear and nonlinear audio in the previous sound design chapters as well as in Chapter 1. These same concepts apply to music. Linear music has a structure that flows predictably from A to B to C *sequentially* each time a track is played (see Figure 5.1). The player has (virtually) no interaction whatsoever with linear music. This is essentially how all film and popular music is heard. It sounds the same every single time from start to finish.

By contrast, nonlinear music flows *non-sequentially* and in no chronological order. So A could flow to C and then back to A again on one playthrough, and in the next it might start on B and flow to C and then to A (see Figure 5.2). This means 100 playthroughs could theoretically have 100 different outcomes musically. In video games, these changes occur when the player takes an action or makes a decision that affects the game in some way, changing the state of the game (or game state). This concept of

FIGURE 5.1 The sequence of linear music flows in order, every time the music is performed.

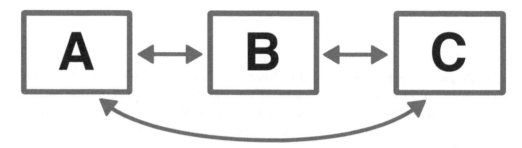

FIGURE 5.2 The sequence of nonlinear music is more like a logic tree; each section of music can move to any other section, or even loop back on itself making the "performance" of the music different every single time.

nonlinearity is the core of video game music. Other media (with some rare exceptions) exclusively uses linear music, but games are unique. Games need music that can change and adapt when the player makes choices, so understanding how to write nonlinear music is absolutely essential for a game composer.

Interactive Music

When we talk about nonlinear music there are actually two different categories that we are referring to: **interactive** and **adaptive music**. These terms are not always well defined and are often used either interchangeably or with ambiguity between definitions. Although some gray areas do exist, there are important differences between them, which we will explore in the following chapters. The defining factor that separates these terms is *how the player interacts with the music*.

Players interact *directly* with the music itself when dealing with interactive music. Games like *Guitar Hero* and *Rock Band* are examples of interactive music. Players are actually influencing the music that is playing in real time. In these games players have agency over individual notes and rhythms. In a sense, the game state becomes a playable musical instrument. In a slightly more distilled sense, mini-games or small events within a game can also be interactive. It is very common for games in the *Super Mario* franchise to have direct methods for players to interact with music. For example, by collecting items players can piece together a melody note by note in *Super Mario Galaxy*. These examples may not be fully interactive in the way *Guitar Hero* is, but they can be considered to have interactive musical components.

Adaptive Music

Adaptive music can be defined as a soundtrack where players have *indirect* influence over the music. In this case the player can affect the game state, which in turn affects the music. Or in other words, the music *adapts* to the game state. Any time we hear a transition from one musical cue to another, we are really hearing adaptive music responding to changes in the game. The player has taken an action, thereby altering the state of the game and the music is adapting to compensate.

Adaptive music will be our primary focus in these next few chapters mainly because in learning the methods behind adaptive scoring you will be capable of writing music for a wide range of game genres and situations. However, the applications for interactive music are far-reaching and continue to be explored by game designers and audio professionals. The overview in "Aesthetic Creativity and Composing Outside the Box," Chapter 9, page 330, will cover some more experimental methods of interactive scoring and provide some ideas for future development.

CHALLENGES OF NONLINEAR MUSIC WRITING

Nonlinear music is at its core very different from the music that you would listen to on the radio, or during a film or television show. Video game scores need to be responsive to the changing states of the game and *must* be created with this in mind. Our challenge then is to write music that functions well in a variety of contexts. As a composer, the priority changes from writing music that fits a specific scene (film music), to writing music that fits the overall *mood* of a particular game state *and* is flexible enough to transition based on unpredictable player input.

Picture yourself playing a game where the object is to solve a puzzle and then defeat an army of zombies. It would be an impossible endeavor to write linear music that synchronizes exactly to the player's actions during the puzzle because every player will take a different path to solve it (for advanced solutions to this problem see the sections on "Vertical Layering" and "Horizontal Resequencing" in Chapter 9, pages 000 and 000). You can however write nonlinear music that supports the mood and emotion of the scene as the player is solving the puzzle. Common styles that fit this situation would be subtle, yet mysterious music that does not distract from the player's problem solving. This music should be written to transition smoothly after the puzzle is solved because the mood will immediately change to something else afterward. It is up to the composer to think ahead and plan the musical pieces to fit well together, and to allow for a cohesive overall player experience.

ESSENTIAL SKILLS FOR THE GAME COMPOSER

Game soundtracks range in genre from electronic to orchestral to ambient and beyond. This makes video game soundtracks an extraordinarily diverse medium, and affords composers the opportunity to create unique and personal sound for each game. However, this also means that new composers usually need to be skilled in a few different musical idioms. The two most common styles in games are cinematic and electronic. Often composers need to work out some combination of the two. This is not to say that composers should always try to write within those frameworks, but it does mean that developers will likely ask for or provide references in these styles. From that, we can ascertain a list of skills that will be helpful in streamlining creativity and workflow (also see "Essential Soft Skills and Tools for Game Audio," Chapter 1, page 16).

Consistency

Although this is not something often discussed, arguably the most important skill for new game composers is the ability to *consistently* and *frequently* write music within a given timeframe. It is a myth that great music comes from a spark of inspiration, manifesting itself as a fully formed *magnum opus*. Inspiration can indeed strike at any time, but in reality that is the first step in a long process which requires intentional and consistent practice to bring great music to fruition.

The best way to develop this essential skill is to set aside a minimum timeframe (e.g. ten minutes), and without any expectations or judgment about what you are doing, write music every single day within that timeframe. At first, much of what you write will be of little value. But if you think of it as practice, and push yourself to write quality music within this timeframe, you will find that you are developing the ability to write high-level music quickly, which is very important for a sustainable career in games. You can then expand this timeframe to 20 minutes, 60 minutes, or more. You can eventually invert it so that you are tasked with writing "X" minutes per day, every day. This is even more challenging because you will be tempted to seek perfection in your own work. The important thing is not to micromanage every detail, but to learn how to harness your creative abilities quickly and consistently to meet deadlines and to raise your output capabilities as high as possible.

Spencer

Basic Loops and Stingers

These are the "hard" technical skills for the new composer, which you will need under your belt if you expect to start scoring games right away. We have discussed loops in depth in "Essential Soft Skills and Tools for Game Audio," Chapter 1, on the companion site, so refer back to that section to refresh your memory. Loops and stingers function in a very similar way musically, but there are some important differences that we will cover in the following sections.

In adaptive music, **loops** are musical tracks whose ending seamlessly blends back into the beginning *without pops, clicks, pauses, or obvious changes in musicality.* The simple way to do this is to plan to copy your first bar into your last bar (or a slight variation of it to keep things from getting boring). When this is not possible, it's important to identify four things: instrumentation, **dynamics** (and expression, modulation, and other relevant MIDI parameters), **density**, and tempo. These four things *must* be the same for the first and last notes (or possibly full bars) of the loop in order for the transition to be seamless. In the Sound Lab (companion site) you'll find an audio example, score, and screenshots that demonstrate composing music with looping in mind.

Stingers are short musical fragments that play *linearly* and are triggered by a player action or game event. These can be thought of as straightforward cue, written as a very short piece of music. The key when writing stingers is to convey the *mood* of whatever is triggering the stinger.

In the Sound Lab (companion site) is an example of a short death stinger, meant to trigger when a player has died. Notice that unlike a loop, this stinger can change dynamics and articulation at the composer's discretion. The dynamics increase steadily throughout, and end in *fortissimo*. We also end with another double stop, which increases the dynamics and density of the stinger. It sounds very "final," which works well when triggered at the death of the player character. If you listen carefully you can also hear the note hold on the last bar. This would not be appropriate with a loop because the rhythmic **rubato** would feel uneven when transitioning back to the start of the cue. Stingers like this are short because they need to match an animation or briefly highlight something in the game. However, stingers can also be longer at which point they function more like a linear cue.

We will come back to loops and stingers later on as we get into implementation in Chapters 8 and 9. For now it's important to know that they are basic examples of

horizontal resequencing. **Horizontal scoring** is essentially a way of creating musical "chunks" or modules that transition when appropriate in the game. It is a way to offer variety to the score so that the emotional arc of the music closely follows the events of the game. For example, one might compose three loops (A, B, and C) and one stinger (a "win" stinger). Using this horizontal method during a battle of some kind, our music can transition along with the enemy's health. Module A moves to module B, which in turn moves to module C when the enemy's health is low. Finally, when the enemy is defeated and the "win" stinger triggers.

Spencer

Critical/Analytical Listening

One absolutely indispensable skill for game composers is the ability to listen to music critically and analytically. This goes hand in hand with aural skills, which help with identification of chord progressions and melodic shapes. Aural skills, however, are not enough. As a game composer you must also be able to play a game and understand how the implementation may have been executed. To do this it is important to play games often, and listen not just to the content of the music, but to exactly how the music *changes* and *transitions*. Are the transitions smooth? Did the composer write a linear transition, or are two cues crossfading together to adjust to the gameplay? Most importantly, ask yourself how *you would approach those same game scenes*.

On top of that, it's also important for composers to understand what's going on technically in a game's soundtrack. Try to familiarize yourself with a broad range of instruments and plugins. Learn how synthetic sounds are made, and have a foundational knowledge of recording techniques. These will all help you listen more critically to a game's score, and in turn will help you be more analytical about your own work.

Arranging and Orchestrating

While it is not necessary to be an expert in all types of arrangement and orchestration when composing for games, it *is* necessary to have a strong knowledge of one particular style at least. You can then use this as a foothold into other styles. Many composers are experts in a niche genre of game music, and don't need to venture too far outside of their comfort zone. However, this can be a difficult path, and for the average person learning a wider variety of musical idioms usually creates work opportunities. Being able to compose flexibly can often lead to more work and a more interesting day-to-day lifestyle. As you build your career and gain more experience, you will most likely be forced to adjust your style as projects can be very different in terms of the mood and the artwork.

EXPLORING MUSIC AS AN IMMERSIVE TOOL

As human beings we perceive music as an inherently emotional medium. This makes it an effective tool for communicating mood and setting in video games. This communication (when successfully done) is an effective way to immerse the player in the game world. Referring back to our previous example, the mysterious and subtle music we created for the puzzle scene has two functions: 1) to encourage player action and 2) to immerse the player in the mystery and intrigue of the game. Our puzzle track adds tension, which compels the player to take action to solve the puzzle. By adding subtle elements of mystery to the track we can convey the primary mood of the scene, thereby focusing our player's attention on the game world and *story events*, rather than focusing attention on the music itself. This is the essence of how music immerses players in a game.

The Role of Music in Immersion

Immersion as it relates to music comes down to a single question: what does this game *need?* In essence we are asking what task our music needs to accomplish in a given scenario. The answer to this can be split into two very broad categories: emotional impact and gameplay support. These are certainly not the only tasks music can accomplish, but they are two very important ones that are used in just about every game.

Emotional impact usually comes very naturally to us as composers. If a game calls for an emotionally charged moment, we usually have the freedom to write out a compelling thematic cue which will draw the player into the story of the game. This is a more overt process, and we are free to score any and all aspects of the scene. A fantastic example of this is the track "All Gone (No Escape)" by Gustavo Santaolalla, heard in the game *The Last of Us*. The instrumentation is very "no frills," comprised of only a small string ensemble. This makes the whole cue feel intimate and vulnerable, which exactly mirrors the fear of loss that Joel is living through.

The track is triggered near the end of the game, where the player is carrying an unconscious Ellie through an army of enemies. This scene is the climax of Joel's character arc, where he decides that Ellie's life is more important than the life of every other survivor on the planet. The music here justifiably takes precedence over almost all other aspects of the game through strong emphasis in the audio mix. This was a bold and intentional decision on the part of the audio team at Naughty Dog, and it pays off. The cue only enhances immersion by pulling players into the emotions of the scene. On top of this, the entire structure of the track heavily leans toward an emotive climax, which adds a heightened sense of tension and build-up to the scene. Ironically, not all thematic cues meant to deliver an emotional impact need to have a strong melody, but in this case the player is left with one that is beautiful and memorable. In that way "All Gone (No Escape)" expertly accomplishes its goal of delivering emotional impact.

The other example of musical immersion is much more subtle, and it is unique to games (as opposed to film and television). This is music written to support the gameplay experience. This is the core of video game music because this type of musical immersion has an effect on what actions the player will take. In turn, it changes the player's experience completely. One example of this is the *speed* of gameplay. If we once again take to our hypothetical puzzle example, we can replace our mysterious music with an action-packed cue set at 200 bpm. This kind of music will make the puzzle scene less cerebral and more of a frantic race to finish in time. Any game from *The Legend of Zelda* franchise is a great example of this. During time trial encounters a cue will play. When the timer gets closer to zero the cue will change to a faster, more active track, highlighting the fact that the player is quickly running out of time. Similarly, most (if not all) infinite runner games apply the same principles to elicit excitement and energy in players. The important takeaway from these examples is that the tempo and rhythmic activity of the gameplay music *indirectly* elicits excitement in players. The faster the track, the more this will take effect.

Gameplay-specific immersion also manifests by adapting to and influencing player action *directly*. When a player enters a room and an ominously low drone triggers, this *feels* immersive because it is adapting to her actions. The ominous music influences how the player *perceives* and *reacts* to the scene. In this case the player might proceed quite cautiously. She might even take out her weapon and ready it for battle. Or she may leave the room entirely! Regardless of the player's subsequent action, it is the music that is immersing the player in the experience and influencing how she interacts with the game.

Mood in Game Music

There are many other ways music can be used for player immersion both directly and indirectly, and composers are thinking of more creative methods every day. However most of these methods are tied to a very important aspect of game music: *mood*. Mood can be a nuanced characteristic of any kind of music, but in games it can be particularly hard to pin down. How do you translate a mood that is slightly anticipatory, yet somber, set in a post-apocalyptic cyberpunk universe? And how do you then turn that into a musical cue with appropriate harmony and instrumentation? And *then* how do you assure that millions of players with differing backgrounds and life experiences will also associate those particular harmonies and instruments with your version of an anticipatory yet somber mood set in a post-apocalyptic cyberpunk universe...? The answer is you can't – at least not fully.

Attempting to produce a universal gameplay experience for players through music actually defeats the purpose of video game interactivity. Every player's experience is unique. As a composer, it is important to keep this in mind and formulate *your own* connection to the game. How do *you* feel during this scene? How would you feel if you were a character in this universe, experiencing these events first hand? Are there any true experiences you've had that resemble these events? Don't just ask these question – answer them. Try to articulate exactly what you're feeling when you play an early build.

Perhaps you don't just feel sad, you feel bittersweet nostalgic – or maybe it's more of an empty tragic feeling. These details will make a difference, and although you can't force players to *feel* the same way that you do about a scene, this line of thought will ensure that the gameplay music you write will be authentic, and therefore it will be more immersive to the player. This is your starting point for using music to create mood.

Determining the mood of a game scene is a very intuitive process for most composers because it is a reaction to early game materials that are presented to us (an early build, concept art, storyboard, etc.). The next step is a bit trickier as we now have to account for the details of the gameplay. We know how we want players to *feel*, but how do we want them to *act* in game? More to the point, how do we get out of the way so that players can *decide for themselves* how to act? The latter is the trickiest to achieve compositionally, yet it is also the most important because it results in a feeling of *agency* for the player.

Composing music to accompany gameplay really needs to begin with actually *playing* the game. If developers are interested in your input on implementation (and if they are not, convince them to be) then you need to play the game to accurately judge what music it needs and how it should react to the player. Playing through a level a few times with no sound will make it clear exactly what is so challenging about this type of scoring. Our sensibilities tell us to compose something extraordinary, yet playing the game tells us that something more ordinary, or even no music at all, may be more appropriate for immersion. This issue can be solved in a variety of ways, but the result is a musical systems that can accommodate a *range of moods*.

The Sound Lab

 To explore this let's head over to the Sound Lab where we will get our first taste of using middleware. We will return to this same example later on with a more technical focus, but for now we will simply draw our attention to how mood can change in a game scene.

DIEGETIC VS. NON-DIEGETIC MUSIC

Before moving onto more intermediate topics it is important to take note of the difference between *diegetic* and *non-diegetic* music. Non-diegetic music is any music whose *source is not present in the scene*. Any time Darth Vader is shown, the "Imperial March" is played. This is *non-diegetic* music because there is no orchestra present on the Star Destroyer, so the source of the music is not in the scene. Most game music is non-diegetic because themes, battle music, and underscore do not have instruments playing on screen.

Diegetic music, by contrast, is music where the source is *present in the scene*. If you are playing a game and you pass a musician on the street playing the violin, this is *diegetic* music because the source of the sound is in the game itself. This is highly relevant to our last topic of immersion. Diegetic music usually makes a scene feel *more immersive*. Non-diegetic music can do the same, but careful consideration must be taken. The music you write should not distract the player or call attention away from the scene unless you want to break the sense of immersion.

Diegetic Music in Games

Diegetic music is a bit less common than non-diegetic music in games, but it offers a unique opportunity for immersion. A fantastic example of diegetic music enhancing immersion is "Zia's Theme," as shown in the game *Bastion*. In this scene Zia is playing guitar and singing. As you walk your character closer to Zia, the sound becomes louder and clearer until she is singing right in front you. In a sense you (as the player) are following the sound of her voice to your destination. This is a scenario where a very high level of immersion can occur because the music is inseparable from the gameplay.

In some cases, examples like this can border the lines between adaptive and interactive music. In *The Legend of Zelda: Ocarina of Time* one of the most iconic scenes is set in The Lost Woods. Here the player follows a mysterious melody through a forest maze. At first this melody seems like non-diegetic underscore. Eventually the player realizes that the music actually has its source in The Forest Temple, which is where the next objective is located. Only by listening carefully to the music is it possible for the player to find her way to the next objective. In this example the music actually *affects player action in the game*. Without the music, the player would be unable to complete this task, so interaction with the source of the music is crucial. Most game scoring is non-diegetic, but this kind of creative diegesis can be breathtaking.

To summarize, diegetic music is music that functions as a sound effect. It is placed in 3D space, and the player can interact with its source in some way. Non-diegetic music can adapt extremely well to player actions, but because the source is not present in the scene it lacks something visual for players to interact directly with. Where immersion is concerned, the high level of interactivity the player has with the game environment, the more immersive the experience. This extra dimension of interactivity is what makes diegetic music feel so much more immersive.

Visiting Artist: Martin Stig Andersen, Composer, Sound Designer

Music, between Diegetic and Non-Diegetic

Dealing with the challenges of game music I often find it helpful to dismiss the traditional dividing line between music and sound design, and instead enter the ambiguous midfield between diegetic and non-diegetic. This allows for

(*Continued*)

anchoring music in the game's world, in this way providing an alibi for its presence. The approach is particularly helpful in dealing with areas of a game in which the player is likely to be stuck for a while. In such cases non-diegetic music easily becomes annoying to the player, acting like a skipping record reminding the player that she's indeed stuck. If, however, the "music" or soundscape somehow appears to emanate from an object within the game's world, say a generator, it's more forgiving to the player. The idea of having music disguised as objects belonging to the game's world also has special relevance in horror games. Where non-diegetic music, regardless of how scary it might be, always runs the risk of providing comforting information to the player, such as when and when not to be afraid (as if taking the player by the hand), music residing somewhere in between diegetic and non-diegetic will appear more ambiguous and unsettling to the player.

PRODUCTION CYCLE AND PLANNING

The production cycle for game music, as in sound design, can be very broad. Because development cycles are so unique to the developer, there really is no "one size fits all" approach to planning and executing a score. However, there are a few general trends to consider.

Time Management for Indie Projects

When working with some developers, composers are often given a surplus amount of time to work on a soundtrack. This is due to indie developers needing extra time for fundraising and marketing. At first glance the extra time may seem helpful, but this is actually one of the most difficult scenarios for composers. It can be extremely challenging to keep yourself active and engaged in the game material when deadlines are stop and go, and when there is little direction or feedback for your work.

The best approach to combat this is to become your own audio director. You will have to adopt organizational responsibilities on top of your creative duties. Lay out an internal schedule for yourself and stick to it. Include time for brainstorming (one of the most important parts of the creative process; see Chapter 6), writing, and review. Do your best to stay on top of your creative point-person and ask for feedback. Maintaining communication with the developer will ensure that you are both moving in the same direction.

Time Management for AAA Projects

At the other end of the spectrum, some developers may put you in the position of writing music under a very tight timeframe. If at all possible, it is best to maintain a consistent work schedule and try to anticipate deadlines with room to spare. This is not always under our control, unfortunately, and the occasional 12-plus-hour workday is something most of us have to deal with on occasion. When this happens make sure not to neglect your mind and your body. Eat regularly and take breaks often to keep your mind sharp. After the deadline has passed, take a day or two off, even if it is mid-week. Overworking yourself is the quickest way to mediocre work and to a potential decline in your health (see "The Pyramid of Sustainability," Chapter 10, page 345).

When planning a soundtrack it is very common to find logical deadlines for composers to deliver a number of tracks. These are called **milestones**. Milestones are important because they keep teams on track and help you plan out reasonable timeframes to write within. Overall, when deciding on these milestones with your developer, it is important to keep in mind that they are very subject to change as issues come up during development. Try to be flexible, but keep your writing as consistent as you can. If you're juggling a few projects at once, find an appropriate amount of time per week to devote to each project. This will help you avoid having to reacquaint yourself with your previously written material each time you sit down to write.

WORKING WITH A TEAM

Especially at the indie level, it is quite common to see game composers responsible for roles like music editing, orchestration, implementation, recording, and even sound design and implementation (refer back to the companion site, Chapter 1, for more roles in game development). However there are times when a composer may end up working as part of a team, or even as a co-writer on a project. These collaborative roles are a great way to gain a foothold in the industry and make some powerful connections. If you are lucky enough to land a project at the AAA level it is unlikely that you will be solely responsible for the above tasks, so it is important to know how to navigate these interactions. We have listed a few guidelines for common game music roles below.

Orchestrator

As an orchestrator it is your job to take the MIDI mockup that the composer has created and use it to create sheet music for the orchestra. This could go one of two ways; it could be a transcription job where you are simply taking notation in one format and delivering it in another; or it could be a very creative process whereby you interpret the composer's work and help the music "speak" in the orchestral medium. These are two very different approaches however, so it is important to clarify at the start how much creative control you have. Either way, be conscientious. The composer has likely spent months or even years working on the score, and may have a difficult time relinquishing control.

Engineer (Recording, Mixing, or Mastering)

As an engineer you are responsible for recording (and possibly mixing) the game soundtrack. There is less ambiguity here in terms of creative control. The composer will usually have an idea already of how the music should sound and your role will be to create a recording environment that suits the imagined sound. If you are mixing or mastering, then your role will be to take the pre-recorded tracks and make the mix as clear as possible while maintaining the emotional intent of the soundtrack. If you are placed in the role of engineer, get to know the music well. The composer may be open to ideas on how to achieve a certain aesthetic.

Performer

Musicians and performers have one of the most fun jobs in the realm of game audio. If you are performing on a soundtrack then you are part of one of the final stages of the entire process, and you are the closest link the composer has to the players. In regards to creative freedom different projects and composers will require different types of performances. Some musical styles require precise adherence to the sheet music. Others require a high degree of improvisation. The most important thing as a performer is to open a line of communication with the composer up front to clarify what she is looking for. Quite often the composer will research a particular performer to find the right sound for the game. If this is the case then the music should already be well suited to your playing style. Even if this is not the case, composers often welcome extra takes with some improvisation or interpretation, so don't be afraid to ask up front what is needed from you.

Composer/Co-Writer

There are two main methods of co-composing. The most common way is to work on separate tracks. Games like *Civilization*, *Detroit: Become Human*, and *Lawbreakers* use locations or characters as an opportunity to call on different composers to write in contrasting musical styles. Another method is to find elements within each cue and split up the work. For example one composer can work on the lyrics and vocal components of a track while the other works out the instrumentals and orchestration. Bonnie Bogovich and Tim Rosko used this method of collaboration for their title theme to *I Expect You to Die*. This method can be difficult, and it requires flexibility and open lines of communication for feedback. It can be very rewarding however, and sometimes it leads to long-lasting partnerships.

PLATFORMS AND DELIVERY

When creating your soundtrack it is important to consider the final listening format of the game. The differences in audio capabilities between consoles and mobile devices are considerable and there is little consistency in home speaker systems. If you are delivering for mobile, it's important not to overcrowd your arrangements. Mixes can become muddy very fast, and frequencies above 15 kHz can sometimes be piercing. Creating a clear arrangement is the first step to a solid mix, but EQ can also be used on the master to declutter some of those problem frequency ranges. If you are composing for console then a good mix will generally transfer well. The key is to produce an effective arrangement, and a clear mix will likely follow. Take the time to check your mixes often on a few different listening devices and make the appropriate changes. We will dive back into some more specifics on mixing your soundtrack in Chapter 9.

Most developers will ask for a specific delivery format so that they can prepare the files for implementation into the game engine. Music usually takes priority in terms of fidelity so most often you will be asked for WAV files in stereo, but there are exceptions. A developer may ask you for mp3s to save space, but these do not loop well and are therefore not adequate choices for music. If space is a concern a better choice would be Ogg Vorbis. This is one example from many, so it is best to familiarize yourself with the majority of file types and compression formats (for more information see Chapter 8).

The Sound Lab

 Head over to the Sound Lab for a wrap-up of the topics discussed in Chapter 5. We will also explore mood and immersion. We will return to this same example later on to dig into some more complex musical techniques.

BIBLIOGRAPHY

Clark, A. (November 17, 2007). "Defining Adaptive Music."Retrieved from www.gamasutra.com/view/feature/129990/defining_adaptive_music.php

6 Composing Music

In this chapter we will dive into the theory and practice of composing music for games. We begin by discussing the creative cycle and some natural starting points for your music. Later in the chapter we will cover, in depth, various approaches to composing melody, harmony, and rhythm. We end the chapter by outlining some tools for designing a musical palette which will help you realize your core compositional ideas and turn them into finalized pieces of music.

GENERATING MUSICAL IDEAS AND THE CREATIVE CYCLE

There is plenty of information about music production, mixing, and recording, but it is unusual for books and tutorials to discuss the actual process of generating musical ideas. This is a shame because it is the foundation of an effective game soundtrack, and it is often the most mysterious part of the whole process. There is no universally successful approach to composition. But there are common stages of the creative cycle that most composers are familiar with. The more familiar you are with this process, the more control you will have over it and the easier it will be to call upon when needed. We have outlined a few of these general phases below.

> The ideas and techniques that follow are completely separate from any decisions on the instrumentation of your music. These methods can be applied to samples, live instruments, or synthesized/electronic instruments with equal effectiveness. In most cases it's best to start with a simple piano timbre so that you have a "blank slate" so to speak. Certain timbres by their nature evoke emotions and this may influence your compositional choices.
>
> Spencer

Preparation

Creating a unique soundtrack for a game is a daunting task, and can be overwhelming at first. It can be helpful to begin with a *preparation phase*. The purpose of this phase is to give yourself a direction for the creative action you are about to take. Similar to steering a ship, it is important to know where you want to go *first*, even though you'll likely make course corrections along the way.

Logistically speaking, this phase should begin with a flexible schedule to plan around any deadlines or milestones. This will help you remain organized and on task. To make the most of your time you can budget about a week (or more depending on the timeline) to let the details of the project "simmer" as you are identifying a direction for your project. In particular, you will need to decide which elements of the game (if any) will need unique themes. These elements could be characters, areas of the game, or even more abstract items such as plot elements.

This part of the process is roughly analogous to **spotting** a film score. Spotting is when a composer and director sit down together to watch the movie in its entirety, making note of what aspects of the film need themes or sonic identifiers, and where and how those identifiers should be developed. It *is not* the time to determine exactly what you will be writing note for note. It *is* the time to determine what kind of music the film needs and where it needs it.

This is how you should approach this phase when working on a game. Your task in this phase is not to write your musical ideas, but to determine *what the game needs and where it needs them*. Make sure you immerse yourself in whatever early development materials you have, and allow your ideas to percolate. Gather artwork from the game, map out storylines, play any and all preliminary builds, and watch gameplay videos. Ask yourself the foundational questions: What instruments *feel right*? What harmonic approaches speak the emotional language of the game? What melodies speak the same language? Does the game feel like it needs the emotional specificity that themes convey? Or does it need something more ambiguous like a musical texture?

Above all, it is important to start listening to music analytically. If the developer has particular reference music in mind, then pick apart the elements that are relevant to the game and think of how you can use similar ideas in your score. If there are no references then come up with your own based on the information given to you. It is sometimes most helpful to listen to music that is completely contrary to the prevailing aesthetic of the game. For instance, if you are scoring an action game, listen to slow and emotive string quartet music. If you are doing a Western shooter, listen to chiptunes. Often you will find that incorporating elements of unexpected musical styles is the perfect way to add a distinctive flare to your soundtrack.

When you have a clear direction for your project, and your brain is brimming with possibilities, it is time to take action and move on to the *brainstorming* phase.

Visiting Artist: Penka Kouneva, Composer

The Pre-Production Process

I receive art (characters, environments), PowerPoint proof-of-concept, or artistic vision-type documents, and prototypes (one or two levels-in-progress). Understanding the genre and the aesthetics is paramount. Getting a sense for the gameplay is also vital, as the games are so diverse. Scoring a platformer (such as *Mario*) would be an entirely different experience to scoring a medieval melee multiplayer (such as *For Honor*). Or scoring a sci-fi dystopian VR game would be approached completely differently to scoring a female-protagonist storytelling indie game like *Her Story*. The conversations about genre, aesthetics, tone, characters, the game's overall feel are the first ones that take place. I usually don't write a note until I discuss with my collaborators all their ideas, background stories, and expectations about the music.

Thoughts on Working with Collaborators

As a collaborative composer working together with other creative people, my #1 priority is to get a clear sense of their musical expectations. I always ask for "music style guides" – mp3s and YouTube links. These are models and inspirations from past game and film scores. Then I ask how these "models" fit their vision (and also what they don't like). The style guides function as a rear-view mirror to help us navigate the scoring journey and move forward. I love researching similar games in the same genre, and the history of the franchise. I listen to a lot of music on a daily basis.

Brainstorming

Many people think that composing music begins with a jolt of inspiration that mysteriously presents itself to "gifted" individuals as a fully formed musical idea. Make no mistake – inspiration can be helpful, but it is virtually useless without *years of practiced experience*. Although it may *feel* like a jolt of inspiration has hit you, creativity is actually the product of our associative memory. When you sit down at the piano to construct a melody, you are actually activating memories of previous melodies (and many, many other memories) that you've heard or written in the past, and combining them to form what will become elements of your musical ideas. Creativity then, is more an act of *synthesis* than of creation from nothing.

The consequence of this definition of creativity is that as a game composer you need to draw from various musical styles and frameworks to create new and interesting music. The more disparate those styles are, the more unique your game will sound. It is then of the utmost importance to give yourself a very broad spectrum of musical experience. If you make it a point to experience a large and diverse range of music you will have a much easier (and more fun) time combining elements to create something unique and meaningful. If you narrow your experience, and only keep your mind open to a tiny portion of the music in the world, then you will likely have less initial material to work with.

> A helpful way to envision associative creativity is with musical genres. rhythm and blues and country are two distinct genres of music. However, if you combine them we arrive at rock 'n' roll, a unique genre with its own particular flavor. By combining two distinct things, something completely new emerges.
>
> Spencer

With the powers of our associative memory in mind we can begin *brainstorming*. We are all probably familiar with brainstorming as it was presented to us as a grade school activity. It is a tragedy that most of us lose touch with this classic art as adolescents because its benefits are immense. This is especially true for newer game composers. People new to a creative task tend to benefit more from experiencing *many iterations of the creative cycle* than they do from belaboring one aspect of it in great detail. In other words, if you are new to composing game music you will benefit most from *writing and finishing* a large amount of music as opposed to working on a single track to perfection. This fits in perfectly with the preparation and brainstorming phases because they prime us to produce a colossal number of ideas to work with. Many of these ideas will be bad, but a few will be usable. It doesn't really matter as long as you start *somewhere*. Once you have this starting point, you can then move on to the *refinement* phase.

Visiting Artist: Wilbert Roget, II

Brainstorming 24/7

Ideas can come from anywhere. When composing a full-length score, one can work actively for hours at a time, but inevitably the mind will continue to find new perspectives, ideas, and musical concepts throughout the day. It's important to be able to capture these ideas whenever they arise, as efficiently as possible.

On the most basic level, I begin every major score with a simple text file. I might write things as specific as individual melodies or chord progressions, or

(Continued)

as broad and abstract as an adjective or a color. Most often it will be instruments, sounds, sonic concepts or performance techniques that I think might work as signature sounds for the score. I'll edit this document continually throughout the entire writing process – as the score evolves, I might cross out some elements that are no longer relevant, or move more successful ideas towards the top of the list, or simply define some concept more specifically with how it will work in game.

Additionally, I make sure that wherever I am, 24 hours a day, I have some way to capture musical ideas. When I'm outside, I have a pocket-sized staff paper notepad that I use for sketching themes or harmonic progressions. Occasionally for rhythmic ideas, I'll sing into my phone's audio recorder app. My bedside has another notepad and a metronome. And even in the shower, I have waterproof post-it notes that I can use to scribble down melodic concepts.

Lastly, to keep everything in order, I'll eventually copy my text file up to a Trello board (an online note-taking software), so that I can add to it on the go from my phone. These managed brainstorming techniques not only help me keep the music consistent with well-thought-out concepts, but also ensure that I can keep writing with fresh ideas throughout the score.

Refinement

In the *refinement* phase you will select your musical ideas from the brainstorming phase and develop them. Where brainstorming was *intuitive*, refinement is *analytical*. It is the most time consuming and technically demanding part of the creative cycle, but it is also the most important. Here, smaller musical elements will become fully realized pieces of music. This is also the time to begin thinking about the ensemble that will sound your melodic, harmonic, and rhythmic ideas. It can be a real challenge to decide which ideas to develop, but we must be bold and merciless. Only the best ideas should make it through in this phase, and there will be many to choose from.

Note that as a composer you have the ability to go *with* or *against* your intuitions. As we were outlining earlier, brainstorming is an intuitive process which allows your associations to help you generate small ideas. Many of these ideas will come from heavily trodden or even clichéd territory – that's perfectly fine, and sometimes necessary. However at this point you may decide that something more novel is necessary for your project. The refinement process will allow you to take a step back, analyze what you have, and work out something that goes *against* your intuition (and potentially the intuition of the player). This doesn't work all the time, but it can be a great way to keep your music sounding fresh.

There are many ways to choose which musical ideas will fit the soundtrack. The most important thing to remember is that every choice you make *must* support the needs of the game. In other genres music composers are free to make choices that suit them and their musical preferences, but this is *not true* of game music. You can certainly push things in one direction or another to sculpt your sound, but the end result must suit the game on some level. The best way to determine if the music suits the game is to play through the game yourself and take note of how the music is affecting gameplay (see "Testing, Debugging, and QA," Chapter 8, page 304). Is your music exciting, or is it slowing the gameplay down? Does it rush the player through important puzzles? Is it overpowering the story or the sound effects? It's crucial at this point to be analytical, but also to really think about how the music *feels* in game because that is exactly how the player will be experiencing your soundtrack.

Another aspect of the refinement phase is development. The way you develop your material is important for delivering an emotional impact to the player. If you begin the game soundtrack on the highest note possible, then that leaves little room to build up again for the ending. At the same time if you start your score too subtly, you may fail to grab the attention of the player. For this reason a deep understanding of the game and how it is to be played is *essential* to the refinement process.

Once you've thought through what your game needs and have decided on your core ideas, it's time to start developing your material. Two important types of development are **motivic** and **structural development**. Motivic development comes from the word *motif*, which is a short fragment of a musical idea. A motif can be a melody, chord progression, or even a particular rhythm. These motifs must then be explored and expanded upon throughout the game. How you expand on a motif is largely determined by the motif itself and by the needs of the game. If you start out with a melodic fragment, then you will likely want to create variations on that melody. You can lengthen it, shorten it, transpose it, invert it, put it in retrograde (reverse it), or use any combination of these operations to achieve a different emotional impact while retaining the core idea. Structural development refers to the form of the music, which we will get to later in the chapter (see "Form and Structure in Game Music," below).

Revisions

Lastly, this is the phase where you will begin *revising* your material. A revision is the act of adjusting one or more elements of a cue based on feedback. Feedback can commonly be in reference to instrumentation ("change out the trumpet for a clarinet"), melody ("I like the chords but I don't like the melody"), or something broader like the overall emotional impact ("this section makes me feel a bit sad, but I should feel happy"). The important thing here is to learn how to translate the feedback that you receive into actual musical changes that better serve your game. This can be difficult at times, and we dive a bit deeper into digesting feedback for audio in Chapter 3 on the companion site.

Revision requests can come from either your client (the game designer/developer) or yourself. Although there are ways to amiably and professionally offer your own opinions on how the music should be written, the needs of the game must always come before your own artistic aspirations. Regardless, the revision process is an important and *inevitable* part of the creative process, so don't shy away from feedback. Welcome it and your music will improve exponentially over time.

It is important to maintain an analytical perspective in the refinement and revision phases, especially when it comes to the implementation of your music. For this reason, many of the revisions you make should be your own. Nobody knows your music better than you do, so make an effort to be part of the process of revision and implementation. Do your best to make sure the music adapts smoothly to all game scenarios and delivers the desired emotional impact to the player. This iterative process is incredibly important toward creativity in general, and especially with regard to achieving the best and most appropriate music possible for your games. When you have approval (from yourself and your client) for all cues, it's time to move on to the *polishing* phase.

The Final Polish

The final phase of the creative process is called *polishing*. In this phase most of the creative work is already done; the focus is on top-notch quality and overall cohesion. This is similar to mastering an album. The writing and recording is done, but the engineer must go through and balance the levels of each track, and ensure the album as a whole sounds compelling. In games this means tweaking any implementation errors, smoothing musical transitions, and balancing cue levels with sound effects and dialogue.

Visiting Artist: Jeanine Cowen, Composer, Sound Designer

Planning Your Music

Even after you've decided on a direction, tonality, arrangement, approach, etc. to the score, it can be a daunting task to decide where to actually start writing within the context of the game. Although music is an art form that we can only experience linearly, there is no need to take a purely linear approach to writing your score. You don't need to start at the beginning of the game and write through your ideas to the end. On the contrary, in many cases it will be better for you to start at the end.

> *(Continued)*
>
> ....
>
> Think about the score and how it will be realized in the final battle scene, or when the player reaches their goal. If you can write the fullest or sparsest amount of music needed for those scenes, then it will be easier to know where your score is aimed. The biggest mistake I see with new composers is that they just start "at the beginning" without truly understanding where the music is going! Some of the best scores have been written backwards ... starting with the end first.

STARTING POINTS FOR MUSICAL COMPOSITION

Now that we are familiar with the creative cycle, let's take a deeper look at the actual process of composing. First, we will outline a few of the many possible starting points for the preparation phase. These are the elements of a game that you can focus on to spark some initial ideas in the preparation phase. The good news is that games are multi-dimensional pieces of art and entertainment, so they come with tons of material for us to draw from. Usually this material will influence one or more musical elements, which in turn help us to work out starting points for these ideas. Looking at game specifics like art and story can influence almost any aspect of a score. However, most composers find that these details will most likely inspire melodic, harmonic, or rhythmic ideas. Equally likely to emerge will be thoughts about **musical palette** (the instrumentation and processing we choose to use). The process of translating game elements to musical ideas is very idiosyncratic, but below we will offer some tips to help guide you through the process.

> The first ideas that come to my mind are always about musical palette. A Spaghetti Western-style shooter might bring to mind a palette that includes electric guitar and whistling. A puzzle game may likewise bring up a chord progression or a mysterious melody. Sometimes these ideas aren't even instruments, they are just electronic or processed textures or ambiences. When these ideas hit you, take note! They can become the foundation of your score.
>
> Spencer

Artwork

Many composers choose to focus on the artistic/visual style when they begin writing. This is a tried and true method. Art and music traditions have always influenced each other, and will continue to do so. Games are a shining example of this. Music plays a large part in supporting the visuals of the game, so focusing on art is a great place to start drawing up musical ideas.

When approaching writing from the perspective of artwork it is important to have a vocabulary to describe what you're seeing. Is the art style *darker* or *lighter*? This will determine how dark or light the *mood* of your music can be. Is the scope large and broad, or small and detailed? This could give you ideas for the *size* of your ensemble. Is the art fantasy oriented, realistic, steampunk? These factors can and should inspire some interesting motivic and instrumental combinations for you. Even conjuring a timbral association could be a really helpful starting place. The main point is to use the artwork as a basis for your creativity to run wild.

> Timbre is the character or quality of a musical sound or voice as distinct from its pitch and intensity. It can also be applied to purely synthetic instruments and textures. In the steampunk example, you may not even need an instrument to communicate a steampunk aesthetic – a gritty, metallic ambient texture may work just as well.
>
> Gina

The Characters and Story

The characters and story of a game are another starting point for generating motifs and musical ideas. For example, in every *Final Fantasy* game there are character themes that play and develop as the game progresses. Starting a project by brainstorming character themes is a very effective approach because these themes will be an important point of emotional connection for the player. They often set the tone for the rest of the soundtrack, especially for incidental and background music. Thematic variations that occur later in the story arc will serve as satisfying musical developments and can help foreshadow or highlight turns in the plot. Music can also influence how a player feels about a certain character, so we hold a great deal of storytelling power as composers.

A simple example of this is to take the basic mood of the story and write something that forms a strong connection to the overarching mood. Similar to the artwork approach, if the story is dark and brooding you can use that element to choose your palette. You may use only low strings as your entire string ensemble. In fact, Jason Graves did exactly this with his score for *The Order: 1886*. The omission of high strings is subtle at first listen, but darkens the quality of the orchestra entirely. The density is also noticeably thicker due to the limited range of the orchestra, thus producing closer voicings. This mirrors the more nefarious and horrific elements of the plot and sets the tone for the events that take place in this alternate history.

When creating character themes look at the character's personality and appearance as well as their overall development arc. A character who is bright and upbeat may have a bittersweet ending, and the music should reflect that in some way. A character that winds up betraying his allies may have a slight note of darkness in his theme. Conversely, an antagonistic character that becomes an ally may have a hopeful note in his theme (or in a variation of that theme). It is important to be aware of what you are and aren't emphasizing as a composer – an obvious emphasis in one direction or another can spoil a twist in a character arc.

The flip side of this method is using the story as a way to generate musical ideas. This is an example of **conceptual music**. Conceptual music takes concepts or narrative themes and realizes them musically. Imagine a scene where a super villain created a doomsday device that reversed time. To turn this into conceptual music we might record glass scraping sounds and reverse them to fit the track as part of the sound palette. Note that with conceptual music the concept translates very literally into music (i.e. the reversed glass is literally a reversal of the sound in time).

Game Mechanics

This one is a bit less obvious, but taking inspiration from the game mechanics can be a great place to generate ideas. You likely won't have any immediately strong associations when taking this route. Nevertheless, game mechanics play a definite role in musical development. Platformers sound quite different than tower defense games, which in turn sound different than role-playing games, and so on. An infinite runner is a great example of this. In these games players must react quickly to avoid obstacles as the environment is procedurally generated. The pace may increases with distance. As a composer it is necessary to match this fast-paced excitement. Usually this translates to exciting rhythmic ideas or a quick catchy melody. Various percussion timbres are often used.

Similarly, a game might be very cerebral and puzzling. In this case something much more subtle and intricate may come to mind. An electronic palette could be employed here, with a somewhat restrained rhythm section. In this case the music needs to stay out of the way so that the player can think through the puzzle strategically. In either case, the game mechanics are an important foundation to the ideas you will be brainstorming.

Visiting Artist: George "The Fat Man" Sanger, Composer, Author

Handling Direction

How do you handle direction when, for example, a client asks for music by referencing Jurassic Park *meets Hans Zimmer's* Call of Duty?

I'm pretty sure Team Fat were first to emulate John Williams in a game, because we were asked to do so for *Wing Commander*. But it so happened that

(*Continued*)

our composer Dave Govett was a huge John Williams fan, that was the music of his heart, and in his high school days he had already mentally composed what we now know as the battle music and theme for *Wing Commander*. It was just a matter of scribbling them down for this game. "Be yourself when you are imitating." [Sideways smiley face.]

Humility and arrogance are best put aside; take your ego out of the equation and look at the work and the mystery that you've been invited to delve into.

These days I try first to find that place in my heart where I am sure I have no idea at all what to do next, or even how to score a game. What feels at first like panic is a sign that the mind is becoming unmoored from routines, habits, and formulae. That is what many great composers do, so I take The Fear as an auspicious sign. Then, because the client has given that direction, I have permission to go back to Composer School, to dip into these great composers' work and see what tools they use, and see if some of their insights can sit in my own toolbox so that I can make something that is new, that is from me, but in some way stands alongside the great movie/game moments in its ability to move the hearts of the players. Because that is really what the client is asking for.

Blues players sometimes say that there is one long note that has been playing since before the beginning and will play for all time. For a while, we get to join in. Who could ask for more?

COMPOSING GAME MUSIC

Once you have familiarized yourself with your new game project it's time to start creating some musical material to work with. Using the game elements above, you should have a basic framework for directing these ideas. There are many musical elements that you can use as a foothold to start writing your soundtrack, but for our purposes we will focus on melody, harmony, rhythm, and musical palette.

In each of the following approaches the goal will be to create *simple ideas* that will eventually form well-developed (and adaptive) musical systems. It's not important that each idea be spectacular, it's only important that you have a large number to work with from the start. This will ensure that you have some solid material when moving on to the refinement and polishing phases. Although we will learn many ways to construct musical ideas in this section, it is (again) important to remember that these ideas are *only* effective if they fit the game. The following examples are all *possible* starting points, but no technique is as useful as harnessing your intuition to understand the needs of a game and how music can fill those needs. Your intuition is a powerful tool. These methods can help to ignite and expand it, but not replace it.

Generating Melodic Ideas

Melody is an extremely important tool in game music because it is the most distinct and "visible" aspect of a soundtrack. Players can easily identify with a memorable melody whether it is presented subtly or in a more obvious manner. We will now go over a few methods for generating melodic motifs. In an actual game, these motifs would become the core of a well-developed thematic soundtrack.

Although this goes against common practice, one of the first considerations to make when writing a melody should be shape. The shape or *contour* is a very general way to conceive of a melody, yet it has an important impact on how that melody is received by a player. A continuously rising melody will usually translate to an *increase* in drama or tension. In contrast, descending melodic motifs are mostly associated with a *decrease* in tension because they sound more resolute and final. An unexpected rise at the end of a static melodic line can also evoke a feeling of hope or optimism while an unexpected descent yields a feeling of failure, or a sudden mood shift from hopeful to dreary. Static melodic lines (lines with limited movement up or down) are quite flexible in that they can be used to propel the melody forward or hold a melody in stasis, halting the forward momentum. The difference here usually lies in the way that rhythms are developed. If a fast rhythmic motif is used the player may feel a sense of forward progression, but if long sustained notes are used then a forward drive is unlikely to be achieved through melodic means alone. In many cases, if a melody remains on the same pitch, regardless of the rhythm used, it is usually perceived as a conclusion. These patterns don't come without qualification of course, and there are exceptions to every rule. It is a helpful guideline though to keep in mind how the shape of a melodic line will likely be perceived by the player.

Another important consideration is the scale or **mode** of your melody. This is usually what composers consider first, but scales can be altered and key signatures can be changed. There are many more ways to create and develop a melody than by using a rigid set of notes. This is why we recommend conceptualizing a melodic shape before boxing yourself into a particular scale or mode as this can sometimes limit creativity.

All that being said, a broad knowledge of scales and modes is a great way to begin creating interesting melodies. It won't be surprising that the most commonly used scales are the traditional major and natural minor scales. These are great starting points because they are used so frequently. In the Western world most of our initial ideas will probably be within the traditional major/minor framework, and for good reason. A very broad range of emotion can be struck using only a major scale. When you already have a melodic shape in mind to combine with the notes of a major or minor scale, your melody can emerge very quickly and naturally. The same can be said of modes, however. Modal melodies can often be even more nuanced and distinctive because of the slight variation on the usual major/minor mode. Each mode has a particular flavor and color, so they can sometimes inspire motifs themselves. If you haven't already, they are absolutely worth exploring.

A great way to explore the sound of these modes is to pick a mode and improvise for some amount of time. This will solidify the unique character of that mode in your ear, and it will likely also offer a few melody fragments for later use. Another way to use these modes melodically is to take a melody that you've already written out in a traditional major or minor scale, and change it to a similar sounding mode. For example if you have written a phrase in C natural minor, try raising the sixth scale degree to transform it into C Dorian. You can also use this tactic during the ascent only, and then during a melodic descent revert back to the natural minor, which would offer a unique sounding **mode mixture**. This occurs when using two or more modes in the same melodic passage. Keep in mind that this works with any mode. So major modes can both transform into, and mix well with Lydian and Mixolydian modes.

Another simple way to add interest to a melody is a technique called **semitone offset**. Take a melody that you have written where the phrase resolves on the tonic (or another foundational chord tone) and change that note by a half step. The melodies that work best with this method are melodies that generally move in one direction toward a "goal" note. To start, take a melody that is mostly ascending by half steps and whole steps and ends on the tonic, then raise the tonic by a semitone. You now have a melody that still has a satisfying resolution, but has propelled us into a completely new area emotionally.

One important note here is that in order for the semitone offset to work, you must rely mostly on step-wise/scalar motion, especially when approaching the goal note. Raising the goal note to create a leap of a minor third will throw off the expectations of the listener and you will lessen the sense of direction in the melody.

When a game scene calls for something more esoteric than the above scales and modes, you also have the option of using **symmetric scales**. Symmetric scales are constructed in a way that the half-step whole-step pattern is symmetrical, as the name implies. Commonly used symmetric scales are the chromatic scale, the whole-tone scale, and the octatonic scale (Figure 6.1 on the companion site). The chromatic scale is probably the most familiar as it includes every note in Western music (excluding microtones). This makes it an ideal candidate for modulation of a melody, and many romantic and late-romantic era composers took full advantage of this.

By contrast, the whole-tone scale is often used as a method for keeping a melody in stasis. This is because the whole-tone scale is a **mode of limited transposition**, meaning that you can only transpose the scale once before you end up with the same notes you started with. Whole-tone melodies, frequently found in impressionist-era music and dream sequences in games, sound exotic but can find it difficult to capture a sense of momentum using only the melody because there are no half steps in this scale.

The octatonic scale is also a mode of limited transposition and it is used often in twentieth-century works by composers like Igor Stravinsky and Olivier Messiaen. Because this scale is created by alternating half steps and whole steps, it has nine notes in it. This pattern also makes it very powerful in terms of modulation because the half step can easily be used as a leading tone moving to the tonic of a new scale. This works best in the context of mode mixture. For example you may base a melody in a natural minor key. When looking to modulate, you could then alter it by introducing the octatonic scale briefly so that the new half-step/whole-step pattern *overrides* the pattern in the original scale. Thus it creates melodic momentum toward a new key. This momentum is achieved by replacing the leading tone of the original scale with a pitch in the octatonic scale, altering the pattern. In general, these symmetric scales allow for some very alien-sounding melodies. They work well for thrillers and horror but don't always work if the goal is to compose a very simple or catchy melody.

Another way to construct a melody is by organizing intervals into a spectrum of consonance and dissonance. The concept of consonance and dissonance can be influenced by many factors including timbre, dynamics, culture, and tradition.[1] Intervals can nonetheless be useful in relation to consonance and dissonance because most of us have subconsciously absorbed comparable harmonic patterns. Thus intervals can help determine the level of dissonance and *density*[2] in a melody based on the ratio of frequencies as compared to the **harmonic series** (Figure 6.1). Check out the Sound Lab (companion site) for more resources on the harmonic series.

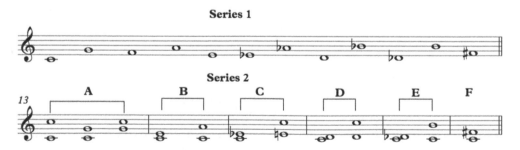

FIGURE 6.1 The harmonic series.

The Harmonic Series

Series 1 shows the order of overtones based on the fundamental (C). Note that the farther we go from the fundamental, the less weak the harmonic relationship is. Series 2 organizes the intervals found in Series 1 based on this relationship. Group A consists of the perfect intervals, sounding pure and open, and offering no real harmonic interest. Group B contains a major third and a major sixth, which are consonant but more harmonically dense that Group A. Next we have Group B, which contains a minor third and minor sixth, which are darker and thus slightly more dissonant than Group B. In Groups D and E we start to see some more tangible dissonance as we now incorporate seconds and sevenths. Finally we reach Group F, which contains only a tritone of C–F#, the least harmonically related interval in the C series. Note also that Series 2 not only gives us a spectrum of consonance to dissonance, it also gives us an order of stability for intervals (notice that thirds always come before sixths). This has important applications in melody and harmony, as well as in orchestration (see Chapter 7). For more information on the harmonic series refer to Sound Lab (companion site) where we provide additional resources.

Using intervals rather than scales or modes as a starting point for writing melodies provides two advantages. The first is that it gives us a more manageable starting point; it is easier to find an appropriate interval than it is to find an appropriate melody for a given scene. Sometimes all you need to really capture the essence of a moment is two notes, and with the organization of consonance and dissonance finding the right notes is not such a difficult task. If a scene calls for a hopeful and romantic melody an ascending leap of a major sixth will often achieve that mood. From there we can easily fill in some incidental flourishes using the major scale and we are left with a melody that hits the core of what we want our players to feel. A classic example of this (although not from a video game) is John Williams's use of the major sixth leap in "Marion's Theme" from *Raiders of the Lost Ark*. It's worth noting here that the wider the interval (as mentioned earlier regarding melodic shape) the more dramatic that melodic line will sound. A leap of a major sixth is wide enough to add some passion to the melody, but small enough that it still sounds lyrical.

The second advantage is that intervals make it easy to modulate the melody later. Intervals respond easily to **transposition, inversion,** and **retrograde** operations as well as traditional tonal modulations. These processes are important to remember when creating melodic ideas as well as developing them. Transposition moves intervals higher or lower while maintaining the same relationship between the notes. This works the same way as transposing a song or a chord progression into another key, but when we use this technique on a melody it creates cohesion and structure. **Sequencing** is a technique where multiple transpositions are applied to small melodic fragments and quick succession.

Inversion of an interval works differently than inversion of a chord. A chordal inversion rearranges the pitches in a chord from bottom to top with respect to the chord tones. Inversion of an interval rearranges two or more pitches with respect to an **axis of symmetry**. Inverting of pitches always maintains the distance to the line of symmetry, but changes the pitches themselves. For example, take the interval C–G (a perfect fifth). It is common to use the lowest note (C in this case) as the axis of symmetry. Think of this as a mirror for our second pitch. The G is a perfect fifth above the axis of symmetry; by inverting it we are left with a perfect fifth *below* our axis of symmetry. So the G above C becomes an F below C.

It isn't necessary to always use the lowest note as our axis of symmetry. If we take E as our axis of symmetry, inverting our C–G interval would yield a different result. We now have to take each note individually with respect to the axis. Where C was a major third *below* E, inverting it will leave us with a G# a major third *above* E. Inverting the G (a minor third *above* E) will yield the C# a minor third *below* E. By inverting the same interval with a different axis of symmetry leaves us with a very different result and therefore gives us more options for motivic development.

Retrograde is actually a very simple operation. Retrograde reverses the order of the notes in an interval sequence. So C–D–G will become G–D–C. This has obvious implications for a melody, especially if you also invert the *rhythm* of the melody. A wonderful use of retrograde is "Ballad of the Goddess" from *The Legend of Zelda: Skyward Sword*. During the middle sections of this track a retrograde version of "Zelda's Lullaby" is played. By playing the track in reverse you can hear a clear rendition of the melody in "Zelda's Lullaby." This is a perfect example of these type of interval operations having a deeply meaningful effect on player experience. The act of simply reversing a melody leaves a subtle but tangible feeling of nostalgia and importance on the listener, making the track emotionally significant. Curious players who actually discover the hidden Easter egg of the reversed melody will find the track has an extra dimension of depth not noticed at the surface level.

Retrograde can also be applied to a melody in tandem with inversion to create a *retrograde inversion*. This is the result of both operations occurring simultaneously. Although these operations can seem overly complex and inconsequential at first, sonically they achieve a very nuanced yet noticeable connection to the original interval sequence or melody.

An underutilized (in our opinion) method for generating melodic ideas is the **twelve-tone method**. This method essentially assigns a number from 0 to 11 to every pitch in the chromatic scale. The idea is to create an order for all twelve pitches into a **tone row** and to use it to compose a melody. The strict rules (as created and refined by Arnold Schoenberg and his students) are to avoid repetition of any note until all twelve tones are iterated. However, much of the fun in life comes from bending and/or breaking the rules. Composers in the past have bent or broken the rules of the twelve-tone method in various ways, and we encourage you to do the same!

As a technical tool, composers use *twelve-tone matrixes*, which visually show a tone row as it undergoes all possible transpositions, inversions, and retrogrades. Taken to the extreme this method can create incredible complexity, with many tone rows iterating simultaneously and many operations taking place at once. Depending on how you treat your melody, the resulting sound can range from terrifying and alienating to novel and strikingly beautiful.

For our purposes the twelve-tone method can be used simply as an idea generator. It is not necessary to adhere strictly to the rules, and in fact can be used quite freely to create a contrast between consonant and dissonant melodic fragments – sometimes within the same melody. It also becomes immensely helpful when the all-too-common "writer's block" hits, and you are feeling stuck on one melodic idea. Often just a small portion of a tone row can create new directions for a melody to take. Many of these directions will be non-tonal as the twelve-tone method was created precisely to *avoid* Western tonality. This works out phenomenally well for many styles of artwork and gameplay. Horror and thrillers work especially well here due to the eerie lack of recognizable motifs. However this method is a tool, and tools can be used creatively. By bending the rules and using accessible rhythmic figures you can create melodies that are beautiful in their tonal ambiguity. By employing simpler rhythmic devices and groove-oriented rhythms it's even possible to achieve a funk or fusion sound.

The twelve-tone method is used to good effect in the independent game *BestLuck*. In a track called "Rain Puzzle" the tone row emerges first in the viola as a transition from the chaotic first section (m. 8 in Figure 6.2). Although the intensity and density of the cue is brought down at this point, the tone row used and the eerie **sul ponticello** (bowed at the bridge) timbre keep the tension high. The row also undergoes a few transformations as it moves into different instruments before transitioning back into tonality in the next section. In this way tonal elements are juxtaposed with the twelve-tone method in an accessible way and aids in developing the tonal material. It also functions as a useful way to transition between sections. Structurally it differentiates the development while building to a satisfying climax.

Another method that can achieve similar results to the twelve-tone method (but is more commonly used) is **aleatory**. Aleatory is the use of randomness in music by degrees or in totality. It is spiritually the opposite of the twelve-tone method but can deliver similar sounding results. Where the twelve-tone method is pre-organized and controlled, aleatory incorporates randomness and relinquishes control. Like tone rows, aleatory is often used in horror-themed games and films to achieve a sense of alienation and uneasiness in an audience. For example, a common aleatoric technique is for composers to notate an upward arrow with instructions for musicians to *"play the highest note possible."* The resulting sound would be entirely dependent on the ensemble performing the technique.

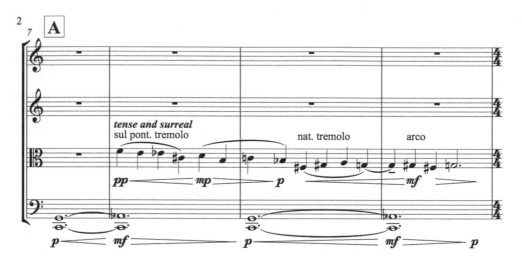

FIGURE 6.2 The tone row as presented in the viola in mm. 8–9 and later handed off to the rest of the quartet.

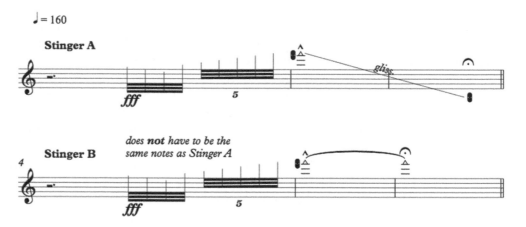

FIGURE 6.3 Aleatoric stingers from the game *Evil Nun* (Keplerians).

Although this technique involves some performer improvisation, the sonic impact of the gesture as a whole can be predicted and exploited. If only the upper strings are told to perform this technique with a straight tone, the sound would be ethereal and possibly even beautiful. In contrast, if a whole orchestra was meant to play this figure with rapid arhythmic tremolos at an *fff* dynamic the result would be pure terror.

Check out the Sound Lab (companion site) for the full version of Figure 6.3 and other aleatoric examples.

FIGURE 6.4 An example of limited aleatory, where the musician improvises a melody based on the instructions and a given set of notes. Note the text above the brackets to play "in order," and then later on "in any order."

Aleatoric techniques have massive consequences in terms of harmony and orchestration (as will be discussed later in Chapter 7 and Chapter 9), but some of the most interesting effects can occur when using aleatory for generating melodic material. One way to approach this is to provide a limited set of pitches, and to define some rules for the player to improvise within a framework (Figure 6.4). Here we have a defined set of pitches marked off in brackets, along with clear instructions for rhythm and duration of improvisation. Although this melody will never be performed the same way twice, we can still control the emotional impact of it by limiting the pitches to be used and by offering clear directions for the performer to adhere to. In the Sound Lab (companion site) we provide the full versions of these excerpts.

Generating Harmonic Ideas

The first consideration when generating harmonic ideas is again to determine what exactly the needs of that particular game scene are. Does it require something somber? Action-packed? Does it need a strong theme, or should it be more textural? If the answer to the latter is that it requires a strong theme, then it may be best to write your melody first and then harmonize it. If the answer is that a textural approach is best, then you can get going right away on writing harmony.

For many composers, the first step in brainstorming is coming up with a chord progression that fits the mood of the game scene. This is a very common approach because most of us grew up listening to popular music, which often utilizes simple, catchy chord progressions that support the vocal melody. However it is worth noting that it can be very easy to get caught up in recycled chord progressions. This can make it difficult to compose interesting melodies. When working with an orchestra this can result in something more like an "orchestrated pop" cue rather than effectively making use of the complexities of the medium. This style is appropriate from time to time, but it is also limiting if this is all you are capable of writing. As game composers we want to have the fluency to write melodic lines that flow seamlessly into new tonal areas without sounding stale or repetitive. We also need the harmonic literacy to write progressions that support more complex melodic lines. For these reasons we encourage you to experiment with writing your melodic ideas first and harmonizing them afterward with an appropriate chordal texture. This usually results in freer melodies, and novel harmonizations.

Spencer

Harmony is similar to melody in that it can have a very strong and memorable impact on listeners. There are few among us who wouldn't recognize the iconic two-chord motif that signifies the presence of Batman in *The Dark Knight*. This shows us that harmony is just as capable of creating a *sonic identity* as melody. But the difference between melody and harmony in game music is that harmony can be somewhat less tangible, and therefore more subtle. Anyone can walk around humming a catchy melody, but it doesn't really work the same way for harmony, no matter how well written the chord progression is. While there are familiar chord progressions that strike up an immediate emotional response (*The Dark Knight*'s I–VI progression included), most players would not be able to articulate *why* it gave them that response. This makes harmony useful for broad musical strokes, like painting the scene with a particular emotion.

Let's ignore genre and stylistic considerations for the time being and jump right into a very fundamental element of harmony: mood. Mood will return again and again in our study of game music because it is one of the most important experiences that game music provides. When brainstorming some harmonic ideas, whether you have already written a melody or not, it is important to stay in line with the intended mood of the particular game scene. Below, we will explore a few ways to generate harmonic ideas that satisfy a few different moods.

By far the quickest way to create a chord progression that elicits a particular mood in a listener is to study other examples and use one of them as a starting point. This may seem like cheating, and it isn't something you should rely on entirely, but it is a great way to build a repertoire of chord progressions that you can eventually tweak or combine to elicit particular moods. Besides that, many of the most effective chord progressions are simple and widely used, making them easy to study. Part of the reason progressions such as the one found in *The Dark Knight* are so successful is precisely *because* they are used so extensively. As moviegoers and players, the more accustomed we become to harmony in certain contexts, the more our brains associate the music with the mood of *that* scene. This influences us to write similar progressions for similar scenes to evoke similar moods (with subtle creative variations of course), thus reinforcing the cycle. Eventually we have a pretty standardized cache of harmonic associates with mood. For composers looking to create emotional gameplay experiences, this is an area ripe for exploitation.

The best way to utilize this technique on a project is to find a variety of similar scenes from other media (games, movies, television shows, etc.) and analyze the harmony that is contributing to the mood. If your game scene requires a high-energy action cue and you're brainstorming a progression to add to this excitement, then play games like *Metal Gear Solid* or *Need for Speed* (or something similar) and take notes on the chords that you hear. If you hear a I–III–VI–VII progression in a few of your study sources, chances are that progression will work well as a starting point for your action scene as well. This works equally well for other moods. Franchises like *Dead Space* and *Silent Hill* are great examples of creepiness and outright terror. Regardless of the source material you choose to study, most games contain a variety of moods – and so will yours. Find a few examples and study the ones that fit with the game you are working on.

The purpose of this exercise is to give yourself one or more templates that you can change, combine, or disassemble completely to generate unique harmony for your game. Starting with game and film music is helpful 99 percent of the time, but drawing from less obvious territory can lead to truly original music. World music, concert music, and experimental music are all fantastic areas for study due to their often esoteric nature. The entirety of the horror music tradition can be traced back to sound experiments in concert and electronic music. This is not to say that game and film composers have not contributed to the tradition (in fact just the opposite is true, they have refined and advanced the tradition into its own idiomatic genre), but it helps quite a bit to have a broad knowledge of these areas in order to create harmony that is fresh and innovative.

Eventually, using this technique, you will have enough examples of harmony to draw from that you can use your intuition directly to pull out new harmonic ideas immediately. However this exercise will be useful to continue throughout your career as music is always changing, and there is no shortage of relevant source material to study.

The simplest way to begin generating harmonic ideas is to explore diatonic chordal permutations. In other words, you can select a scale or mode that fits the mood and try reordering chords that are natural to that key. You'll find that some chord orderings will have strong **cadences** or resolutions. A V–I progression is quite often used because of its momentum back to the tonic. It creates a clear sense of anticipation and expectation in the listeners because we are expecting the major V chord to resolve solidly back to the tonic (major I or minor i). Sometimes these stronger cadences will be right for your score. However, more frequently it will be necessary to avoid these expectations. Because the V–I cadence is so strong, it can become obvious or even distracting during a video game. The sense of finality could imply events that simply aren't in the scene. In these cases it can be more appropriate to end your chord progressions with something more ambiguous, like a vi–I or iii–I. These chords flow nicely into each other, and usually won't stick out as a heavy resolution. On the opposite end of the spectrum, to actually subvert the expected resolution a V–vi cadence works well. This is called a **deceptive cadence** and it is relatively common in games. Listeners will be expecting a strong V chord to lead back into the tonic, and by ending the cadence with a minor vi chord we can take the mood from resolute and triumphant to disappointing or even tragic. You'll find that there is a broad emotional expanse to explore using only the limited set of diatonic chords. However once the basics of **diatonic harmony** have been learned, other forms of harmony can be interesting as well as satisfying.

> There is a whole tradition of functional Western harmony that can be studied. A basic understanding of this is very helpful for game music – but don't get too bogged down by it. Remember that the cadences mentioned above are meant to be the structural *end* of musical phrases. With game music, beginnings and endings take on a different meaning; usually the audio engine handles the structure based on player actions. Most often, the best way to compose a cue is to find a progression that works with the intended mood and stick to it. If the mood is meant to change, then that can be done by composing a new cue and implementing a transition later on.
>
> Spencer

Like melodies, the shape of a chord progression can affect mood. A chord progression in which the chord tones rise usually leaves listeners with a sense of *increased tension*, or even yearning (especially if the *root notes* of each chord are rising). Chordal descent usually gives listeners a sense of *decreased tension*. This is more subtle than the way melodic shape changes the effect of a cue because chord tones don't always move in the same direction. The overall effect on mood is obfuscated, but it is still a useful technique when the scene calls for something less overt.

The concept of harmonic shape also applies to the *quality* of the chords themselves. For example, the progression I–V–vi–IV is a very common progression in many genres of music. However, if we swap out the major IV chord with a minor iv chord, *flattening* the third of the chord, the result is something sadder and more reflective than the original. In essence, we are changing the shape (and quality) of the chords themselves to change the mood. Swapping out the major IV chord with the minor iv chord is an example of harmonic mode mixture (see previous section) because we have borrowed a chord outside the key (in this case we have borrowed the iv chord from the parallel minor key). This is an extremely useful way to spice up or personalize commonly used harmonic progressions.

A beautiful example of mode mixture is "Aerith's Theme" from *Final Fantasy VII*, written by Nobuo Uematsu. The first two chords of the track are D Major–A Minor (I–v). The left hand in the piano plunks out the chords and the right hand plays the melody, which is an inverted arpeggiation of the progression. The theme itself is almost entirely harmonic in nature. Yet (due to the mode mixture) the track is highly effective and emotionally poignant. If the progression were written diatonically those first two chords would be a standard I–V cadence. The minor v chord adds emotional depth and a note of tragic hopefulness to the theme, perfectly outlining Aerith's entire character arc in just two chords.

Mode mixture can be taken further by introducing **chromaticism** into harmony. Chromaticism was used famously by romantic composers such as Gustav Mahler and Frédéric Chopin, as well as impressionist composers like Claude Debussy and Maurice Ravel. Chromatic harmony essentially opens up the diatonic language outside of the seven notes in a scale, to all twelve notes in the Western chromatic scale. Sometimes chromaticism can be added just as a brief flourish or passing harmony. This can be a very effective way to "flavor" your harmony. Other times a heavy focus on chromaticism can create a sonic atmosphere that wanders through tonalities. This can be useful for all kinds of game contexts including magic and fantasy exploration.

Added note harmony is another method of composing that pushes beyond diatonic harmony. With this technique we can take any chord (or series of chords) and add a new chord tone based on the harmonic series (see "Harmonic Series" in the Sound Lab companion site, Chapter 6). If a pure sonority is desired, add a note that results in an interval from Group A (Figure 6.1). If a dissonant or harsh sonority is desired, add a note that results in an interval from Group E or Group F. It's even possible to stack up new chords one note at a time with this method to create completely ignoring diatonic harmony. Figure 6.5 shows a few examples of added note harmony based on the harmonic series.

With added note harmony we are mostly concerned with 1) the interval between the added note and the root, and 2) the most dissonant interval that results from the added note. In Figure 6.5, example A incorporates intervals in Group A and Group D, resulting in a mostly consonant, open sonority. Example B includes intervals from Groups B and E, resulting in a major seventh chord, which is a richer and more complex consonance than example A. Example C has intervals from Group C and the more dissonant Group E, combining into a more obviously discordant sonority. Example D is interesting because with the added note it's technically two triads superimposed: a C major triad and a C augmented triad. Regardless, the addition of the Ab creates a more dissonant and dense sonority than either the C major or C augmented triad alone. Example E, with the addition of the triton from Group F, makes the sonority tonally ambiguous as well as more dissonant and dense (with the resulting minor second from Group E).

Added Note Harmony

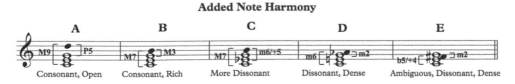

FIGURE 6.5 Five examples of added note harmony.

Another useful way to create unique harmony is to stack intervals *other than thirds*. This process is similar to the way we used added note harmony. Diatonically speaking, we make chords by stacking major and minor thirds up as the key signature dictates. But this is not the only way to stack chords and to create harmony. Due to the diverse nature of video games, many situations will call for harmonic progressions that sound more **idiosyncratic** than that which diatonic harmony can provide. In these scenarios it can be really useful to create your harmony through other intervals. As the composer you are free to stack chords based on any interval (sevenths, sixths, ninths), but be careful not to lose track of your stacking! Using intervals larger than a fifth can quickly revert back to tertian harmony (for example chord stacks in sixths can sound like traditional tertian harmony, but inverted).

Two very common forms of **non-tertian harmony** (harmony based on intervals other than thirds) are **quartal** and **quintal** harmonies. These are chords stacked in fourths and fifths. Chords of this nature sound striking due to the pure quality of perfect fourths and fifths. This also makes these chords more ambiguous in terms of tonality. When considering the harmonic series, the perfect intervals (octaves, unisons, fourths, and fifths) have a very fundamental relationship, and therefore have limited harmonic content compared to thirds and sixths. If used effectively, these chords can be beautiful and resonant in a very physical way. For this reason, quartal and quintal harmonies are wonderful for adding emphasis and novelty to a cadence or climax.

Another consideration with quartal and quintal harmonies (and by extension *any* type of non-tertian harmony) is that it is possible to stack these chords *diatonically* as well as chromatically (as we have been). It may seem contradictory to bring back an element of diatonicism, but using any diatonic scale for non-tertian harmony can actually sound *more* esoteric and complex than stacking intervals chromatically. Stacking fourths on top of each other chromatically will result in a series of perfect fourths, sounding pure and resonant. However, if you move up any major scale in fourths you will hear a tritone when you hit scale degree 4. This is because the seventh scale degree is exactly seven semitones (or an *augmented fourth*) away. The tritone does *not* sound pure, and completely changes the mood of the harmony.

Besides quartal and quintal harmony, secundal harmonies are also common. Secundal chords are sometimes referred to as cluster chords because the chord tones are stacked so close together. Cluster chords are actually a very simple device and can be used to elicit feelings of dream-like nostalgia in a player, as well as feelings of abject terror. As simple as this is, if you sit down at a piano and allow your fist to drop down onto any of the white keys, you will have a cluster chord.

Clusters basically work the same way quartal and quintal harmonies do. Starting with any major scale, you can draw a series of three- (or four-) note diatonic cluster chords. These clusters can then be used in any common chord progression. For example a I–IV–V diatonic progression becomes a I–IV–V cluster progression. The roots (C–F–G) remain the same, but the chord tones are stacked in seconds (C–D–E would be the I chord, etc.).

Cluster chords in the upper register usually sound fluffy and tonally ambivalent. Because there are only two points in the major scale where we would normally find minor seconds, the lighter and more consonant major seconds dominate this progression. These diatonic clusters can be used in many ways, from simply spicing up traditional harmony to creating entire textures.

On the other hand, thinking back to our intervallic consonance and dissonance spectrum, we can stack our clusters up in chromatic secundals instead and use predominantly minor seconds. Because minor seconds are more harmonically dissonant than major seconds, these types of chromatic clusters are best used when trying to scare the daylights out of players. Going back to our I–IV–V progression, the I chord now consists of two minor seconds stacked (C–Db–D). Try playing this at the piano. The fluffy chords have grown teeth and now sound aggressive and dark if played in lower octaves. This is important to note. Because cluster chords can potentially blur any sense of tonality, range has a large effect on the emotional impact. Up in the higher register, even minor seconds can sometimes sound dreamlike and appealing. But lower register clusters will always sound dark and ominous.

Finally, since we are on the topic of harmonic *generation*, it would be an oversight not to include the uses of the twelve-tone method. As we have learned in the previous section, the twelve-tone method can be used to create great complexity, or as a tool to quickly generate novel motifs. Although there is an exception to every rule, the twelve-tone method is best used in the context of game scenes that call for more esoteric sounds. That being said, using a tone row to invent harmony can be fun and rewarding. It can also lead to more creative harmonic ideas that need not be constrained by the "rules" of the method.

Using the twelve-tone method to create harmony begins the same way we used it to create melody. First, order all twelve pitches to create a tone row. Then (instead of imposing rhythm on this tone row to create a melody) stack three or more notes on top of each other to create a few chords to work with. In this scenario it is up to the composer to choose how strictly to adhere to the rules. In some cases composers will stack chords in the order of the tone row with the register of each chord tone being entirely up to creative choice. In other cases every chord tone has its own register in which it *always* repeats (this is sometimes called total or **integral serialism**). In our experience the best way to use this method for games is to let the tone row dictate simple three- or four-note chords, and then leave the rest up to your own intuition and musical sensibilities. Check out the Sound Lab for some resources on the twelve tone method.

Generating Rhythm

Generating rhythmic motifs can be an immensely productive starting point when generating musical material for your game score. For starters rhythmic material is very physically intuitive, so most of us are capable of improvising rhythmic ideas with little or no musical training. Another plus is that it can be liberating and satisfying to step away from the piano and just bang on something! Seriously, try it. For certain game scenes (particularly anything with action or urgency) it just makes sense to forget about the nuances of melody and harmony and instead start air-drumming some interesting rhythmic variations.

Because we are still in the brainstorming phase, before we cover how to *develop* rhythms we must first focus on some strategies for *generating* them. The most basic tool we can use to direct and analyze our efforts is **rhythmic density**. Rhythmic density is an intuitive measure of the number of notes to be played within a given sequence. This is a simple task, but it is often overlooked in the composition process. The tendency is usually to add *too much* rhythmic density, thinking that this will automatically add excitement. In reality, heavy-handed percussion parts can *reduce* excitement if not properly orchestrated. Parts like this leave no room for other musical elements. Rests are an absolutely crucial element of rhythm, and without them we would be left with nothing more than a metronomic pulse. Rests are to rhythms what pitch is to melody. A melody with only one pitch is not much of a melody, and a rhythm without rests is just a metronome. Knowing this, let's now take a look at how we can accurately assess rhythmic density for a few game scenarios, and then come up with some rhythmic material to match them.

As always, the most relevant aspect of our game scene for rhythmic material is going to be the mood of the scene. In essence, we are now trying to match the mood with a corresponding rhythmic density. This is similar to using harmony to impact mood, but with rhythmic devices we are limited in the emotional nuance. Instead of focusing on whether a game scene should feel "nostalgic" or "emotionally ambiguous," we are really going to ask ourselves "what is the *intensity* of the scene?" The answer to this question will inform our rhythmic ideas as well as our *tempo*. In addition, our rhythms must be genre-specific. If your intended genre for the game is rock 'n' roll, then you'll be cycling through rhythms idiomatic to that tradition. The same goes for jazz or pop music. If your goal is cinematic scoring you'll have a bit more freedom in choosing rhythms because the devices you use won't be as categorical.

Let us say that our first example will be of a very high intensity. Since it is focused on action we should create a very rhythmically dense texture. There are basically three steps to create rhythmic ideas from scratch. For step one, let's create a metronomic base rhythm. Try tapping eighth notes in a bpm (beats per minute) of 140 or higher. Ask yourself if this tempo matches the intensity that you want. If it's too low, bump it up a few clicks. This initial tempo is crucial because it sets the stage for all subsequent rhythmic activity. Don't rush this part! Make sure the tempo is right and rhythmic ideas will naturally flow.

Next, start randomly replacing "taps" with rests. Even by arbitrarily choosing a few places to add rests you will hear some clear rhythmic motifs begin to emerge. You may even branch off with your own ideas to develop after a minute or two. Keep in mind that when generating rhythmic ideas in this manner you can choose to add rests on strong beats or weak beats. Simple rhythms often use rests on weak beats since this keeps the rhythm very predictable. However, don't underestimate the power of **syncopation**. Especially during heavy action cues, adding rests on strong beats can keep the listener slightly off-balance and add to the tension and excitement.

We are now left with an actual rhythmic motif, although it is a somewhat boring one. This is because all the note values are exactly the same. Step three then is to zoom into our individual note values and replace some of them with different durations to create a more interesting sequence. Try taking your current rhythmic figure and replacing some of the eighth notes with sixteenth notes to spice things up. Remember that replacing notes with shorter durations will be *increasing* the rhythmic density. Here, adding two sixteenth notes in place of a few eighth notes works well because we are aiming for excitement and tension. Doing the opposite (adding a quarter note and taking away two eighth notes) is also a possibility for step three, but it will only work if 1) you are trying to *decrease* rhythmic density, and 2) you are using percussion instruments that are capable of sustained notes (timpani, cymbals, tubular bells, etc.). You can repeat these three steps in any order as many times as you like to create and fine-tune your rhythmic ideas.

Rhythmic density is only one way to create and develop rhythms. **Augmentation** and **diminution** are two useful techniques for rhythmic development. They can definitely help generate ideas as well once you have a few ideas. Augmentation *proportionally increases* rhythmic values. For example a quarter note and two eighth notes can be augmented to a half note and two quarters. Diminution *proportionally reduces* rhythmic values. That same quarter and two eighth notes would now become an eighth and two sixteenths. This kind of development can be thought of as displacing the register of a chord or melody line by an octave. It's still the same melody, but in a different register.

Finally, like melodies, rhythms can also be put in retrograde. This is an amazing technique for creating variations on rhythms. Like melodic retrograde, the core of the rhythmic idea remains intact but subtly changed. To make things really interesting you can use it on pitched material. Try retrograding the rhythm but not the pitches, or vice versa. This will keep you busy for hours if not days.

Generating a Musical Palette

Because workflow is different for every composer, generating a musical palette takes place partially in the brainstorming phase and partially in the refinement phase. Some composers prefer to start brainstorming at the piano, while others like to create a template of sounds and instruments before they write a single note. Still others incorporate bits of both depending on the situation. To put it very broadly, a musical palette is really two things: the acoustic or electronic instruments you use, and the methods you use to process them. It's pretty safe to say that we as composers have never had as many palette options as we do today. The oldest instrument ever found was a flute made of bird bones discovered in a cave. This gives us roughly 40,000 years of human musical experience to draw from – and that's *only* considering acoustic instruments. Technology has given us all of these musical instruments, plus generations of electronic instruments and processes to draw from as well – all at our fingertips for a relatively small price. This can be overwhelming, especially considering the fact that when we take digital or analogue processing into account we can make an orchestra sound like a distorted hive of bees, or a sound file of waves breaking on a beach sound like an ambient masterpiece. This universe of possibilities is what we like to call a musical palette.

Designing your palette is a crucial step in writing your game soundtrack. For one thing, it is important to know what instrument or ensemble you are writing for when you start brainstorming. Beyond that, your choice of instrumentation has an immense impact on player experience. Can you imagine what retro-style games would be like using a brass band instead of chiptunes, or what *Call of Duty* would sound like without a huge orchestra? The type of sounds you use and the *number* of sounds you use are both tied deeply to how your score will impact the player.

> It can be helpful to layer other soundtracks over video capture of your game. This offers a real sense for how other instrumentation and style choices can change the narrative.
>
> Gina

A simple way to decide on what instruments to use for your game is by starting with commonly used ensembles and then adjusting as needed. An important consideration at this point is to determine what *size* your ensemble will need to be. If your game is an epic fantasy RPG, a good starting point would be to use a symphonic orchestra as a template. Here the scope of the game is quite large, so having 60-plus musicians at your disposal will allow for a large emotional impact. If your project is an intimate puzzle game, then a good alternative might be a quartet or chamber ensemble. String quartets are highly emotive and very flexible in terms of their ability to play different styles, so this is a safe choice for many soundtrack needs.

In addition to the size of your ensemble you will need to pick a *timbral* palette as well. In other words, what *kinds* of sounds would fit the game? Will you be using traditional acoustic instruments like the above examples? Will you need an array of synthesized sounds? Found sounds and ambiences? Electric guitars? Will you be using some combination of all of these? Just as the size of the ensemble has an impact on the player, the timbres you choose will also have an effect.

Choosing your timbres will usually be up to you unless the developer provides specific references. Choose wisely because palette more than any other aspect of the score has the potential to make your game unique and memorable. For example, composer Austin Wintory goes out of his way to use unorthodox groups of musicians. *The*

Banner Saga makes fantastic use of a wind band score, which is very uncommon among game soundtracks. Jessica Curry wrote for church choir in *Everybody's Gone to the Rapture*. Kristofer Maddigan used an equally uncommon jazz ensemble in *Cuphead* to amazing effect. Sometimes the best ensemble for a game is a mix of many different styles as is the case with *Bastion*. In this game Darren Korb uses electronic elements, acoustic string instruments, and vocals in a variety of disparate styles to insert his own unique aesthetic into the game.

Despite the uniqueness of these ensembles, the important point is that these composers carefully considered the artwork and style of these games and chose an ensemble that fit within that framework. We would encourage you to experiment and push the boundaries of your musical comfort zone to find new instruments and new ensembles that fit with your game projects. The only real limitations are budget, and whether or not the ensemble actually fits within the paradigm of the game. We will cover "non-traditional" music ensembles in detail in the next chapter.

> Plenty of games have an orchestral soundtrack, but when you ship a game with a didgeridoo solo, players are gonna remember it.
>
> Spencer

Developing Your Material

Although most of the creative process occurs during the preparation and brainstorming phases, the actual development of your material occurs during the refinement and revision phase. It can be a daunting task to stretch out a few themes into an hour or so of music, so it is crucial to plan ahead and make it a priority to bring back recurring motifs. This will offer two advantages to your score: 1) The player will recognize motivic elements in your music and the consequent structure will enhance the emotional experience, and 2) the music will provide a sense of journey and adventure to mirror the gameplay. These two elements are an important part of the gameplay experience, so don't get lazy once you have a few motifs. Keep your mind sharp so the ideas keep flowing and your creativity does not lose its edge. Look for moments to inject something new into the score, and likewise look for moments where older themes will have an impact.

Any aspect of your music can and should be developed fully. This means that melodies can be changed and brought back in different ways, chord progressions can have variations and recurrences, and even textures or instrument groups can be expanded on (see Chapter 7). We have already discussed in detail methods of developing and varying melody, harmony, and rhythms (i.e. transposition, inversion, retrograde, chord progressions and variations, etc.). Other key areas primed for development are arrangement/orchestration and implementation. Developing the arrangement can be as simple as changing the instruments. For instance, if a character has a theme, then this theme may be brought back each time the character makes an appearance. Perhaps this character is a world traveler. It might be necessary to write a version of the character theme using traditional African instruments, or using an Irish folk ensemble, or maybe multiple variations of instruments and ensembles. The game context will often dictate (or at least inform) this kind of musical development.

Implementation is something that we will cover in depth in Chapters 8 and 9, so we will only touch on it briefly here. For now, suffice it to say that to develop your score using implementation just means that you are changing *how the music acts* in game. If a cue is set to trigger during an alien attack, then triggering it again later on will likely leave the player expecting another attack. This is something that can be used to add nuance and context to the gameplay. *How* you decide to trigger your music, and how you decide to adapt it to the player's actions, can change as the game unfolds. Similarly, if a player is expecting a battle theme to trigger when an enemy attacks, then *not triggering* it is an effective way to develop the score. The lack of music would change the experience.

There are many directions you can take to develop your musical material, but the goal is to balance recurring themes with novel ideas. This is like balancing on a razor blade because if you bring your motifs back too often and without enough variation then the result can sound repetitious. On the other hand, if you alter motifs too much the impact can be lost on the player. It is up to the composer and the developer to decide where that balance is.

FORM AND STRUCTURE IN GAME MUSIC

Let's now backup a bit and take a look at the **form** of our music. If you can think back to your high school or college music class, form refers to how a composition is structured or organized temporally. Although some forms have specific names (sonata form, rondo, etc.), we can also use letters and apostrophes to define sections and variants (A, a, B, b, A', a'). Capital letters stand for larger sections of music, and lowercase letters stand for the subsections that larger sections are comprised of. Apostrophes denote a variation on an original section. For example A' signifies a variant on a larger section A, and b' signifies a variation on a previous variation of a subsection b. This is a simple and effective way to define and analyze form.

While this method works well as a labeling system, the meaning behind the term "form" is lost to a certain degree. This is because "form" assumes a particular musical structure from start to finish. But in games, not everything functions in precise chronological order. Some events happen out of order, or not at all, or over and over again, which makes the form of game music tricky to pin down and apply. For the purposes of analyzing adaptive musical structure here on out we will still use the lettering system, but we will abandon the term "form" in lieu of the terms microstructure and macrostructure. Microstructure will refer to the basic structure of a section or "chunk" of adaptive music. For example, if a loop is in **binary form** (A/B) or **ternary form** (A/B/A) then we are discussing the microstructure. Macrostructure will refer to the overall sequence of musical events as presented in the entirety of the gameplay experience. In other words, macrostructure is the game soundtrack as a whole.

Microstructure

Microstructure is very similar to the term "form" in linear music, except for one key distinction: the musical modules in a game are assumed to be repeated or skipped or jumbled up in some way in the greater context of the macrostructure. This fundamentally changes how we look at the music structurally. As mentioned above, a common form in linear music is A/B/A or ternary form. This is effective in linear music because we are presented with a musical idea in section A that we can latch on to. Then section B moves away from this musical idea, either by contrasting it with a new idea, or developing the idea by transforming it somehow (modulation, rhythmic manipulation, etc.). Finally we arrive back at the A section, which is satisfying because we finish where we started. As the listener we remember where we came from, and having departed from this idea, returning to it then feels like we have made it home from a meaningful journey. But does this work as well when this structure is made for adaptive game music?

Take a simple loop in ternary form as an example. The first time through, the loop has the exact same effect described above. The musical idea is presented, developed, or contrasted, and then we have a "return home." *But then it loops.* Our return home is now doubled again, making our microstructure A/B/A/A/B/A. Every time the cue loops, we hear the A section twice in a row. This might not make much of a difference if the style is ambient or textural, but for heavily thematic cues this could get very repetitive. Unless there is significant variation in the second A section (A/B/A') we are essentially adding an extra section, which will inevitably be repeated anyway. A possible alternative would be to cut the final A section entirely and bring the module down to binary form (A/B). This leaves us with a more balanced track structurally speaking. Because we know that the A section is bound to loop back around, the result is a track that oscillates back and forth between development and return, which usually proves much more sustainable over long-term listening.

Another way to take advantage of nonlinearity is to stack up simple microstructures to form a more complex macrostructure. For example, in linear music **rondo form** is relatively complex. The primary theme in rondo form alternates with at least one contrasting theme. An example of rondo form would be A/B/A/C/A/B/A, or A/B/A/C/A/D/A and so on, yielding a complex weave of musical ideas. But if we take microstructure into account this is actually pretty simple to approximate in an adaptive

Adaptive Rondo Form

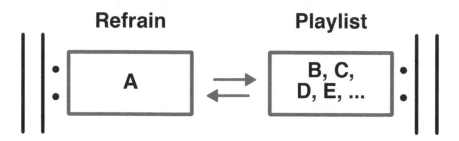

FIGURE 6.6 An example of how to take a traditional musical form, in this case rondo, and make it adaptive. This will result in significant structural development of your score, and increase playability of the game. What are some other ways to make traditional musical forms adaptive?.

setting. We would need four loops, one for each larger section (A, B, C, and D). Then we can set the A section (in this case our refrain) to loop in game. Next we can implement a transition from A into a **playlist** (we'll cover this later in Chapter 9, page 323) containing each of the contrasting themes. The game engine can then cycle through (in order or randomly) B, C, and D. The interesting part is that each of these contrasting sections can be set to play *only once*. This way our track will always move back to the refrain after a contrasting section is triggered. The setup here is minimal, and it is a very effective way to showcase a high degree of contrast, which always resolves back to a thematic return.

There are many other ways to use microstructure adaptively. However, in some cases it won't have a huge effect on the overall tone of your music. It will likely only be through many playthroughs that the negative repetitious aspects will come to light. But by being aware of these pitfalls, and by keeping your eyes open for opportunities to exploit microstructure in your adaptive systems, your music will be more resilient when looping, and much more effective due to adequate contrast and development.

Macrostructure

Since macrostructure is the organization of your soundtrack as a whole, you will never actually be able to control the exact presentation of your music. This is neither a good thing nor a bad thing, but it must be addressed. It has less to do with the specifics of structural organization and more to do with how the soundtrack as a whole feels to the player on an emotional level. Does it resonate fully throughout?

Does the music start off too strong, spoiling its development and possibly the plot? Does it have enough contrast that themes are still meaningful and poignant rather than repetitive? Your macrostructure should take players on a journey as much as the game itself does.

When evaluating the macrostructure of your soundtrack, you are basically deciding how the themes/textures you have written will develop as the player advances through the game. Sometimes themes, or **leitmotifs**, can be expanded on and fortified throughout a game. A leitmotif may first be presented with the vulnerability of a solo violin, but the final iteration at the game's climax may be orchestrated fully, with the power of over a hundred musicians. This in itself shows meaningful development.

When adaptivity is taken into consideration, the most controllable aspect of the macrostructure that we have is deciding *when* to trigger our leitmotifs and recurring thematic material. This is the essence of structural development, as mentioned earlier in the chapter (see page 184). When we decide that a theme should retrigger here over an ambient background cue, it is a clue to the player that something important is either happening or about to happen. It also makes a significant emotional impact because the player will respond to themes that she has already heard, and will respond in some way based on the previous context in which the theme was triggered.

The key to making this kind of development effective is ensuring that it relates very specifically to some relevant aspect of the game. This could mean the story, gameplay, visuals, or anything else you can think of. With the theme mentioned above, perhaps it is presented with full orchestra because the ending of the game is a climactic battle against the forces of evil. Seeing an army of a thousand enemies would certainly warrant a full orchestra. An equally valid reason would be if the gameplay changed significantly. Perhaps the original iteration of the theme occurred during the opening credits where the player had nothing to do but sit back and watch. The end of the game might be an insanely difficult challenge which needs an orchestra to fully capture the enormity of the task. The possibilities are limitless, but the bottom line is that there needs to be justification from the game in order to decide how to appropriately develop your ideas, either through arrangement or some compositional variation.

The Sound Lab

 Before moving onto Chapter 7, head over to the Sound Lab for numerous practical exercises and additional resources for the topics covered here in Chapter 6.

NOTES

1 A. Farnell, *Designing Sound*.
2 O. Legname, *Density Degree of Intervals and Chords*.

BIBLIOGRAPHY

Farnell, A. (2010). *Designing Sound*. Cambridge, MA: MIT Press.
Legname, O. (1998). *Density Degree of Intervals and Chords*.Retrieved from www.scribd.com/document/287087067/Density-Degree-of-Intervals-and-Chords

Arranging and Orchestration for Games

In this chapter we will take what we learned of composing musical ideas, and explore ways to arrange them for various ensembles. We will begin by covering basic textures often found in game music, and then we will move on to an overview of the orchestral instruments. Once we have a better idea of the orchestral instruments and how they function we will cover template building. We will then analyze some practical orchestration techniques for acoustic and synthetic instruments. Finally, we will share a brief overview of some alternative arranging styles found in games.

GENRE AND STYLE CONSIDERATIONS

Before we embark on our journey into arranging and orchestrating for games it is important to understand a few things about musical genre. There are some clear defining lines between genres that allow us to categorize musical styles. For example, the "rock 'n' roll" style commonly makes use of electric guitar, drums, bass guitar, and a vocalist. Jazz usually has piano, drums, and bass as well, among other instruments. Using the terms "rock 'n' roll" and "jazz" are thus helpful in defining *instrumentation* as well as the groove of a musical cue. But genre can also be too broad to be useful. One person might be thinking of "rock" with respect to Led Zeppelin, while another is thinking of Coldplay. These bands have vastly different instrumentation and emotional direction. In the case of the jazz example, a jazz trio could have piano, bass, and drums, but what if the developer was actually referring to big band jazz? This can potentially mean anything from a horn section up to a full orchestra! These types of discrepancies can make genre sometimes unhelpful when speaking about musical direction. With regards to instrumentation, it is usually better to speak to your developer in terms of their *desired emotional impact* than in terms of genre. If the developer is musically literate or has preferences about particular groups of instruments, it can be beneficial to discuss this, but try to avoid getting bogged down in too many technical details before working up a sample of your own.

Breaking Down References with Genre

As far as the actual process of arranging goes, it is important to keep yourself open to different genres and styles of music because you will likely receive references from developers in unfamiliar styles. When this occurs, put yourself into analysis mode and try to dissect the requested style. Start by researching what a traditional ensemble would look like. Learn the instruments and how they function. Next, try to break down the rhythmic activity. Is this style groove oriented? Is there a commonly used time signature? It is crucial to research particular scales and chord progressions in that style. Certain genres of "world music" can be especially difficult if you grew up without much exposure. If you pay attention to the details of that style and do your research, you will be more likely to produce an authentic score.

In games you are equally if not more likely to write in a hybrid style as you are to write in a traditional one. Because games offer unique experiences to players, particular genres are often insufficient to address the needs of the game. For example, if you are scoring an iPhone app and the reference music is a song by Janelle Monáe, it is highly probable that what the developer is really asking for is a similar *groove*, minus the vocals. Popular music is tailored for people to sing along, so catchy melodies and repetitive lyrics are common. In video games prominent lyrics can be distracting, so it is necessary to take specific *elements* from the reference style and leave out the elements that do not work well with the game. This is true across the board. You may decide that an EDM synth bass works well for your game, but the melody and chords of that style don't fit. Perhaps the melody and chords from a heavy metal track fit much better. In this way, mixing and matching genres can be a good idea in games. Sometimes the most unlikely pairing of styles can be the best fit.

Games are unique, and while they can encompass just about any style of music, careful thought needs to be put into determining which styles to use and to hybridize. The most common hybrids are orchestral, and any pop or electronic style. Orchestras are flexible and can fit quite smoothly with almost any other genre. They also allow for a range of emotional impact and mood. Learning orchestration well also makes it easier to quickly learn how to arrange in other styles. For these reasons we will focus primarily on orchestral instruments in the next sections. Keep in mind that orchestral music is just one of many possible ensembles. Your imagination is really the only limitation when it comes to the virtually infinite selection of musical palettes that we have at our disposal.

FOUNDATIONS OF MUSICAL TEXTURE IN GAMES

For most people, musical texture is a very marginal consideration. Many of us talk about music that we like in terms of the melody and whether it's "catchy." The standard melody and accompaniment has become a sort of a catch-all for musical texture. But as game composers there are many situations where other textures can be equally if not more effective than a melody and accompaniment. For our purposes there are three basic textures in music.

Monophony

Monophony is the simplest musical texture. A monophonic arrangement consists of a single melodic line and no accompaniment. Obviously this includes solos, which are very well represented in games. But it also includes passages where musicians are playing or singing in unison or octaves. This can be very effective in a variety of situations, and it's definitely a tool you should keep at your disposal. It provides a stark contrast to the other musical textures, so don't underestimate the value of monophony.

A close cousin to monophony is **heterophony**. In a heterophonic texture a main melody is still the prominent focus, but slight variations are added simultaneously. Heterophony is prevalent in Eastern cultures and can be heard in traditional Arabic music as well as Indonesian gamelan music. It was also employed by classical composers such as J. S. Bach and Mozart, and more recently by a few twentieth-century composers. Heterophony offers a bit more complexity than monophony, but not as much as polyphony.

Homophony

Homophony is very familiar to our ears, even if the word isn't recognizable. The homophonic texture is the standard melody and chordal accompaniment that we regularly hear in just about every genre of music, from pop to symphonic. In homophonic textures the accompaniment often has the same or similar rhythm to the melody, but it can also have slightly different rhythmic activity. The accompaniment never reaches the point where it has enough *weight* to be considered a counter-melody as this would move the texture into the realm of polyphony.

Polyphony

Polyphony is the most complex of the three textures, and it is usually the most difficult to write. Fugues and renaissance motets are examples of the polyphonic aesthetic. These works employ musicians and singers that balance with each other in terms of compositional importance; every voice has equal weight. Usually this means that each part is written in a perfect balance (as is the case with motets) or that various unequal melodies and timbres are crashing into each other as is the case with contrapuntal symphonic works. In either case, each "voice" (or *part*) is distinct from the others, but they work together to create an overarching sense of movement and musicality.

The complexity of polyphonic textures is a powerful tool for a game composer, but it must also be used with discretion. A cue that is heavily polyphonic can easily overpower a scene. Likewise a minor error with voicing or dynamics can cause major issues with transitions between cues. Genres like horror and action can actually handle polyphony quite easily. The organized chaos fits perfectly well in a frightening or action-oriented scene.

> These three terms – monophony, homophony, and polyphony – are used loosely here. Each has its own framework, rules, and musical traditions from around the world; we use them here specifically to denote the general texture of game music. The rules of each are well worth studying, especially polyphony, which is a rigorous endeavor in its own right, but it will leave you able to expertly navigate contrapuntal voice leading and chordal movement.
>
> Spencer

ARRANGING AND ORCHESTRATING USING SAMPLES

The technology of sampling has improved immensely in the last few decades. As composers we now have the high-fidelity sound of a 60-piece orchestra (or more) at our fingertips. We also have the power to record that orchestra while drinking coffee in our underwear (seriously, try it). Samples certainly do not replace the need for actual human beings who have devoted their lives to their instruments, but they do afford us a few important advantages. It is because of these advantages that we have decided to explore sample-based approaches to arranging before covering arranging for actual musicians.

By far the biggest advantage of using samples is the price point. It may cost $1,000 for a boutique orchestral string library, but once you have paid you can use the library for years before it will become outdated. By contrast $1,000 will only get you about one minute of music recorded by a full orchestra, at most. This advantage really dominoes into other perks as well. With the cost being so low you can afford to experiment, which will in turn teach you a great deal about orchestration and mixing. Most new composers can't afford to learn orchestration by hiring an orchestra to read through their music, and we lack the mentorship framework that was prevalent in the days of court composers. On the bright side, new composers usually don't have a problem purchasing a basic orchestral sample library and learning as they go. Samples offer us the ability to sketch out orchestration and immediately hear how they will sound without going through the expensive and time-consuming process of producing parts and booking rehearsal and recording time.

In order to thoroughly cover this topic, we first need to address a long-standing war that has been raging between composers for millennia.

TEMPLATES

A template is a blank session (in Logic or Pro Tools, or whatever DAW you happen to be using) that has pre-loaded and pre-mixed sample libraries and synthesizers. Many composers swear by templates, while others eschew them. Despite our biases, templates are not *always* necessary. Some genres require very few instruments, which do not need to be organized in advance. However, when you are working with a large or frequently used ensemble, templates are absolutely essential. Without using a template you will waste countless hours loading in ensembles that you have already loaded in many times before. You will then spend even more time mixing those same ensembles over and over again to achieve the same results. With a carefully designed template you can skip those steps. You will be able to open up a session and start composing, and your music will already sound polished. In addition to saving time, templates are also a means of finding your unique "sound." Over the years you will rework your templates to fine-tune mix issues and shape the way your virtual orchestra is used. All of this will speed up your workflow, and point you towards more distinctive arrangements.

Template Types

Large symphonic and Hollywood ensembles are the main targets for template creation due to their sheer size. They also always have a similar core group of instruments. Even if you decide to add or take away instruments, you will almost always end up using the main orchestral groupings: strings, winds, brass, and percussion. A "Hollywood-style" template would also likely have synthesizers and a rock ensemble.

There may be a need for other kinds of templates as well. Depending on the projects that come your way it may be beneficial to have a "world" music template (or even multiple templates depending on the musical traditions you most often need to draw from), which would include samples of commonly used string, wind, and percussion instruments from around the world. If you are a musician yourself it is also highly advisable to have a template prepared and pre-mixed for the instruments that you play. This will ensure that you waste no time on setup when you need to record quickly.

Remember that aside from the time spent on creating one or more templates, there are no downsides. You will save time in the long run on setup and mixing, and you will have a much smoother workflow when using a template.

TEMPLATE PLANNING

The first thing you must consider before you start on your template is how it will run on your computer. This informs how many instruments you can load into this template. The unfortunate truth is that it's unlikely that all of your orchestral samples will run smoothly if they are all loaded at once. Most professional orchestral setups are too much for your standard laptop, so refer back to Chapter 1 on the Sound Lab (companion site) to brush up on computer specifications. Despite that, you probably won't need everything you own on every single project. The best approach is to plan ahead and decide which samples will 1) allow you to compose quickly and efficiently without overloading your computer, and 2) save you time by loading the core group of instruments that you *most frequently* use in your compositions. Keep in mind that this latter point will change over time. Templates are never finished, so it's best to save versions of them as your production skills and tastes improve. It may even be beneficial to take projects that you have worked intensely on and convert them to templates afterward.

The next thing to consider is exactly *what* to put in your template. The short answer to this question is that your template should consist of the instruments you use consistently from project to project. This will save you the hassle of loading up and mixing each instrument every time you start a new session. In addition to your "bread and butter" instruments it is highly advisable to make the standard orchestra the core of your template, even if you don't use the full orchestra all the time. The reason for this is that unlike instruments used in pop and electronic music, orchestral instruments are usually used in the same or similar ways from project to project. For example, a guitar could be used in a variety of mix settings from style to style. An oboe, however, will usually have the same or similar volume relative to the rest of the orchestra, with the same or similar reverb settings, and so on. In addition to this is the fact that orchestras (usually) only differ from one another in terms of size and a few specialized instruments. They (almost) always contain the same core instrument groups. Your standard orchestral template should include the main sections of the orchestra: strings, woodwinds, brass, pitched percussion, and unpitched percussion. Then you may also want to add choir, commonly used synthesizers, and scoring libraries like Omnisphere or Symphobia. These aren't necessary for every composer, but it can be helpful to have them at the ready if you use them often. By creating a balanced template that adequately covers your basic instrument groups and reverb settings pre-mixed and pre-loaded, you will have 90 percent of your orchestral needs at your fingertips. With the addition of the hybrid samples you will easily be able to load up your template and get to writing quickly in any style.

A trick to keeping your template light and speedy for Kontakt users is to set your libraries to purge and dynamically load samples as needed. You can also use networking software like Vienna Ensemble Pro to connect multiple computers and/ or run samples from external drives without hassle.

Spencer

Finally (and this is the time-consuming part), you must decide *how* you're going to organize and mix all of these samples. This is a more difficult question than it seems. Libraries all have different methods of articulation switching. For example, LA Scoring Strings uses the modulation wheel to control dynamics while EastWest Hollywood Strings uses expression. These discrepancies can be awkward and waste time later on. A bigger issue is that many libraries use keyswitches while others don't. Cinebrass uses an elegant velocity control to switch between long and short notes. This is a great method for brass, but strings have so many articulation possibilities that the velocity method doesn't work as well.

On top of all of that, sample libraries are almost always recorded in different places. The Sample Modeling libraries are among the most realistic and playable libraries that money can buy, yet they were recorded in an anechoic chamber with no reverb whatsoever. This makes them incredibly flexible, but sometimes impossible to mix in with libraries that were recorded *wet*, or with reverb baked into the samples. For this reason some composers sometimes choose one library manufacturer and use all of their products to make it easier to mix.

Without getting too bogged down in the details of library selection, the most important thing to consider when choosing which libraries you purchase and add to your template is that *they fit efficiently within your workflow*. The best sounding library in the world isn't worth a dime if it constantly inhibits your ability to use it to compose the music in your head.

In summary, be sure to listen to mockups and do your research before purchasing. If the library developers offer a demo make sure to try before you buy. In the Sound Lab (companion site) we present a brief list of some of our favorite industry-standard sampling companies.

THE ORCHESTRAL INSTRUMENTS

Getting to know the orchestral instruments and how they work together is an important part of game music composing. Even if you work primarily with synthesizers or electronic instruments, odds are that you will use orchestral samples at some point. Knowing how to write idiomatically for each instrument can massively improve the sound of your score. Below we have laid out a common setup of instruments within a basic orchestral ensemble and a brief description of their function within a sampled orchestra. This is not an exhaustive list. It is meant to cover only the core orchestral instruments. We highly encourage you to study a dedicated orchestration textbook on your own.

Samuel Adler's *The Study of Orchestration*[1] has some indispensable information on other instruments and techniques, and combines it with listening examples. This can provide a quality foundation for any game composer. Likewise, Thomas Goss's *100 Orchestration Tips*[2] is a quick study and lays out the most important points. As we are primarily concerned with how orchestration relates to video game music, we will assume a basic familiarity with the orchestral setup and move on relatively quickly.

Strings

- Violin I
- Violin II
- Viola
- Cello
- Double bass
- Harp*

Note: * indicates that this instrument should really be in its own category in terms of technique and use, as it varies greatly from the instruments in the rest of its group.

Orchestral strings, particularly the bowed violin, viola, cello, and bass have earned the nickname "the backbone of the orchestra" for good reason. String players don't rely on their lung capacity to make sound, so it's not uncommon to hear a string section playing consistently through an entire cue. This doesn't mean that they *should*, but they are more than capable of it if the orchestrator calls for a pervasive string timbre. The standard **arco**, or bowed articulation, is incredibly flexible, allowing for a wide dynamic range on long notes and short notes. This bowing also allows subtle or overt sliding between notes. In sample libraries a subtle slide between two notes is called **legato** and is activated by overlapping MIDI notes (Figure 7.1). In libraries like LA Scoring Strings it is possible to take this to the extreme and glide long distances to a destination note. This is called **portamento** or **glissando**. In your career you will undoubtedly write many sweeping string melodies, so it is important to remember that a little bit of portamento goes a long way and can make a melody truly come alive.

Other articulations to be aware of are **staccato** and **spiccato**. These are bowed attacks with a short and *very* short duration respectively. Some libraries even separate these into different *kinds* of spiccato, which can be useful. If string samples have one strength, it is the spiccato articulation. Due to their short duration it is very easy to sample. The technique itself is used ubiquitously in film and trailer music especially, having been pioneered by composers like Hans Zimmer and Brian Tyler. Although the spiccato articulation usually gets all the glory, there are many many lesser used articulations that are commonly sampled in string libraries. Sul ponticello is a technique where the player bows near the bridge to get an airy, nasal sound. It works quite well

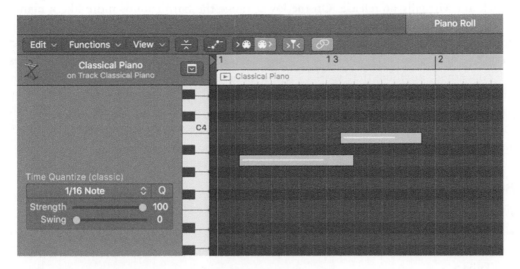

FIGURE 7.1 Overlapping MIDI notes triggering a legato sample.

in horror scores or in combination with **tremolo,** wherein the player will bow many times in rapid succession. **Harmonics** are also commonly sampled and can offer a very distinct flute-like and airy timbre for strings.

A go-to string library is likely the most important purchase you will make with regards to your orchestral sound. By far the most important factor is playability. This means that the library should allow you to sit and write out your ideas with very minimal tweaking. If it needs endless tweaking for you to make it sound polished, it is probably not the right library for you. Instead look for a library that you like the sound of "out of the box." Also try to choose a library that includes many articulations and allows for a very wide dynamic range. On top of that, it can be nice to have options for **divisi,** or split string sections. This will allow you to add extra parts, or split chords up, or even opt for a thinner, more intimate chamber sound.

> We are quite fond of LA Scoring Strings as a mainstay string library supplemented with Orchestral String Runs by Orchestral Tools. EastWest Hollywood Strings is also an industry favorite due to its large number of articulation choices.
>
> Spencer

It is worth noting that the harp is somewhat of an outlier in this section. It is technically a string instrument, but it produces sound by plucking rather than bowing strings. Some (but not all) string libraries include harp. In terms of timbre it can be thought of as two instruments in one. In the upper range a harp sounds like bright **pizzicato,** the sound that orchestral strings make when plucked instead of bowed. It has a short attack and virtually no release. On the lower range the harp sounds more like a piano, with a rich warm timbre and a long release. The harp uses pedals to sharpen or flatten notes, and thus to modulate scales. In short, the harp is a completely different beast from the orchestral strings (and really from most other instruments). Where samples are concerned, anything that works for the game will work as a harp part. However, the harp is very idiosyncratic, so it's best to consult with a harp player about any parts that need to be recorded prior to an important session. We also recommend carefully studying a dedicated orchestral text, as well as the work of composers like Ravel and Debussy for good harp-writing etiquette.

Brass

- Horn
- Trumpet
- Trombone/Bass trombone
- Tuba

If the string section is the backbone of the orchestra, the brass section is the muscle. Brass is by far the loudest section of the orchestra and can be a formidable ensemble even on its own. The typical "trailer" sound is a combination of string ostinati with heavy brass blasts for emphasis. And it works! The biggest strength of brass samples is that it is quick and easy to assemble unrealistically huge numbers of brass musicians to blast away. It's not uncommon to see a twelve-French-horn ensemble patch in a basic brass library (for reference the classical setup is usually only four). This makes brass samples a huge hit for any cue that needs to pack an epic punch.

There are two common functions for brass instruments: 1) to play loud and heroic melodies, and 2) to produce stabs and blasts during climactic moments, or moments of tension. French horns are usually used for the former and trumpets and trombones for the latter. The tuba can (more or less) achieve either effect. It sounds great when doubling French horns an octave below on melodies, but it can also be used to round out the low end of chords underneath trumpets and trombones.

The brass section brings to light an interesting issue that we don't really have to deal with regarding string sections. This is the issue of using solo instruments vs. full sections. With strings, even when using divisi you are still playing with the sound of six or more musicians. Because the dynamics of strings are so flexible, we never really have to worry about balance when using samples. But brass instruments are so loud that it is an important consideration when arranging. Often one trumpet is enough to add weight and density to a phrase, but three or four trumpets might overpower the other instruments playing in unison. A good rule of thumb is to use unison brass on big melodies that you want to emphasize, and to split the section up in "chorale" style when playing chord tones. This will ensure that the chords are balanced and that the melody will be the main focus.

Of course brass instruments are capable of a wide range of emotion, even when using samples. Although brass instruments will overpower other sections at *mf* and *f* dynamics, at *p* a trombone trio or quartet can be quite soft and beautiful. In addition many brass libraries come with muted patches which change the timbre and overall volume of each brass instrument. This offers more options for orchestration and can be really handy when blending brass with other sections of the orchestra.

When choosing brass instruments the main goal is again to find a library that is playable quickly and efficiently for you. The library should sound good at all dynamic ranges and not too tinny or metallic. You're aiming for a library that is full sounding, but not "boomy." A very important selling point for a library is its ability to perform fast successive articulations in the style of John Williams. Real brass players can double and triple tongue, so these idiomatic techniques should be available from your library as well. Two other huge pluses to a brass library are extra muted patches, and solo patches as mentioned above.

There are many high-quality brass libraries out there. Berlin Brass from Orchestral Tools is absolutely top notch but incredibly expensive, especially if you want the muted patches. Sample Modeling Brass is easy to use and probably the most flexible brass library available, but it can be extremely difficult to mix into an orchestra. Cinebrass and Cinebrass Pro are a great halfway between the other two libraries, and not too expensive either. Especially when deciding on French horn samples, make sure you use a library that includes **true legato** samples. This means that the transitions *between* legato notes are recorded, just as in string portamento and glissandos.

Woodwinds

- Piccolo
- Flute (alto, bass)
- Clarinets (Bb, bass)
- Oboe
- English horn
- Bassoon
- Contrabassoon
- Saxophone (soprano, alto, tenor, baritone)*

Woodwinds are in some ways the forgotten stepchildren of the orchestra. It's not uncommon to hear cues entirely devoted to strings and brass (often with synthesizers adding color and interest) and no woodwinds whatsoever. This may be because woodwinds provide a softer dynamic range than brass in many cases, and are thus difficult to hear over the blasts of horns and trombones. Adding to this is the fact that woodwinds (apart from percussion) are the most eclectic section of the orchestra timbrally. There are at least two distinct timbres within the woodwind section, and individual instruments often display *both* timbres depending on the range they are playing in. In light of all this it seems many composers forego the myriad colors of the woodwind section in lieu of the louder and more homogenous brass section.

Despite occasionally being undervalued, when orchestrated properly woodwinds can have an unparalleled effect on the orchestra. The disparate timbres offer the experienced arranger a nearly infinite palette of sounds to work with. In most cases, the primary function of woodwinds is to add color to string parts. Doubling a string melody with a flute line is a very common example of adding color by combining timbres (more on this in later chapters). Combining bassoons with low strings and brass is quite possibly the most powerful combination in the orchestra in terms of volume and punch.

Another important role that woodwinds play is to *contrast* string (and brass) timbres. Voicing a few chords in the string section followed by the same chords orchestrated for woodwinds can be striking to say the least. This trick adds more interest than you would expect, and it can make a frequently repeated motif sound new again. Finally, the most notable strength that woodwinds have is their ability to perform fast and climactic runs. Strings are capable of this as well, but due to the physics of the instruments, woodwind runs offer more style and flare. This is especially true when using pre-recorded flute and piccolo runs and rips. In their lower register they can sound slightly dull, but as a run ascends it naturally picks up volume and brightness due to the extra air moving through the instrument. This can be emulated by using the Mod Wheel (CC1) and Expression (CC11) controllers, but it doesn't usually beat fully recorded phrases.

Woodwind samples can be tricky to decide on, and it seems that every composer has libraries that they love and libraries that they hate. Part of the issue is that woodwinds (as mentioned earlier) don't have a homogenous sound. A clarinet sample might sound great but the flute sound in that same library might sound horrible. It may be necessary to pick a few different libraries to mix and match. For starters, make sure to find a core library that has good-quality legato sounds. You'll be using this articulation quite a bit, especially with the flutes. We have found staccato wind samples to be all over the place in regards to dynamic range, so also try to choose a library that allows for short and long notes in both *p* and *f* dynamics. You may have to find wind runs in another library altogether, but if you can find a core library that includes them then that is a huge bonus.

Berlin Woodwinds is a fantastic wind library which offers a good deal of microphone positions to help these instruments sit well in the mix. The Vienna Symphonic Woodwind libraries are some of the oldest libraries on the market, but still hold their own due to their intimate sound and flexibility. Cinewinds has pretty solid legato articulations, but we've found their dynamic range to be limited. However, it is a great price for all of the basic wind instruments. Finally, you may want to grab Hollywoodwinds at some point because it is a very fairly priced library that includes dozens of runs, rips, and wind textures.

The saxophone instruments are technically part of the woodwind family, although timbrally they fit somewhere between a clarinet, a bassoon, and a French horn depending on the circumstance. High-quality saxophone libraries are somewhat elusive on the market, likely because of their extreme flexibility regarding genre and their lack of standardized orchestral treatment. They are absolutely essential however if you are doing big band, jazz, or horn arrangements. We recommend the Audio Modeling Saxophone library for the same reasons as the SM Brass library mentioned earlier. The downside is that these can be difficult to mix into an orchestral setting due to their dryness. Regardless, saxophones are often (and tragically) forgotten about, but can really help blend the woodwind and brass timbres when orchestrated properly.

Unpitched Percussion

- Snare
- Tom-toms
- Bass drum
- Cymbals

Pitched Percussion

- Xylophone
- Marimba
- Glockenspiel
- Chimes (Tubular bells)
- Timpani
- Piano*

If woodwinds are the forgotten stepchildren of the orchestra, percussion instruments are the cool older cousins. Percussion instruments are even more disparate timbrally than woodwinds. There are literally hundreds of percussion instruments to choose from and thousands of libraries available for purchase. Percussion instruments are actually very well suited to the process of sampling. This is because by nature percussion instruments sound for a very short duration. This **transient** characteristic means that the nuances of note transitions and legato articulations aren't necessary to record. The keyboard itself is a percussion instrument, so after the recordings have been made, mapping the instrument range of, say, a Marimba to a MIDI keyboard is a relatively straightforward process.

Percussion samples work so well and can be used so effectively that it is extremely common for large orchestral scores to be supplemented by potentially hundreds of sampled percussion tracks. There are very few downsides to using percussion samples vs. recording percussion in a live orchestral setting. Certainly the core unpitched percussion instruments (snare, timpani, etc.) and tonal percussion (piano, marimba, etc.) need the experience of live players to bring the music to life, but having a large auxiliary percussion library will come in extremely handy.

For our purposes, the percussion section can be split into two groups: pitched and unpitched (see above). Pitched (or *tuned*) percussion instruments like the piano and marimba have the ability to add melodic and harmonic layers to an arrangement, but the sustain differs from instrument to instrument. Unpitched (or *untuned*) instruments, like the snare and bass drum, do have timbral and register qualities, but cannot add any sort of melody or harmony. The timpani, although it is tuned by the player and can be used for a basic melody of sorts, is generally considered separate from other percussion instruments due to the specialization required of timpanists. It *does* have the ability to play up to four pitches simultaneously or in direct succession, but usually it lingers on the root or the fifth of the key. For our purposes with samples, it can be considered a bridge between pitched and unpitched percussion, offering rhythmic support in climactic moments, and reinforcement of chord tones when needed.

Apart from the notation and part production of percussion sheet music, writing for percussion samples and live instruments are remarkably similar. We ascribe to a "less is more" approach for a few reasons. Most composers go overboard when they download their first percussion library and realize they have access to drums the size of a small house. "Epic" drum ensembles are necessary for some genres of course. But they don't leave much room for arranging and orchestrating. In fact, when properly balanced just a few percussion instruments can overpower an entire orchestra. So when it comes to percussion arranging, save the "big guns" for the right moment, and use smaller detailed instruments and rhythms to get the job done. A little bit of percussion goes a long way in most situations. Try thinking of percussion as "punctuation" more than as the content of a sentence.

In terms of function, unpitched percussion is usually used in one of two ways: rhythmic ostinatos and added layers of color. Both methods of percussion writing can be used to intensify transitions and climaxes. Think of the snare drum in a marching band. The rhythmic loops *push* the energy forward during the piece. It works exactly the same way with video game cues. Denser rhythmic activity (see Chapter 6, page 179) *adds* energy. This is why it's common to hear tom-toms or trailer percussion hammering away during battle scenes, very often heard in tandem with string ostinatos.

An example of adding intensity to transitions and climaxes is the classic cymbal roll. Cymbals are so complex in terms of their overtones that they can't be used as a true "harmonic" instrument. However, this complexity makes it a great addition when transitioning between sections of a cue. A cymbal roll can start at the end of one section, and peak at the beginning of the next to effectively *smooth* the transition. It also heightens the drama momentarily.

Pitched percussion instruments are functionally capable of doing almost anything the other sections of the orchestra are capable of. Sometimes pitched percussion instruments are used as a focal point for a cue. The marimba and vibraphone are both more than capable of expressing complex melodic and harmonic material. Alternatively, pitched percussion can be a fantastic source of added color when paired with strings and woodwinds. It is very common to hear flute or piccolo in unison with xylophone. John Williams uses this technique often in his score for *Jurassic Park*, and the effect is dazzling. This pairing works because it combines the bright timbre of the flute/piccolo with the percussive strikes of the xylophone to emphasize the *attack* of every note. This is an example of **complementary orchestration** (combining colors/timbres in a way that "fills in the gaps" like a puzzle piece). The strong attack of the xylophone is used to strengthen the rounder sustains of the strings and woodwinds. Without the percussion, the attack wouldn't cut through the mix as sharply; without the woodwinds and strings the duration of each note would be too short to make as much of an impact.

When buying percussion libraries it is important to find a library that has as many auxiliary percussion instruments (world instruments, esoteric instruments, etc.) *in addition* to the core orchestral percussion instruments as possible. It is often wise to find a dedicated core percussion library first to take care of the bass drum, snare, cymbals, and timpani. Then you can find another library to supplement the pitched percussion as they can have idiomatic articulations (like mallet rolls on a marimba, or a bowed glockenspiel) that would be too much for a basic percussion package to include. Often the number of "round robin" (RR) samples is a good indicator of a strong percussion library. The higher the RR number, the more repeated notes you can sequence without the phrase sounding mechanical or unnatural. Additionally the choice of mallets (soft or hard) can be very useful in orchestration. Eventually you can supplement your percussion library with various world percussion libraries as well.

Here again we have an outlier – the piano. The piano is technically a percussion instrument, but its history in the Western music tradition is all over the place. It can be used as a solo instrument, in conjunction with the orchestra, as a chamber instrument, as an accompaniment to a soloist *or* orchestra – it can even be prepared ahead of time and made to sound completely specialized. Like the harp, the piano has its own framework for performance and pedagogy. Keyboard players are *used* to reading parts a certain way, so piano scores should be studied in depth before delivering parts. On top of that, the piano repertoire spans just about every genre you can think of: jazz, big band, rock, pop, and everything in between. The fact that the piano is usually placed near the other instruments on this list (toward the back on the stage) is about the only thing it has in common with the other percussion instruments – and even that is not set in stone! Needless to say, the piano samples you choose need to either be very flexible, or very specific depending on how you are most likely to use them. We've found that piano libraries are among the most difficult to sound "right" due to the extreme contrast with the way pianos are mixed from style to style. You will likely have to purchase multiple libraries to cover all your bases.

Expanding Your Template with Synthetic Timbres

There are no hard and fast rules for including synthesized sounds in your orchestral template. A good starting point would be to have a basic patch setup with each of the above instrument types. It is also common to see blank synthesizer patches loaded into a template to make it easy to build a new sound from scratch if it is needed. The setup really depends on your aesthetic preferences and workflow. Often synthesis is so personal that composers choose *not* to add it to their templates, leaving it as part of the composition and arranging process.

Regardless of what you choose to add to your template, keep in mind that software synthesizers are usually quite loud, and often need to be brought down a bit to avoid covering up the quieter orchestral instruments. For more information on mixing with synthetic timbres visit the Sound Lab (companion site).

Template Setup

 Now that we are familiar with the basic orachestrl instruments we can move on to the process of organizing our template. This is a crucial aspect of producing compelling music for games. Before moving on, head over to the Sound Lab for a practical walkthrough on template setup and mixing for realism.

Your template will never truly be "finalized." You will likely expand with new libraries, and tweak the template as your production skills increase. For this reason your template will always be more of a work in progress, and we recommend saving versions of your template for reference.

Spencer

WRITING FOR STRING SAMPLES

Writing and orchestrating for string samples is an absolute necessity as a video game composer. Out of all of the sections of the orchestra, strings are the most ubiquitous in game music. But most game projects simply don't have the budget to hire a 30- to 60-piece string section for every cue, so composers must opt to use string samples in such cases. The good news is that string libraries today sound spectacular, and are often very intuitive to use. However, no string library is going to sound professional and polished without a competent orchestrator making the decisions.

The typical attitude towards strings, especially for new composers proficient with a piano, is to improvise a few chords with a strings patch loaded up. Then maybe make a few tweaks, and *voilà* – we have orchestrated string parts. This *may* work well in some cases, but more often playing keyboard parts with string sounds is a very inadequate way to write for strings.

This scenario gets to the heart of what orchestration actually *is*. Orchestration is the *idiomatic writing of parts such that the resulting sound is balanced and achieves the desired effect*. It is the application of a thorough understanding of how orchestral instruments produce sound, and how to best present that sound. In this case, a keyboard is not a string instrument, and so parts easily playable on the keyboard are at times not playable with a string section. They are usually not even *desirable* either, because they do not allow the strengths of the stringed instruments to come through in the music.

So what *are* we looking for? When writing for strings the best approach is to visualize each of the string instruments – violins I, violins II, violas, cellos, and basses – and imagine that an actual human being is playing each part that you write. This means that your parts need to be interesting enough to hold the attention of an actual professional string player. It also means that your voice leading *has* to make sense. If a part is to be playable, it should smoothly move from chord to chord without crossing voices (unintentionally) or making ridiculously large jumps. The result should sound almost melodic in many cases.

Another thing to incorporate into your string writing is a variety of articulations. Many novices will mistakenly use *only* legato strings, or *only* spiccato strings. But strings are possibly the most flexible section of the orchestra when it comes to timbre and articulations. Most sample libraries offer a range of articulations to choose from, so take advantage of them.

Entire classical works are written with just strings. The possibilities are high. In the Sound Lab (companion site) we offer additional information on TwoSet violin as a primer on what is possible with the instrument.

Lastly, we are looking for a range of *textures* as well. We will dive into texture a bit later in the chapter, but for now keep in mind that a string section is not limited to playing chords and melodies. *Independence of parts* is an important concept, and strings are so homogenous that even the most complex polyphonic material can be easily balanced. Effectively composing parts that are independent is the opposite of our keyboard approach from earlier. There, we were limited by the size of our hands on the keyboard and our ability to play contrasting simultaneous parts. In a string section, whether live or sampled, we have the full range of notes from the low E in the bass section (or lower with extensions) all the way up to (usually) a high D in the first violins and beyond.

So let's summarize the main points that foster effective string writing. The first is more of a guidance tool for inspiration, but the last three translate directly to the quality of your string orchestrations. In some cases these elements can be the difference between strings that work for your game and strings that need to be cut altogether.

1. Part interest (imagine that a human is playing each part)
2. Smooth voice-leading
3. Utilization of available timbres and articulations (pizzicato, sul ponticello, etc.)
4. Utilization of available textures (part independence)

Many novice orchestrators begin their studies by practicing four-part string writing before even touching the rest of the orchestra. See the Sound Lab for an exercise in four-part string writing.

Potential Problem Areas and Workarounds for Samples

String samples currently on the market have some wonderful strengths. Most libraries have smooth, buttery legatos and exciting, natural-sounding spiccatos. There are even some great libraries with interesting special textures and effects for strings which are extremely useful for horror games and trailers. Despite all this, there are a few idiosyncrasies most string libraries share.

Articulation Switching

Switching quickly between articulations can be a bit sloppy. Actual string players can move back and forth smoothly from long to short notes and vice versa, but samples often sound "jumpy" in terms of volume and transitions. The best workaround for this is to load *both* short and long articulations on separate channels and use expression and modulation to clean up the transitions. This helps compensate for any volume discrepancies between articulation samples.

Repeated Notes

Repeated notes can sometimes produce the "machine gun" effect. This occurs when it is obvious to the listener that the same samples are being fired over and over. The best way to avoid this is to buy a library with *railroad staccato* or *spiccato*, or *round robin* sampling. This means that multiple samples are recorded of each short articulation, and the library is capable of alternating them with some randomness.

Bowing

Even the best sample libraries cannot capture all the possible ways that a string player can use the bow to create sounds. Buying a few different libraries can help. There are a number of specialty effects such as sul ponticello, col legno, and bartok pizzicato that can fill the gaps in the string timbres of your template. This still pales in comparison to what a live player can do, and we will explore ways to exploit this in later sections.

Tricks to Make the Most of Your String Samples

Use MIDI CCs and Velocity

Nine times out of ten, if your string parts aren't coming through in a compelling way it's because of a lack of attention to your MIDI data. Velocity plays a large role in the quality of your sample writing with all instruments, not just strings (see Figure 7.2). Velocity curves should be varied enough to sound human, but not so much that notes are dropped or out of place. In most sample libraries expression will be a fine volume control, working off of a percentage of your track volume. This can be helpful for balance.

Modulation usually controls the actual *sample* itself. For example starting a modulation of zero, a cello sample at *ppp* will be triggered. Moving the modulation all the way up will crossfade to a *fff* sample. This is audibly different from the way expression works. The *fff* sample will include artifacts of heavy bow pressure among other things. By contrast, changing expression will *not* change the sample, it will only change the volume. In short, modulation control usually changes the sample itself, which can be used to make your orchestrations more detailed.

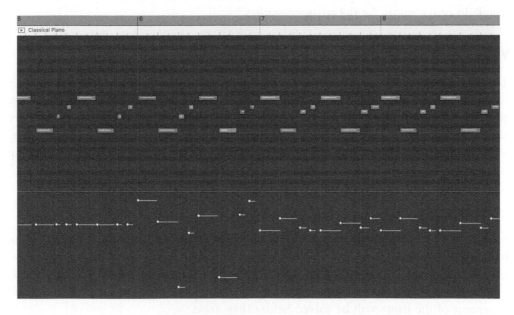

FIGURE 7.2 The "Goldilocks Zone" of velocity curves. Bar 5 is too narrow, bar 6 is too random, and bars 7 and 8 are just right. Bonus: Can you guess what theme these MIDI notes are referencing?.

Focus on Strong Articulations

Most string sample libraries have very well-recorded legato and spiccato articulations, and composers make good use of these. Without omitting the other articulations, it can be helpful to use legato strings as a foundation for your melodic writing. Spiccato articulations are great for energy and momentum. It's important not to rely too heavily on these articulations, but they are a staple element in the composer's toolkit for good reason.

Portamento and Trills

One of the best ways to add realism to string samples is to use a bit of portamento. When used sparingly this can bring out some life in a string part. It can also make your writing sound very natural and emotive. Likewise, trills can be a great way to spice up chordal movement. Trills don't often switch smoothly into other articulations the way a live player can, but the cautious inclusion of a trill now and then can make a dull cue sound magical.

Use Other Orchestral Sections

String samples are flexible and reliable, so they are understandably a go-to for game composers. But over-reliance on strings can be monotonous in a game soundtrack. It can be very effective to bring other sections to the foreground as a contrast to the strings. In other words, let the strings *breathe*.

Writing for Live Strings

Modern string samples are workhorses, and in some cases can sound *better* than live recordings. However, when the time comes to record a professional string ensemble, mock-up orchestrations don't always translate to idiomatic string parts. Remember that players are usually *sight-reading* their parts at the session. It is in your best interest to make these parts as idiomatic and playable. Below are a few tips to making your parts as playable as possible in a session.

Scalar Melodies

String players eat, breathe, and live scales. The best way to get a string session running efficiently is to write scalar or stepwise parts. Because orchestral strings don't have frets, writing large leaps and arpeggios can cause tuning issues unless the fingering is completely idiomatic. Keep the accidentals consistent, and write scalar melodies, and 90 percent of the issues will be solved before they arise.

Write in an "Open" Key

This goes for all instruments (not just strings), but prepare your parts in an "open" key signature. This means write your score with no sharps and no flats as the key signature, and write in all of your accidentals. This will save time in the session, avoiding questions like "is this still a G-sharp, or is it a G-natural?"

Buy an Orchestral String Instrument

It sounds like overkill, but purchasing a cheap string instrument can really help with your orchestrations. You don't need to spend a lot of time with it before you are able to plunk through parts with your left hand. The goal isn't to be a virtuoso, it's to evaluate how awkward leaps are. In many cases you will also be able to determine whether or not certain figures are even physically possible. It will also help you with planning double and triple stops. As a rule of thumb, if you can slowly finger through a part with your left hand, then a seasoned player shouldn't have a problem with it.

> In *all* cases of translating sampled parts to live instruments, the best way to ensure that your orchestration is idiomatic is to have an actual player take a look at the part in advance. This is true *regardless of the instrument*. Our advice is to make tons of friends and seek out their opinion on parts for instruments you don't play yourself. You can practice by writing a few bars for a solo instrument, and then asking someone to record it for you. Listen to how it sounds in reality and compare it to how it sounded in your head. Understanding the difference between the two is what will teach you how to orchestrate/arrange.
>
> Spencer

Examples of String Writing in Games

 In the Sound Lab we offer a few examples of string writing for games.

WRITING FOR WOODWIND AND BRASS SAMPLES

Although woodwinds and brass are actually two distinct sections of the orchestra, they function very similarly when using sample libraries. For this reason we have lumped them into one category. More detail on the idiosyncrasies of each instrument group will be discussed in later sections (see "Advanced Orchestration and Arranging for Live Ensembles" below).

Brass and woodwind instruments are very important to consider when orchestrating. For one thing, they offer a contrasting color palette. Compared to the relatively homogenous sound of the strings, woodwinds and brass are far more diverse timbrally. This is both a strength and a weakness. An effective arrangement that uses brass and woodwinds will offer sonic variety and add interest to a game soundtrack. Unfortunately, due to this contrast between instrument timbre and mechanics, balance and blending in an orchestral setting can be a challenge.

In general, the four main points for string writing also apply here. Part interest, smooth voice leading, use of available timbres, and use of available texture are all important aspects of brass and woodwind orchestrations. However we will now add a fifth point.

5. Timbre and volume balance between voices

Timbral balance is important with string writing as well, but it is usually much simpler to achieve. As mentioned in the case of divisi strings, a volume balance can easily be achieved in a chord progression by splitting all instrument groups into divisi. On top of that, the timbre of the string section is so similar between instruments that a timbral balance can be assumed from the start. This is not the case with woodwinds or brass instruments. Woodwinds alone have a variety of instrument designs and mechanics, which means that care must be taken to place each instrument in its proper register and dynamic when voicing a chord. Similarly, brass instruments have a variety of mutes, which are often included in sample libraries. These mutes can subtly change the volume of brass instruments, but their main function is actually to change the timbre. This leads to some of the same issues that woodwinds have with overall balance.

Solo Instrument Patches vs. Ensemble Patches

Due to a lack of homogeneity in wind and brass samples, it will be necessary to split a brass or wind section into multiple parts similar to the string divisi. In these scenarios our preference is to use solo instrument patches rather than ensemble instrument patches. This results in a much more balanced sound within the section. Your parts will blend better, and you will avoid any single instrument overpowering another. For example, if you are trying to split the trombone section into a three-voice chord progression, using an ensemble patch would *triple* the number of trombones that were sampled. This can sound gigantic and might bury another section playing within that frequency range. By using solo trombone patches instead, the voicing will be balanced with the rest of the orchestra. This voicing will also translate very well to live performances since you will likely not have nine trombones at your disposal.

However, the opposite is true as well. If, for example, you are trying to blend the first violins with a flute on the same part at a *forte* dynamic you might run the risk of overpowering the flute. This is especially true if the flute is playing in its middle or low register. In this case, copying the flute part to each of your solo flute channels would result in an undesirable *phasing* effect, which sounds tinny and unnatural. A better option is to switch back to an ensemble patch of two or more flutes and use it in unison with the first violins. This will sound like a nice blend of the two timbres, neither overpowering the other.

Melodic Lines

Melodic writing for woodwinds and brass differs slightly, but it does have one thing in common – it takes *breath* for these instruments to make sound. It is important to keep this in mind for all your wind and brass orchestrations, but particularly for melodies. Run-on melodies are simply not possible for woodwinds or brass. Allowing breathing room can add realism to your parts. Some sample libraries actually won't allow absurdly long held notes, which can be frustrating for composers who aren't aware of the physical limitations of these instruments. Don't ignore this! Keep breath in mind when writing your melodies and you will avoid over-writing for wind instruments.

Apart from breathing, woodwinds are a great choice for melodic lines. Flute samples tend to be quite reliable from library to library, and are thus a great choice. Legato phrases sound fantastic, but staccato articulations speak just as well. For various reasons flute, piccolo, oboe, English horn, and bassoon samples *usually* switch between legato and staccato quite smoothly, but clarinet and saxophone samples don't always switch articulations as smoothly, so keep this in mind when writing your parts.

Commonly you'll hear a flute melody doubled with the first violins. This is a very effective pairing. Woodwinds blend very well with strings, and can bring out a brightness in a melody line that would be absent in the strings alone. At lower string dynamics (or at the top of the flute's range), one flute will balance well against a string part. At *forte* or louder, it can help to use an ensemble patch of two or more flutes so that the woodwind timbre is not lost in the strings. This balance is typical of every woodwind and can be used as a general rule of thumb, even for live settings.

> Other useful pairings like this are bassoons and cellos, baritone or bass saxophone and double bass (arco), and alto flute and violas.
>
> Gina

Brass instruments are fantastic for strong or punchy melodic lines. In games brass is often used for heroic melodies due to their volume and power. The number of players can range from one to twelve for an epic melody. Horns blend well with strings or woodwinds as well, so you'll often hear a horn melody blended with the cellos and doubled an octave below by the tuba (and/or string bass).

Trumpet and trombone samples are also capable of effectively voicing melodic lines. Using solo patches these melodies can be heroic, triumphant, and intimate. Using ensemble patches these melodies can be heroic, triumphant, and *loud*. Take care to decide which patch is fitting for your arrangement.

> Another useful tip for brass writing is to voice a melody (or chords) in the horns, and add a single solo trumpet in unison with the top horn line. This adds subtle brightness and clarity to the part that might otherwise sound dull depending on the range.
>
> Spencer

Unlike woodwinds and French horns, trumpet and trombone samples don't blend all that well with other instruments. Usually the volume and power overshadows other instruments. That being said, with some modulation and expression adjustments, trombones (especially muted) paired with low strings and bassoons can be effective for big moments in an arrangement. Trombones, cellos, and bassoons are often paired together for stabs and ostinatos. You may need to use ensemble patches for the bassoon timbre to be heard, but this pairing is a powerhouse for action and adventure cues.

Chordal Movement

When using a woodwind or brass ensemble to voice chords the most important thing to consider (aside from breathing) is balance. By this we mean balancing the *weight* of each chord tone. For woodwinds a 1:1 ratio of virtual instruments is usually satisfactory. If the top note in a chord is voiced by a solo flute patch, using solo instrument patches to voice every note in the chord will be preferred to maintain balance. But if the top note in the chord is voiced by two woodwinds (say two flutes, or a flute and a clarinet), then use *two* solo instrument patches for every note in the chord to maintain balance. In short, maintain a 1:1 ratio of musicians for woodwind chords and the weight of each chord tone will balance approximately.

Similarly, brass patches need to be balanced in this way as well. However, in the case of brass patches, usually one (un-muted) trumpet or trombone is equivalent in volume to two French horns. In other words, for every solo trumpet or trombone designated to a chord tone, there should be two French horns to maintain the dynamic balance. Table 7.1 shows the basic dynamic ratios of each orchestral instrument based on Nikolai Rimsky-Korsakov's orchestration method. In simplest terms, this method assumes that at *piano* all instruments are approximately equivalent in volume, but they start differentiating at *forte*, as outlined in the table.

TABLE 7.1 All *piano* dynamics are relatively the same; at
forte they begin to diverge

Section	piano dynamic	forte dynamic
Strings	1	2
Solo Woodwind	1	1
Solo Horn	1	1
Solo Brass	1	4

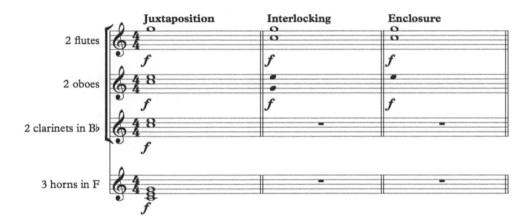

FIGURE 7.3 Bar 1 is a typical juxtaposed voicing of woodwinds and horns.

Aside from balancing the *volume* of your voices, it's also important to balance your chords *timbrally*. The simplest method of achieving this balance is by separating timbres, also called **juxtaposition**. For example, in Figure 7.3 you can see that a C major chord is split into two octaves. This is a typical voicing for brass and winds in an orchestral setting. Using these ratios as a *starting point* can help achieve a balanced chord. The woodwinds here are voicing C–E–G an octave *above* the horns. For every solo horn there are two woodwind instruments to compensate for volume (in terms of volume, trumpets and trombones share an approximate 2:1 balance with French horns; likewise, French horns share an approximate 2:1 balance with woodwinds resulting in a 4:1 balance between trumpets/trombones and woodwinds at a *forte* dynamic). If the winds are in their upper register, this adjustment may not be necessary (especially for a piccolo, which almost never needs doubling for support to cut through). However, there is also *fullness* to consider. Even if balanced with a brass instrument in volume, a solo woodwind will always sound *thinner*.

Note the 2:1 ratio of instruments between woodwinds and horns to compensate for volume and fullness. Bar 2 is an example of interlocking, where the flute and oboe timbres alternate up the chord tones. Bar 3 shows a similar voicing without the second oboe part, thus enclosing the oboe timbre inside the flute timbre. These are all valid ways to voice chords, and you should strive to employ them appropriately in various game scenarios.

> Although we have offered a basic equation for balance here, note that it is a vast over-simplification. It's great for newer orchestrators, but other factors like register, dynamics, "fullness," and musical context play a large role in how an orchestration sounds. When using samples, the mix can completely change the impact of an orchestration as well. We will dive a bit deeper into these concepts later, but for now use this as a tool to train your ear to listen for balance in a chord. Once you get some practice you will be able to use your own judgment to balance chords, but it takes time and a critical ear.
>
> Spencer

Another method of voicing chords is **interlocking** (Figure 7.3). Instead of splitting chords up by their timbre we would interlock and alternate timbres. In this example of a chord voiced by woodwinds, we have split up the first and second flute, and inserted the oboes above and below the second flute. If we were to take away the second oboe, we would have an example of **enclosure** because the flutes will have encapsulated the oboe timbre (Figure 7.3).

The above techniques work in many cases, but they are not quite as effective for achieving chordal balance as juxtaposition is in a homophonic texture. However, as game composers, interlocking and enclosure are sometimes necessary to maintain adaptivity in a score. For example, in a scenario where a cue must loop, using inter-locking timbres to maintain smooth voice leading is preferable to a perfect balance that forces an awkward leap in one voice. Interlocking can be a workaround for this. Use this technique cautiously, because some timbres will stick in more than others. Interlocking a flute timbre and the more nasal oboe timbre works well, but interlock-ing a trumpet timbre and a flute timbre could cause imbalance in your chord tones.

Textural Effects

Textural effects are certainly a strength of woodwind and brass samples. In most cases sample libraries will designate a single note on the keyboard for a particular effect, making it extremely easy and flexible for use in games. See the examples below for common effects.

Runs

Woodwinds are capable of performing scales quickly, and we call these **runs**. String (and brass) samples are capable of this as well, but due to their brightness and dynamic range runs are very idiomatic for most woodwind instruments, and the effect can be dazzling. For this reason, as mentioned earlier, most libraries will either include a "runs" patch, or include pre-recorded runs that are typical of film and game scores. A quality runs (sometimes called "runs transition" patch) will allow you to piece together your own custom run, with the only limitation being that it must fit a particular tempo range. Pre-recorded runs and scales patches will sound very natural, but are further limited by offering only a few options of scales and gesture lengths.

Rips

Brass **rips** are non-tonal glissandi that are performed quickly, and at a loud dynamic. French horns are famous for these effects, and they work quite well for moments of surprise or shock in games. By contrast, these effects can't truly be simulated with a legato or staccato articulation the way that runs can. Most of these effects will be found as pre-recorded patches.

Clusters

Clusters are actually possible with all instrumental groups (see Chapter 6), but brass samples voice clusters especially well. A cluster is essentially a chord with tones stacked into major and minor seconds. These tones are "clustered" together, hence the name. In some libraries clusters are included as pre-recorded patches, but because clusters require no special playing instructions, they can be very realistically simulated by normal legato or staccato patches.

Stabs

Stabs is a term typically used with regard to brass instruments. Due to their power and dynamic capacity brass instruments are sometimes asked to play a loud accented chord of an extremely short duration. Sometimes stabs are used in conjunction with clusters to create a sense of shock in the listener, or to highlight an urgent moment in a film or game. Like clusters, stabs can be fully realized using staccato or staccatissimo patches, as well as with pre-recorded patches.

Potential Problem Areas and Workarounds

For the most part, brass and wind libraries are reliable and consistent. The trickiest aspect will be balancing the powerful brass with the slightly more delicate woodwinds. Make sure not to overpower the woodwinds too, but don't neglect other timbres altogether. Remember the previous lessons on balancing your template (refer to the

Sound Lab for detailed instructions on mixing your template). Remember that brass instruments at a *forte* dynamic should naturally be louder than the other sections. If you adjust the mix so that they are *balanced* with the winds at that dynamic then you will lose the natural sound of the orchestra.

Tricks to Make the Most of Your Brass and Woodwind Samples

MIDI CCs

As always MIDI CCs (specifically modulation CC 01 and expression CC 11) are your first line of defense for balancing woodwinds and brass. Additionally, some libraries may even use the breath controller (CC 02) to adjust vibrato or other parameters.

Layering Articulations

It may be helpful to organize your tracks by splitting the long and short articulations. This can save you the time and frustration of managing keyswitches, and it can also offer some creative solutions down the road. For example, many brass libraries have no middle ground between sharp accented staccatos and lush legato. A common solution is to actually copy and paste the MIDI region exactly for a particular phrase from the staccato track to your legato track. This effectively layers both articulations on top of one another. This will allow you to easily add attach sharpness to legato phrases by increasing CC 11 on the staccato track, or you could add some controlled sustain to staccato phrases at your discretion.

> In our experience 99 percent of the time a brass figure sounds awkward, the solution is to layer articulations in this way.
>
> Spencer

Writing for Live Winds

Breathing Room

The importance of allowing players to breathe cannot be understated when writing idiomatic parts for woodwinds and brass. Woodwinds are incredibly flexible instruments capable of churning out tons of notes in very short periods of time, but they cannot play long drawn out phrases for an entire cue with no room to breathe. If you are dead set on writing pervasive ostinati for woodwinds or brass, make sure to include instructions to "stagger breathe," and don't expect an incredibly sharp performance.

Be Aware of Transpositions

Many woodwinds and brass instruments are transposing, which means that the note you *write* is different than the note you will *hear*. An important part of orchestration is understanding this and delivering parts *in the proper key for the player*. Being comfortable transposing in your head can be immensely helpful in this regard.

Watch the Breaks

When writing for actual woodwind players a common mistake is to write trills or ostinatos that bounce back and forth between two notes. These figures are easy to write, but can be devilishly difficult to play. All woodwind instruments have "breaks," where the fingerings change drastically. In some cases trills can be unplayable even by professional musicians at certain tempi. We recommend using a dedicated instrumentation text (like Blatter's Instrumentation and Orchestration) as a reference for these breaks to make sure your parts are playable.

Trombone Slide Positions

Similar to the breaks in woodwinds, trombones have slide positions that can be awkward or impossible to play fast. Generally, large leaps at fast tempi can be at best sloppy, at worst physically impossible. Scalar motion will usually solve most of these issues, but it can be helpful to keep a slide chart for reference. Similarly, some glissandi are not possible as well, depending on the range. To water this issue down, an open trombone can only play glissandi ranging a tritone or less. In the Sound Lab (companion site) we provide some additional resources on trombone slide positions.

Examples of Brass and Woodwinds in Games

 Despite the differences in the sonic properties of wind instruments, they are still extremely versatile and useful when effectively orchestrated. In game scores, composers often use woodwind and brass instruments in a few ways: melodic lines, chordal movement, and idiomatic textural effects. Check out the Sound Lab for some examples.

WRITING FOR CHOIR SAMPLES

The choir isn't always considered a section of the orchestra per se, but it nonetheless can be used very effectively in games. The choir is similar to strings in that it has a very homogenous sound and can be used in a variety of ways. For these reasons, many basic techniques of good string writing listed above also apply here. The choir can be used effectively when moving from chord to chord homophonically, but it also balances very well when counter-melodies or polyphony is introduced. The obvious difference when working with a choir library is that care must be taken when choosing lyrics or vocalizations.

As far as choosing a choir library, there are many options – from basic libraries that offer simple vocalizations ("oohs" and "aahs") to boutique packages which focus on sampling specific vocal traditions practiced by soloists and choirs all over the world. Still others offer the option to use pre-recorded phrases, or allow you to build your own phrases from scratch. These libraries all come with their own benefits and downsides. It's best to try before you buy, and stick to libraries that fit into your workflow.

Potential Problem Areas and Workarounds

With choir libraries achieving realism can be tricky. This is because our ears are very attuned to the human voice, and it is relatively easy for us to identify anything unnatural when it comes to vocal samples and speech. Choir libraries can be effective when using the basic vowel sounds to add drama or intimacy to a cue, but phrases and word builders can take an immense amount of work to sound natural.

The solution to this issue is to know your library, and know its strengths and weaknesses. Some phrases may sound better than others, so avoid the awkward samples and stick to phrases that sound clear and human. If you absolutely *must* use phrases or words that don't sound quite right, the general rule with samples is to double that part with other instruments. The bigger the arrangement is, and the more effectively you orchestrate around the problem areas, the less noticeable the offending part will be. The opposite is also true – the more exposed the part, the more noticeable it will be. This applies to virtually any samples that you use, not just choir libraries.

Tricks to Make the Most of Your Choir Samples

Reverb

Reverb is again a hugely important factor in achieving realism with choir libraries. If too much reverb is applied, the phrases can be drowned out and your singers will almost sound like they are singing from a cavern miles away. However, an appropriate amount of reverb is entirely necessary to hear a smooth legato sound, and to emulate the proper space. Microphone positions can be used to balance the choir mix. Try playing with the close and far microphones if your library offers it. The goal is to introduce enough stage and room sound for the choir to sound big and dramatic without negatively impacting the clarity of the sound.

Layering Live Vocals

This is somewhat of a cheat, but the best way to add realism to choir samples is to overdub live vocals. This approach can make a mediocre choir library sound absolutely fantastic, especially when lyrics are involved. The listener's ear will naturally be drawn to the live recording when mixed well, and the samples will act as reinforcement. Since almost everybody knows a singer, this can be a really easy trick to pull off.

Writing for Live Choir

Lyrics

Writing the *notes* for choir is really no more difficult than it is for any other instrument group. In fact, it's much easier due to the aptitude for singers to blend together. Unlike woodwinds, it's even common for sections to stagger breath, making longer phrases less of an issue. The trick to writing for choir is notating lyrics without confusing the singers. This is a sizable topic on its own, but in general sticking to "oohs" and "aahs" is a tried and true method for adding choir color to a game cue. As mentioned in earlier chapters, lyrics can sometimes take the player out of an immersive experience. That said, lyrics (even Latin or gibberish lyrics) can add depth and detail to a cue that simple phonations can't. The nuances of notating text in a choral work are beyond the scope of this book, but we recommend checking out Elaine Gould's *Behind Bars*[3] and using it as a reference (along with literally any choral scores from the classical period or later).

Aleatoric Effects

We mentioned aleatory in Chapter 6, but let's bring it back briefly with regard to choral writing. Aleatory is the technique of adding "chance" or improvisation into a cue. With live choir writing, you get the added option to improvise *lyrics* as well as notes and rhythms. You also have the option of using spoken word rather than sung notes, which can be surprisingly affective. This can sound like anything from powerful to nightmarish. The simplest way to achieve this effect is to give the singers a "bag of notes" or words, and have them improvise to simple directions. Check out the Sound Lab (companion website) for a short snippet of whispered text that would create eerie murmuring in a horror cue.

Examples of Choir Writing in Games

In the Sound Lab we take a look at a few examples of basic choir writing.

WRITING FOR PERCUSSION SAMPLES

Percussion samples are diverse in terms of timbre and function. This makes them both useful and at times tricky to write for. Percussion instruments of indefinite pitch (cymbals, drums, etc.) are relatively intuitive. Depending on your needs these instruments can be used to accentuate important moments in a game. Unpitched percussion are very effective for stingers because they overpower the rest of the arrangement and smooth over transitions or loop points regardless of harmony or voice leading. This is useful to keep in mind if you are struggling to make an adaptive music system sound cohesive.

Unpitched percussion instruments are also used quite often as a "groove" layer in vertical scoring. In many cases percussion acts as an effective foundation for a loop, and other sections of the orchestra can be stacked in vertical layers to change the mood of the cue. We'll take a look at more complex examples of this later (see "Advanced Vertical Techniques," Chapter 9, page 311).

Pitched percussion can be used to great effect as solo instruments, ensembles, or combined with other orchestral sections. Marimba, glockenspiel, and xylophone are all percussion instruments that are more than capable of holding their own when it comes to melody and harmony. Pitched percussion sample libraries are quite versatile, and transfer easily to the MIDI keyboard so don't neglect them.

Potential Problem Areas and Workarounds

The only real problem area with percussion instruments is overuse. Many new composers will purchase an epic battle drum library and write a complex part at a fast tempo thinking it will blend with a huge Hollywood-style orchestral arrangement. This may *sometimes* work, but usually the battle drums will either overpower the rest of the arrangement, or they will be lost to the brass and strings. Often simple parts with orchestration that *leaves room for percussion* is the best way to achieve an epic sound. For drums to sound "big," there needs to be room in the mix for the listener to hear them.

Tricks to Make the Most of Your Percussion Samples

Complimentary Scoring

The best way to avoid the issues above, and to make the most of your percussion samples is to use them sparingly and intentionally. Don't just pull up the fattest drum patch you can find and hammer away. For action-packed grooves to work there has to be samples that complement each other rather than compete. For example, when writing a drum part for an action cue, try adding a shaker rather than layering more battle drums. The shaker is a higher-frequency instrument and has a sharper attack, so it *complements* the part. Likewise, using higher-frequency instruments like bamboo sticks can act as a fantastic counter-melody of sorts.

Utilizing Range

When choosing percussion instruments focus on a broad range of instruments rather than overusing the "big guns." It will make the moment when you bring out the ten-piece bass drum ensemble more memorable. When range is taken into account, percussion instruments are capable of producing an incredible amount of complexity and intensity.

Managing Velocity

Aside from proper choice of percussion instruments, it's incredibly important to fine-tune your velocity curves. Percussion instruments are transient (excluding cymbals and gongs), so they usually don't hold their sound for long. This means that velocity is for the most part your *only* means of shaping the dynamics of each phrase. Listen carefully to your percussion parts alone and with the orchestra to ensure that notes aren't unnaturally loud or disappearing in the texture.

Writing for Live Percussion

There is a surprising amount of nuance to writing percussion parts. Percussionists (especially unpitched percussion specialists) are capable of reading extraordinarily difficult and complex rhythmic figures. This doesn't mean it's always best to *write* extraordinarily complex rhythmic figures, but it doesn't hurt to add in something fun for your percussionist to play. Beyond that, any given cue could employ dozens of percussion instruments or more. Below are some tips to make this easier.

Write for Humans

Generally speaking, the most important thing to consider when writing for an actual percussion ensemble (or section) is to make sure you know 1) how many percussionists will be at your disposal, and 2) allow enough time for each of them to transition from instrument to instrument logically. This means you should always write to the *number of percussionists in the ensemble*. If you have two percussionists, don't write a part for bass drum, cymbals, and tam-tam simultaneously. That said, in an orchestral setting timpani are usually assumed to have their own player at all times who will not switch to another instrument. However, pitched percussion instruments like marimba and vibraphone often do require percussionists to switch on and off if the part allows for it. The bottom line is that you should write for the number of people you have. By extension, you will most likely have to consolidate the parts, including instrument changes, for percussionists to read.

Pay Attention to Notation Etiquette

Again, notation etiquette is important. For example, all rolls are notated with three-slash tremolos for percussion instruments regardless of speed. We would again refer you to Elaine Gould's *Behind Bars*[4] for specifics.

Playing Techniques

Writing for percussion samples can be somewhat of a hit-or-miss task. Most composers plunk notes on the keyboard until they hit a sound that works, and then they record it. When writing for a live ensemble you can actually instruct players to produce sounds in any way you want. The standard example is to direct a player to use hard mallets vs. soft mallets. This is just scratching the surface though. Most percussion instruments including cymbals, gongs, crotales, and even vibraphones can be bowed. The resulting sound can be anything from glassy and beautiful to horrifying. Similar to string writing, the cue for percussionists to bow an instrument is to write *arco* in the score. The best resources for creative playing techniques like this are twentieth-century concert scores. Anything from ratchets to whips were commonly used, and to great effect. These techniques have since become standard repertoire for modern film and game composers.

Improvisation

Don't underestimate percussionists' ability to improvise grooves, especially when working with a drum set. If you aren't familiar with percussion instruments, or are still new to a style, sometimes the best route is to simply write "swing beat" or "funk groove" in the score, and leave it up to the drummer. At turning points or climactic moments you can then write the **composite rhythm** using slashes, and the drummer will perform some variation on it.

Examples of Percussion Writing in Games

 On the Sound Lab we take a look at a few example of pitched and unpitched percussion writing.

WRITING FOR FULL SAMPLED ORCHESTRA

Now it's time to put it all together. With a full orchestral template you will have so much versatility that your options as a composer will be virtually limitless. Keep in mind that the goal is to arrange and orchestrate *to the needs of the game*, not necessarily to your tastes. All sections of the orchestra do not need to be playing at all times. In fact, doing so will drastically limit your timbral palette. The orchestra is used so often in games because of the range of colors and emotions that good orchestration can yield. It is important to be mindful of the strengths and weaknesses of each section so that you are able to orchestrate appropriately for any given situation.

Basic Orchestration Technique

If you have assimilated the information on each section, then writing for the full orchestra will be relatively simple. The basic guidelines for orchestration apply here as well. Part interest, smooth voice-leading, utilization of timbres and textures, and balance in timbre and volume are all essential for full orchestral arrangements as well as sectional arrangements. The difference when orchestrating for full orchestra is that you'll have to be aware of how the sections are interacting with *each other* as well as themselves.

A very helpful way to think about orchestration is in terms of the harmonic series, as mentioned in Chapter 6 (page 168). In the harmonic series the lower frequencies are spaced out widely (in octaves, fifths, and fourths), while the higher frequencies are spaced closer together. This is a very general way to voice chords in a full orchestral setting. By voicing the roots of chords (and sometimes the fifths) in the lower-ranged instruments it keeps your arrangement from getting muddy and unclear. It also reinforces the most *harmonically related* tones in each chord. Unless you're aiming for a special effect, stick to octaves and fifths in the lower register at first.

In Figure 7.4 we can see a D major triad voiced across the entire orchestra. Notice that the lower instruments are spaced no closer than a fifth apart, and most are paired in octaves. Looking at the mid- and high-ranged instruments we can see that the spacing closes into thirds and fourths. This example makes use of juxtaposition applied to brass

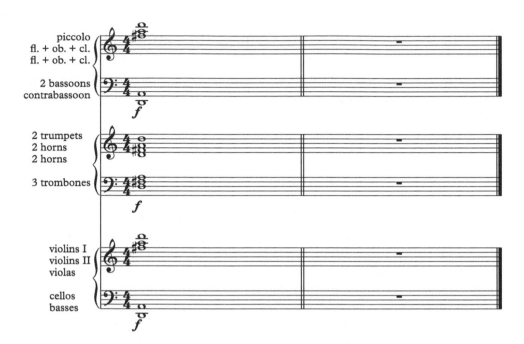

FIGURE 7.4 A typical example of a balanced D major chord, voiced throughout the orchestra.

instruments. They are the only instruments in their range. Notice that the woodwinds are paired with the strings both above and below the brass in various combinations, making this somewhat of an enclosure example. The timbre grouping and **open voicings** (chords voiced across more than an octave) here are extremely helpful when orchestrating for games in a homophonic or chordal setting. It makes it easy to keep track of the voice leading on every instrument, and therefore yields smooth loops and transitions. It also allows a wide variety of flexibility in the case of vertical implementation (see Chapter 9, page 311).

Note the ratios between winds, strings, and brass. The only exceptions are the piccolo, whose volume is perfectly adequate in that range, and the contrabassoon, which is also in a strong range. Check out the Sound Lab for the full score rather than the reduction we've presented here.

Note also the *forte* dynamic and the doublings. The brass instruments are in balance because there are two French horns for every trumpet or trombone part, as mentioned earlier (see "Writing for Woodwind and Brass Samples," page 209). An octave above that, the woodwinds are paired with the strings at a ratio of 2:1 woodwinds and strings. The exception is the piccolo, which is on the top chord tone and can easily be heard.

To achieve this level of balance and clarity it's important to be meticulous with every single part that you write. It's exceedingly easy to lose track of a part, and if too many voices become disorganized, especially in the low end, your chord will not sound clear and the result will be a sloppy mess. It might be tempting to quickly fill out chord tones with any instrument patch you have, but taking the time to conscientiously voice your chords will ensure that your musical intentions are conveyed clearly. It also sets you up for a very efficient recording session when you get the opportunity to work with a live orchestra.

Another important point for orchestral writing is to avoid doing *too much*. Especially with video games, writing overly complex arrangements can really pull the player out of an immersive experience. There are certainly times when this will actually support gameplay, but it must be intentional and appropriate for the scene. When too many parts are competing for the player's attention it will sound sloppy and unprofessional. Our advice is to keep your orchestrations simple and controlled at first, and then branch off into more complex textures after you are comfortable using the entire orchestra as a cohesive unit.

Similarly, do your best not to overuse timbral pairings and blends. The temptation for new orchestrators is to combine every instrument at all times for that "big" orchestral sound. But the result always sounds static and monotonous. Why? Because there is no room for variation or dynamics; there is nowhere to go when you use all of your colors right from the start. A painter that mixes every color at her disposal is likely to end up with something dark and ambiguous. The same applies here, by blending every timbre at our disposal we are lessening the impact of those timbres. When composing and orchestrating try using one color at a time, and be *intentional* when you add anything new. This will make every contrasting color sound fresh and exciting. By holding your colors back as long as you can you will maximize the impact of your orchestrations.

Writing *for* Samples vs. Writing *with* Samples

The difference between writing *for* samples and writing *with* samples is subtle but important. Samples have come such a long way that they have their own set of strengths and weaknesses, and should be considered a robust instrument in their own right. If you are writing to those strengths and making the most out of the samples you have then you are writing *for* samples. If you are using samples as a means to an end (i.e. a mockup for an orchestral recording session) then you are simply writing *with* samples. Both of these approaches will be necessary at one point or another during your career, and here we will cover some tips to make the most of a full samples orchestra.

Samples offer a virtually limitless range of instruments and timbres to work with (even more if you also add in synthetic timbres and less "traditional" instrumentation and processing). The ability to experiment with samples is a huge bonus for composers still learning the craft, as well as seasoned composers who are interested in developing more esoteric and interesting sounds. To add to this, even with traditional orchestral instruments, samples are fantastic when looking to achieve a "huge" sound. Twelve-piece French horn section? No problem. A string section made up of 16 cellos and 12 basses? You got it! Samples just can't be beat in terms of quickly and efficiently getting gigantic sounds for your score. We briefly touched on this earlier, but this technique is called **layering**. Layering occurs when composers use multiple patches (of *different* libraries or articulations) on top of one another to achieve a richer, denser sound. Layering is very common in cinematic cues and in trailers and it works exceedingly well with string melodies and epic brass figures.

Of course, samples have some drawbacks as well. Samples cover the vast majority of orchestral articulations and effects if you're willing to pay for it, but they can never truly capture every sound orchestral instruments can make. In some cases the sound you're looking may not even exist as a sample. Listen to any contemporary concert piece and you'll likely hear sounds you never thought were possible coming from a standard orchestral instrument. Many of these sounds may be of limited use in games, but there are plenty of game composers that take full advantage of sonic experimentation and indeed make it part of their unique voice as game composers. Can you imagine Jessica Curry's score for *Everybody's Gone to the Rapture* as samples? It seems impossible to even imagine.

In our experience the most universal issue with samples is articulation switching. Real players are capable of switching articulations smoothly and can even blend them together given proper instructions. Listen to pretty much any Haydn symphony and you'll commonly hear a figure where a slurred phrase incorporates one or more staccato notes. This is standard musicianship, yet most sample libraries force you to choose between *either* a legato articulation *or* a staccato articulation. Often there is no way to transition a legato note *into* a staccato note without some kind of layering or mixing trickery (see "Tricks to Make the Most of Your Brass and Woodwind Samples," above). The most common way to switch articulations is through keyswitching, which can be clunky and offers no blending options. Our preferred method is to load up each articulation as a separate patch in the template. This also has its issues, but it will allow you to approximate articulation blends. It will force you to quadruple the size of your template however.

Sometimes writing *for* samples means avoiding their weaknesses in addition to playing to their strengths. Start by choosing articulations that you know your template can handle. This is partially why trailer composers don't often need to hire large orchestras to get the job done. The trailer style, rife with piercing staccatos and giant horn sections blasting triumphantly, has already been recorded in many sample libraries. And these libraries sound great! Do your best to choose libraries carefully and get to know them inside and out.

Although we have laid out numerous guidelines for orchestration, sometimes it's necessary to throw them out the window. "Safe" scoring can only take you so far, then it gets boring. If every measure of your orchestration were perfectly balanced in all the ways we mentioned, there would be no room for creativity or tension. If you balance one section of a cue perfectly, try making the next section intentionally unbalanced and listen to the effect. Contrasts like this make the difference between good and great orchestrations. The key is to know exactly when and how to break the rules, and to do it tastefully as it is needed in the game.

Spencer

Examples of Full Sampled Orchestra in Games

 In the Sound Lab you'll find some clips of game music that utilizes the full orchestra.

ADVANCED ORCHESTRATION AND ARRANGING FOR LIVE ENSEMBLES

We've clearly shown that a fully sampled orchestral template is a powerful tool for a game composer. A properly mixed template can allow composers to jump right into the action and begin their game score without worrying about production aspects until much later on. It offers incredible flexibility regarding style and mood, and it also makes the creative process flow smoothly.

But with samples, a composer must sometimes make compromises and choose to write *for* the samples themselves rather than what's in her head. This brings us to a very important point – including live instruments into your score. Incorporating live players into your game soundtrack can be extremely helpful in terms of setting the proper mood, as well as conveying as much emotion as possible. Musicians have trained their entire lives with their instruments, and as such their expertise can never be fully replaced by samples. By working with a musician you are essentially hiring a collaborator, specialized in one or more instruments. Even adding one live instrument track to your sampled ensemble can make a drastic improvement to the overall quality. Below we have outlined some advanced concepts for orchestration and arranging.

Layering Solo Instruments

Overall it is relatively simple to layer in a solo instrument to a sampled orchestra. In some cases musicians will be able to take a part and record it by ear using only your mockup and the MIDI file. Other times it's necessary to take the MIDI and translate it into a professional-looking part (see below). Regardless of how you get the part recorded, there are two main goals: 1) to produce a solo as a focal point for the cue, or 2) to support and add realism to the samples. Either of these options will add detail and depth to your music. See Chapter 9 for information on recording and mixing solo instruments into your track.

However, game scores are not linear and a sloppy recording can completely throw off an otherwise smooth loop or transition. For this reason it is sometimes prudent to write loop points that work with or without the live instrument part. Virtuosic melodies can be difficult to loop due to rhythmic discrepancies, so avoid this style of writing when working with tight transitions. Another issue is reverb. Particularly when working with large ensembles in concert halls or scoring stages, attention must be paid to reverb tails as these tails will need to be chopped and transferred to the start of any loop you are recording (see Chapter 2).

Register

We've discussed register a bit in our chapter about samples, but with live instruments it is hugely important to have a fluent understanding of instrument registral dynamics. Each instrument has a different **dynamic curve** depending on its shape and physics. For example oboes and bassoons are double reeds with a **conical bore**. This refers to the actual shape of the instrument itself, and the physics of air being pushed through a cone. Because of this, oboes and bassoons have different ranges, but similar dynamic curves. For an oboe a good amount of air needs to be pushed through in order for low notes to sound. The dynamic curve of an oboe then looks like a long triangle pointing to the right. In other words, notes in the lower register of the oboe are naturally louder than notes in the higher register.

By contrast flutes have a **cylindrical bore**. The air pushing through the instrument is in a straight column. This means that the dynamic curve is almost exactly the opposite of the oboe and bassoon. Lower notes are very weak and dull in the flute, but the higher notes are loud and brilliant because that is its **tessitura**, in other words its strongest range. If you were orchestrating a tutti chord, a flute or a piccolo in its tessitura would audibly cut through the entire orchestra. However an oboe at that range would certainly not, because in its higher register oboes are much weaker. Every instrument has multiple registers and a tessitura, where the instrument "speaks" most clearly. It is the job of the orchestrator to advantageously choose which register each instrument should be in to accomplish the goals of the cue.

You can find additional resources for orchestral instruments, their ranges, and their dynamic curves in the Sound Lab (companion site). This is a helpful starting point in getting to know each instrument well enough to select the appropriate register for your orchestration. If an instrument or section is getting buried by other instruments in your mockup (or at a live session), there is a good chance you are overlooking register. However, as we have seen, register is just one of the major tools an orchestrator has to work with.

Idiomatic Writing and Instrumental Idiosyncrasies

Instruments all have their own idiosyncrasies that the astute orchestrator can exploit. We have all heard trombone slides and harp glissandos. These are examples of idiomatic writing. They work well because they are unique to their specific instrument. If approached with care and written idiomatically, they can be an excellent way to show the strengths of the orchestra. Learning how each instrument functions and how sound is produced is critical in order to take advantage of these effects.

The best way to gain this experience is simply to write music for musicians to play through and have them offer feedback. If this is absolutely not possible, the next best thing is to find some scores to solo works by classical or contemporary concert composers. Unlike video game and film music, these scores are either free or widely available. Most solo pieces from contemporary concert composers are like dictionaries to interesting musical techniques and idiomatic playing. These scores can show you notation that is standard for an instrument, which is highly valuable when you need to get a piece recorded on a tight budget and in a short timeframe.

In some cases these idiosyncrasies are actually things to avoid. Woodwinds, for example, all have an awkward "break" (see "Writing for Woodwind and Brass Samples," above). This occurs when switching between registers and the fingerings can be treacherous. For a recording session to go smoothly, trills and repeated oscillations between notes around the "break" should be avoided. In some cases certain trills can be literally unplayable. It's best to do your research into the instruments you are using and make sure the parts are at least possible, if not smooth and idiomatic. It's easy to get carried away when using samples, but the challenge is to use samples to write parts that not only work, but work well in a studio environment.

Advanced Timbral Groupings

We've already gone through in some detail methods of orchestration and timbral pairings for sampled instruments. We relied mostly on Rimsky-Korsakov's general equations for dynamics. In large part, this framework transfers very well to live instruments. Using the split-timbre/juxtaposition method still works best when trying to balance homophonic textures and create clear harmony during tutti chords and climax points. The interlocking and enclosure methods also have an important place in orchestration for games.

However the biggest difference between samples and live instruments is that samples are flexible, even after recording parts in. This means that we as composers can "cheat" a bit and adjust dynamics and timbre as we see fit. This is, of course, not the case with live ensembles. There may be some flexibility when recording instruments separately, but recording an ensemble all together can limit the effect of post-recording adjustments. This issue is especially dangerous when you factor in session fees for large ensemble recordings.

In this section we will review the timbral groupings and expand on the Rimsky-Korsakov balance equations. We will also show some more complex examples of tutti chord voicings where timbres are combined and balanced effectively.

Table 7.2 is a helpful tool when trying to balance the timbres and dynamics of an orchestra. The timbral groupings come from Henry Brant's orchestration handbook called *Textures and Timbres*.[5] Brant is not entirely consistent, but Table 7.2 is a good representation of his ideas. He recommends using these equations to adjust 1) volume, 2) fullness, 3) timbre (tone-quality), and 4) articulation, so that chords can be a) balanced (as in homophony), or b) contrasted (as in polyphony).

We will outline the basics here because it is supremely useful to have balance and control when orchestrating adaptive music systems. We encourage you to read through *Textures and Timbres* as well because it contains valuable information outside of the scope of this textbook. Keep in mind that most orchestration books introduce orchestral instruments based on their *section* and not their particular timbre. This can lead to some sloppy writing, which is exacerbated by adaptive transitions in games, so familiarity with these timbral groups can save you a ton of time and frustration down the road.

> These ratios are approximations, and don't always hold true in every possible situation. In some scenarios they may not be ideal for game scoring, and priority must be given to other elements of the orchestration beside the balance. An important point to remember is that the simpler and more familiar the harmony is, the clearer it will be regardless of how you orchestrate it. A C major triad will almost always sound like a C major triad. But change that to a C major 7 flat 9 and the harmony might not be quite as clear if attention to balance is not paid. The best way to think of this chart is a foundation – a starting point – for developing your own style of orchestration, and for developing your ear to begin listening for balanced voicings.
>
> Spencer

The basic idea here is that each group makes up a fundamental orchestral timbre which has its own properties, and can be combined in different ways with the other groups. Brant's point is that by understanding the properties of each group you can then adjust *volume, timbre, thickness*, and *articulation* such that homophonic textures and chord voicings are balanced and clearly represent the intended harmony. We would also like to add *register* to this list. Some of these recommended doublings might be overkill for instruments in their stronger registers while others might be insufficient for instruments in their weaker registers. Adding the extra register consideration helps tighten up the orchestration a bit. This will also aid with choosing timbral pairings for melodies. For example, if you want a melody to be a brilliant flute timbre, then it should be written in the upper range for the flute, not in the lower range, especially if the high strings are playing in unison. However if you want it to be

TABLE 7.2 Henry Brant dynamics

Timbral Group	Example Instruments in Group	Dynamic Units at piano	Dynamic Units at forte	Fullness Units	Special Notes
Wind Group I (flute)	Flute family, clarinet family, bassoon (top octave only), strings (harmonics only), horn (fiber mute)	1	1	1	
Wind Group II (oboe)	All double reeds, clarinet family (bottom fifth only), horn (hand-stopped), muted trumpet and trombone	1	1	1	***Non-blending
Wind Group III (1/2 brass, or horn)	Open horn, trumpet and trombone (open in hat), muted tuba (restricted range), all saxophones (top two octaves)	1-2	2	2	
Wind Group IV (full brass, or trumpet)	Open trumpet, flugelhorn, trombone, tuba, horn (high range), all saxophones	1-2	4	4	***Flugelhorn = 1 trumpet + 4 Bb clarinets in terms of *fullness* (same for trombones and euphonium)
Strings Group I (arco)	Normal arco	1	2-3	2	***One unison bowed string section = 2-3 *identical* woodwind instruments
Strings Group II (pont)	Special arco, ponticello, fingerboard, mute	1	2-3	2	***One unison bowed string section = 1 saxophone or horn in dynamics + fullness
Strings Group III (pizz)	Pizzicato, harp, harpsichord, banjo, etc.	1	2-3	2	
Strings Group IV (legno)	Col legno strings, piano with felts removed	1	2-3	2	
Piano	Percussion of definite pitch				
Perc Group I (semi-pitched)	Timpani, timbales, tom-toms etc.				***For all powerful percussion instruments
Perc Group II (wood)	Xylophone, marimba				2 combined parts at *f* or *ff* can overpower
Perc Group III (metal)	Glockenspiel, vibraphones, chimes, etc.				The entire orchestra!
Perc Group IV (non-pitched)	Snare drum, tambourine, ratchet, etc.				

dull and soft, then the lower range might work coloristically, but since the flute is weak in that register it might be better to orchestrate *very* minimally around it.

These groupings are not just helpful for creating balance, they are also fantastic for orchestration *contrast*. If all you need for a cue is a balanced chordal texture, then follow this chart and you'll be on your way. But if you want to create a contrast, then you can use disparate groups to emphasize the differences in color. When dealing with polyphonic textures, especially orchestrating with this kind of contrast, it will create clarity for the listener. If your goal is to bring attention to distinct melodies and countermelodies, write each part in a *contrasting* timbral group. If your goal is to bring attention to *the same* melody in canon, then it's helpful to write all parts within the same timbral group. We would encourage you to try thinking about the orchestral colors by timbre group rather than orchestral section as much as possible as it will translate very smoothly into orchestrating for adaptive systems (see Chapter 9).

Part-Prep

The importance of part preparation *cannot be overstated*. In some sense, part prep *is* orchestration. It is the process of taking your musical ideas from the DAW and translating them into parts that musicians can (ideally) sight-read. This requires knowledge of the instrument, notation, and implementation. Not only must these parts be clear and idiomatic to each instrument, but they also must take into account the function of the cue within the context of the game. If it is a looping cue, the part needs to reflect that. If it is a vertically layered cue, the part needs to make it clear exactly when and where the musician needs to play and when not to. Having a piccolo accidentally play while a percussion layer is recording will either break the cue or waste valuable session time.

To put it simply, MIDI files are *not* acceptable parts to deliver to an orchestra. Before the recording session the MIDI files *must* be exported from your DAW and imported into a dedicated notation program like Sibelius, Finale, or even MuseScore (if the price of the others is too high). At this point the parts *will* need to be organized and cleaned up. Keyswitches will need to be deleted, tempo and technique markings will need to be added, and notes will need to be adjusted to fit the framework that each instrument is used to seeing. This means things like harp and piano pedaling will need to be added, string divisi will need to be decided on, and any transposing instruments will need to be transposed. In many cases it's actually faster to use the MIDI in your DAW as a reference and type in the notations from scratch.

Sometimes it is necessary to adjust or rearrange parts because they are simply unplayable with a live musician. This occurs when composers write *for* samples. This kind of orchestration can sound great, but it won't transfer well to a live ensemble. Make sure your parts satisfy all of the points mentioned in this chapter regarding each instrument section and all of their idiosyncrasies.

FIGURE 7.5 An excerpt of a polyphonic cue written for various adaptive applications.

We'll cover the process of taking parts from a DAW into the hands of a player in the Sound Lab. We will also take a look at writing homophonically, thematically and polyphonically, and provide basic guidelines on part preparation from DAW to score.

You will also find a full orchestrational analysis of the excerpt in Figure 7.5, and other examples. In Chapter 9 we will discuss the highlights in the image and the various adaptive methods employed.

WORKING WITH SYNTHETIC TIMBRES

Although orchestral timbres are infinitely useful, *they do not constitute all of the instruments found in games*. Synthesis is a topic as wide and complex as orchestration itself, and much of that will be beyond the scope of this textbook. We will however attempt to provide some of the basics in the Sound Lab that can be used in conjunction with the previous timbral groupings to get you experimenting with synthesis alongside your orchestral palette as quickly as possible. We will also provide a resource for adding synths to your template.

Visiting Artist: Atsushi Suganuma, Audio Director

Thoughts on In-Game Mixing

The easiest way I find to begin in-game mixing is to have a clear picture of how music should be layered, along with dialogue and sound-effect elements, in the game before the music implementation process takes place. If you spend a decent time to conceptualize how the game should sound without any music, it becomes a much easier transition in your mind to work out how music elements should be implemented to tell the story in your game.

You will be surprised how effective this process can be and how easy it is to be applied to any genre of game titles without having a huge academic and technical knowledge.

OTHER STYLES OF ARRANGING IN GAME MUSIC

Until this point we have spent most of our time discussing the orchestra. This is because orchestration is extremely difficult and it is very prevalent in games. Knowing how to effectively orchestrate will help you hone many valuable skills. However orchestral music is certainly not the only type of arranging you'll be asked to do. The good news is that (with some adjustments) many of the voice-leading and arranging principles we have discussed can be used with other musical styles as well. Below is a list of some of the more common or interesting styles found in games.

Chamber Music

Common Instruments

Chamber music consists of small ensembles of orchestral instruments, more as soloists rather than sections. For example, instead of having a violin section, you'll use a single first violin on a part, a single second violin on another part, one viola on another, and so on. In games these instruments can often be processed or extended sonically with electronics. We would *highly* encourage you to explore chamber music composition in depth at some point. It offers a unique opportunity to focus very narrowly on composition and color, which can be a breath of fresh air compared to a full orchestra. You'll come away from it an expert in every instrument in the ensemble.

Musical Texture

Chamber music is capable of just about any texture a full orchestra is capable of. The main difference is that much more of the detail and complexity in the timbres of each instrument is present. Where an orchestra is very full sounding and "big," a chamber group is smaller and more *present*. The smaller size does not in any way relate to a lack of emotive power. In fact, chamber ensembles can sometimes sound *more* aggressive and powerful because of the detail that can be clearly heard in every articulation. It's like having a microscope and pointing it at a particular group within the orchestra. The nuances in timbre and performance are all of a sudden more clear and prominent. Chamber groups are equally capable of sounding harsh and edgy.

Defining Characteristics

The defining quality of most chamber groups is the intimacy that they convey. Orchestras convey size and breadth, but chamber ensembles are very close and personal. This could translate to a beautifully exposed and vulnerable solo violin part, or to an aggressive and unsettling bassoon multiphonic. The flexibility in emotion is still there, but it is amplified by the lack of other elements happening at once.

In video games some of the most effective creative uses of a chamber group are when composers juxtapose them, or layer them on top of a full orchestra.

Hybrid Styles

Common Instruments

Orchestra, choir, and electronic or heavily processed elements.

Musical Texture

Hybrid styles can incorporate any kind of texture, but usually with this palette composers tend to lean toward homophonic chordal movement and heavy thematic melodies. Polyphony in hybrid styles definitely exists, but is a bit less common due to the already complex range of musical timbres.

Defining Characteristics

This style often sounds like a massive build-up. Composers writing hybrid music are masters at layering sounds on top of one another to add depth and intensity to a track. Usually there is some type of "motor" that keeps the piece moving forward – a plucky synth ostinato or a repeated figure of some kind. Synth pads usually fill in extra space between orchestral instruments and provide some interesting timbres. Usually the main goal of hybrid music like this is to achieve a massive "wall of sound," which makes the player feel a sense of intense drama.

On the other end of the spectrum, hybrid styles can also be fantastically quirky and fun. Orchestral elements are playfully juxtaposed with harmonically pleasing synthetic sounds, which can be both catchy and aesthetically interesting. Often genres like this are for games aimed at younger audiences.

Electronica/Dance

Common Instruments

The instrumental focus is on anything synthetic with the possible addition of drum samples or voice (heavily processed).

Musical Texture

The texture is usually homophonic with a heavy amount of rhythmic and timbral complexity. Heavy, catchy melodies are also quite common here.

Defining Characteristics

Aside from the electronic instruments themselves, this style has an enormous range of emotionality. Electronica can be danceable, aggressive, action-packed, nostalgic, or even ironic. The defining elements are the methods of obtaining sounds through synthesizers. This process is highly creative and personal because often composers develop these sounds from scratch rather than working with the established framework of the orchestra. Stylistically speaking, electronica can range from atmospheric, to action, to dance, and pop. In a way, every early game soundtrack is electronic in nature due to the lack of sampling power. Some of these soundtracks clearly *modelled* orchestral ensembles, but many of them explored other styles as well. Using retro sounds similar to this is now a huge chunk of the electronic soundtrack being produced under the genre "chiptunes," (referring to the chips found in classic game consoles like NES and SNES).

Rock and Pop

Common Instruments

Acoustic and electric guitar, bass guitar, drums and percussion, voice (with lyrics), horns, electronics.

Musical Texture

This style of music is usually a mixture of homophony or monophony. Often rock and pop music can be riff-based as well, resulting in chords and an ostinato which acts as the "motor" for the track.

Defining Characteristics

This is the kind of music most of us are familiar with growing up listening to the radio and attending concerts or playing in bands. The songwriting can often be distilled into chord progression, vocal melody, and rhythm section groove. The attraction to these styles of music in games is the familiarity players have with rock and pop; it can be highly engaging and/or nostalgic for listeners. There is also remarkable flexibility with the timbre palette for these tracks. With the same core group of instruments you can achieve a light and airy upbeat track, or a dark, distorted, intense track that's filled with foreboding.

Jazz and Big Band

Common Instruments

Brass (solo and section), saxophones (solo and section), orchestral woodwinds (solo and section), electric guitar, upright or electric bass, drums and percussion, voice (with lyrics).

Musical Texture

Homophony or polyphony, sometimes groove oriented.

Defining Characteristics

The defining characteristics of any kind of jazz are usually 1) the chord types and voicings, and 2) complex groove-based rhythms. The chord types are almost always seventh or ninth chords, possibly including many other upper extensions. Chord progressions commonly feature ii–V–I cadences and frequently modulate into other keys. Jazz harmony is very complex, as are jazz melodies. But this complexity is somewhat hidden by the fast-moving harmonic rhythm and the groove, which is usually swung.

"World" Music

Common Instruments

Traditional instruments from non-Western European cultures all over the world.

Musical Texture

Mostly homophonic, monophonic, or heterophonic (two or more voices simultaneously performing variations on a particular melody).

Defining Characteristics

"World" music is not actually a real musical style. It is a catch-all for music that is not based in Western European musical traditions. It is a shame that we often think in this way because it waters down (and somewhat invalidates) a creative universe of musical history and aesthetic. Western harmony is prevalent in game music, but we encourage you to explore other styles of harmony and instrumentation as much as possible.

The challenge then, is that every culture has a different defining characteristic so this is hard to pin down into a generalized category. Most often the history of a culture plays a huge role in the traditional music created by it. For example, the tools and resources available to a population usually factors into the kinds of instruments that are produced, and *how* they are produced. That being said, many cultures do have their own versions of instruments we commonly use, such as guitar, violin, percussion, etc.

Finding commonly used instruments and how to write for them is a great first step to writing world music. The next step is to listen to as much of that culture's *traditional* music as possible so that you can internalize the feel of it. The most important thing to remember when writing elements of world music is to do your research, and try to account as fully as you can for that culture's idiomatic writing, instrumentation, and aesthetic. The best way to do this is to hire a musician fluent in that style of music and collaborate with them on your tracks.

Electroacoustic Music

Common Instruments

Found sounds, heavily processed sounds, and possible hybridization of sound design elements with traditional instruments.

Musical Texture

Non-tonal, often rhythm-based.*

Defining Characteristics

Electroacoustic music is the technical term for art music which includes elements of electronics. We use the term here to broadly define game music in which the foundation is sound and timbral palette rather than any form of harmony. It sounds extremely complicated but in reality music like this is extremely suitable for many scenarios in video games. With current technology it's easy to sample sounds and mangle them in a DAW, and this often leads to an interesting and distinctive palette. Horror games in particular have a wonderful relationship with electroacoustic music because it can very quickly become unsettling. There is a fantastic gray area in this genre between sound design and music. In some cases there is absolutely no difference, which make these soundtracks all the more immersive. With all that said, the defining characteristics here is *sound-oriented* music, and total immersion with the game as if the music itself was diegetic.

> Note that "rhythm-based" is not the same as "groove-based." A groove implies somewhat predictable rhythmic patterns, which is not always the case with acousmatic music.
>
> Spencer

The Sound Lab

Before moving onto Part III, head over to the Sound Lab to wrap up Chapter 7. We will cover some techniques and analysis for orchestration, and link some examples of the orchestra as used in popular games. We will also share and analyze various examples of full orchestral scores.

* * *

PART II REVIEW

Don't forget to head over to the Sound Lab (companion site) if you haven't done so already for each of the chapters in Part II. In addition to the resources we called out in each chapter, we will offer a Score Study section.

In *Chapter 5* we defined linear music as music whose structure proceeds sequentially from A to B to C in order every single time it is performed. We defined nonlinear music as music that flows non-sequentially and in no chronological order. Later in Chapter 5 we listed some essential tools and skills for game composers. Then we followed up by exploring an example of using an adaptive music system for a hypothetical game. Finally, we outlined the basic production cycle of a game, and listed some core roles you might find yourself in as part of the music team. We also discussed the differences between diegetic and non-diegetic music. Visiting Artist Martin Stig Andersen also gave us his thoughts on diegetic vs. non-diegetic music.

In *Chapter 6* we explored the creative cycle and how to start writing music. We offered some strategies for generating musical ideas from scratch and then outlined some key stages in the creative cycle. These stages are preparation, brainstorming, refinement and revisions, and the final polish. Next we explored a few helpful "footholds" like artwork, characters, and gameplay mechanics to latch onto when generating musical ideas.

Next we took a deep dive into music theory and compositional practices and looked at ways to construct appropriate melodic and harmonic ideas for your game. We also looked at melodic shape and gesture, and harmony from basic progressions through to chromaticism and serialism. Then we covered ways to similarly generate rhythmic motifs and a musical palette. Finally we looked at some basic ways to develop the ideas you've created with these techniques. Visiting artist Penka Kouneva, composer, discussed working with collaborators and the pro-production process, Wilbert Roget, II imparted his wisdom on brainstorming, Jeanine Cowen talked about planning your music, while George "The Fat Man" Sanger explained his process for handling direction from a client.

Chapter 7 was a rigorous study of orchestration and arranging. We learned how to take our basic compositional ideas and transform them into full cues that will be triggered in the game. We covered the musical textures: monophony, homophony, and polyphony, and discussed their characteristics. Later on in Chapter 7 we covered arranging and orchestration using samples very broadly. This included planning, preparation, and execution of a full orchestral template. On our way to a fully functional virtual template we covered topics like mixing and reverb, as well as a basic overview of the orchestral instruments and families.

We then launched into a deep study of writing for techniques for each orchestral instrument and how to get the best sound of samples. Then we followed with a look into writing for live ensembles where we covered topics like register and timbral groupings. Next we took a hard look at synthetic timbres and how to incorporate them into your template. Finally we went over some basic characteristics and textures of common non-orchestral styles of game music. Atsushi Suganuma, audio director, also revealed his thoughts on in-game mixing.

Chapter 8 begins Part III, the audio implementation portion of our textbook. If your focus is music for games we recommend reading Chapter 8 followed by Chapter 9. From there we recommend reading Part IV, Chapters 10 through 12 for the business side of game audio.

* * *

NOTES

1 S. Adler, *The Study of Orchestration*.
2 T. Goss, *100 Orchestration Tips*.
3 E. Gould, *Behind Bars*.
4 Ibid.
5 H. Brant, *Textures and Timbres*.

BIBLIOGRAPHY

Adler, S. (2016). *The Study of Orchestration*, 4th ed. New York: W. W. Norton & Company.
Brant, H. (2009). *Textures and Timbres: An Orchestrator's Handbook*. New York: Carl Fischer Music.
Goss, T. (2017). *100 Orchestration Tips*. Available from https://orchestrationonline.com/product/100-orchestration-tips/
Gould, E. (2011). *Behind Bars: The Definitive Guide to Music Notation*. London: Faber Music.

Part III
IMPLEMENTATION

8

Audio Implementation

In this chapter we will we explore game audio implementation. Armed with an understanding of the theory and practice of game audio asset creation from previous chapters, we will explore planning for **audio implementation** by considering engine capabilities, target platforms, and resources. The core of this chapter will cover topics from asset preparation, implementation, **dynamic mixing**, and finally onto testing, optimization, and reiteration.

IMPLEMENTATION BASICS

In previous chapters we explored the challenges presented in nonlinear media and how the player's input affects the game state, leaving audio designers unable to determine *exactly* when a particular audio event will trigger. However, we can plan ahead to answer these challenges during the implementation process. Implementation is essentially the process of assimilating audio into the game engine so that audio can trigger synchronously with other game events. Using a variety of implementation methods we can create dynamic audio systems that adapt seamlessly to gameplay events.

> Game development is a multifaceted and collaborative process in which game audio meets at the crossroad of creativity and technology. Technical and creative skills go hand in hand in shaping a game's sonic experience. Implementation, more than any other aspect of game development, proves this.
>
> Gina

Implementation Methods

There are several foundational methods within the process of audio implementation. As an audio designer you will work with one or more programmers to determine the best tools and techniques for implementation. The budget, schedule, as well as your team's work capacity and level of knowledge/experience will all be factors in deciding these tools and techniques. Implementation tools include the game engine[1] (Unity, Unreal, proprietary engines), a software development environment for building games, and game audio **middleware** (Wwise, FMOD), which is a third-party audio-focused software environment that sits between the game engine and audio designer. The term middleware in the tech world means software that acts as a bridge or connection between other software or systems. Essentially, audio middleware puts more control into the audio designer's hands, with less scripting and programming to deal with.

Let's break down a few typical scenarios in which audio assets are implemented into games. The **"asset cannon"** approach is the simplest form as it consists of the audio designer delivering assets on spec (concerning file type and compression) to the programmer, who will then implement the audio into the game engine natively. While this approach may seem like a good choice for those who feel a bit uncomfortable with the more technical side of game audio, the downside is a lack of control over assets once delivered. It also places the burden of finalizing audio on the programmer. Not all programmers have a deep understanding of audio concepts and theory, so by working in this way the team's expertise is not being fully exploited. This approach can also lead to many more iterative reviews and adjustments as the programmer becomes the primary work horse for completing modifications and updates as advised and requested by the sound team.

This method of implementation can also result in the wrong real-time effects being applied, or inconsistencies in compression and volume. With this delivery process you should *insist on testing builds* once the sounds are in place. Also be sure to build time into the schedule to make changes if necessary. Testing and iteration are the best way to deliver effective game audio using the asset cannon approach.

Implementation of audio natively into an engine is a step above the asset cannon approach. With this approach the audio designer has direct access to the game engine and can import assets and tweak components herself. Compared to the asset cannon approach, direct **native implementation** saves a lot of the back-and-forth time with the programmer when mixing and making revisions. However, an audio designer not familiar with the engine or uncomfortable with scripting might find it a bit more difficult to achieve all of their audio goals.

Audio implementation natively into the engine is limited to the resources of said engine, and heavily dependent on scripting for audio behaviours beyond *play* and *loop*. However, there are plugins like Taz-man, Audio Fabric, and Dark Tonic's Master Audio which extend the game engine's native audio functionality, allowing for complex and rich audio behaviours while reducing time spent scripting.

When the programmer agrees to use an audio middleware solution such as Audiokinetic Wwise (Audiokinetic), FMOD (Firelight Technologies), Criware ADX2 (CRI Middleware), or Elias Studio (Elias), it can be the best-case scenario for both the programming team and the audio designer. It's important to note that these middleware solutions have licensing costs which are indie-dev friendly but should be considered in the process.

Middleware solutions are built to better utilize a game development team's expertise. With any middleware, whether it is for audio, physics, or anything else, the build allow the experts in each discipline to do what they do best. An audio middleware solution will allow you to control the creation of the sound and music as well as the implementation and mixing of those assets. During development you will spend more time working on audio (as you should) and less time trying to explain to the programmer how the audio should work. The development team will then have more time to spend on other programming tasks and the audio team will have more capacity to make the game sound great!

Later in the chapter we will explore the methods discussed above in greater detail. At times we may generalize the game engine's audio function and/or audio middleware with the term **audio engine**.

> In situations where the sound designer is delivering assets to the programmer for implementation (asset cannon), a video mockup which demonstrates audio behavior in game can be very helpful in relaying to the (non-audio) programmer how sound integration should be approached. The next best option is to create an asset sheet with notes and comments on each sound. Both will go a long way in creating solid communication between the programmer team and yourself.
>
> Gina

Integration and Testing

Later in the chapter we will break down the specific functions of implementation and testing, but for now let's take a macro look at what is involved.

Regardless of the implementation method utilized, the game engine sits at the core of development. A considerable amount of scripting will be required, regardless of who is implementing the audio assets. Audio assets imported into the game engine can be easily attached to **game objects** – empty containers placed in a game engine that "hold" gameplay instructions (see "Game Engines" later in this chapter for more details). **Scripts** are then attached as a **component** onto each game object. A script is basically a set of instructions which tell the engine what to do with the audio assets. At the time of writing this text, game engines are "smart," but do not operate with much common sense. The future of AI and machine learning might tip that scale, but for now we can assume we need to provide *specific instructions* to the engine for proper triggering of audio events. Later in the chapter we will break down the key features and functions that make up native engine implementation. We will then take you through a tutorial on the companion site to tie it all together.

As we mentioned previously, audio middleware offers the audio designer control over many functions like randomization of pitch and volume, playlists, beat sync, and more. These features are made easily accessible without scripting in a graphical user interface. Later on we will discuss in detail the various features available in middleware, but for now let's talk about how it is integrated into the game engine.

Integration is the process of synchronizing a middleware tool with a game engine. Some middleware developers offer integration packages for supported game engines while others use the comprehensive **API**. Either approach will configure the two pieces of software so they can communicate and share data. A programmer typically completes the integration before sharing the project with the team. After an audio event is created and configured in a middleware program the audio designer can export what is called a **sound bank** (or bank) to be shared along with the engine. The sound bank provides the audio events and configuration information, which can be used to hook them into the game engine.

During the integration package installation, a specific set of file paths are set up to tell the game engine where to find these sound banks once exported. The sound bank information shares all of the necessary information for the game engine to trigger the audio events as instructed in middleware. The programmer will need to do some light scripting to trigger the events and connect game parameters. Middleware offers users precise control over specific audio event properties such as **real-time parameter control** (**RTPC**), which feeds on data values generated by the game engine. RTPC is fundamentally a way to automate multiple audio parameters in real time, and can be used to create fades, volume and effects automation, and mixes that change on the fly.

It's important to note that native audio implementation into a game engine offers some control of audio based on parameter values, but typically offers limited flexibility and requires additional scripting.

Implementation Cycle

Implementation is a very important step, and (like game audio itself) it is not linear. Your audio assets will be created in a DAW, then imported directly into the game engine or middleware tool for testing. The process is circular in that you will continue to move between these software applications as you create, implement, and test, and then create again. Testing often leaves the audio designer with revisions or re-works, which takes the process back to the audio engine or DAW, and so on.

Visiting Artist: Jeanine Cowen, Composer, Sound Designer

Thoughts on Implementation

Implementation, when it's done right, opens up new opportunities that the audio designer hadn't imagined at the outset. Finding a balance between the creativity

(Continued)

of sound design and music and the problem solving of implementation is often delicate. Working in game audio implementation is both a technical and a creative job. There is no wrong or right way, there is only, how do we get this audio to sound great and help to enhance the game. Rather than view it as a struggle, the good implementers I know see it as controlled guidance. Like the painter who only had the natural dyes found in the real world, they work with the materials they have rather than trying to force their resources to be something they aren't. If you find yourself struggling with getting something to work, it might be time to step back and re-ask yourself the question "what is it that I want the player to experience right now?" This question should guide you and allow you to get out of the weeds that can sometimes consume our focus as implementation specialists.

Implementation for Immersion

The level of sonic immersion (or aural cohesion) in a game comes down to the level of detail and planning in the audio implementation. In Chapter 2 as an exercise we asked you to close your eyes and listen. Sound is all around us. Since the days when we heard our very first sounds we understood these details to be cues with valuable information. The same is true of the sonic experience in games. Sounds give players details about the environment, immersing them in the game world.

While sound is clearly used to create affective response in films, in games there is a level of interactivity that also influences the player's emotional state. In a game scene the player will experience a different level of intensity of music and sound when they are exploring than they will when experiencing an enemy attack. Sound will act as a warning of danger to the player; it prepares them for this transition in gameplay intensity. Sound reacts to the player's choices and in turn provides the player with information which influence actions. In this way audio creates momentum for the narrative, and reinforces it with aural feedback. This idea is supported by the Gamasutra article, "How Does In-Game Audio Affect Players?"[2] Here, Raymond Usher examines the results of a study in which players were monitored while playing games with and without sound. The results proved higher heart rates and slightly higher respiration rates in the audio group. This suggests that audio is a crucial part in our integration and immersion into game scenes.

Playlists and Randomization

Without variation and a certain amount of **generative** aspects in the sound, games would feel static and lifeless. Gone are the days of the *Super Mario Bros.* footstep bleeps that repeat over and over again. Variety in even the smallest sonic details can go a long way to making the experience more believable. In film and other linear media we can add variation manually as we build each sequence and it stays synced to the visuals. With interactive media we must create dynamic systems and implement them into the audio engine to vary the soundscape. Sound effects are typically delivered with a few slight audio variations and then grouped into playlists. These playlists can then be randomized in the engine. In addition, the audio engine can also apply randomization to certain properties like pitch and volume. But this is just the tip of the iceberg.

By breaking down a single sound into *multiple* playlists (sometimes called **random containers** or **multi-instruments**) we can add variety and dynamism. For example, an explosion sound could be broken down into attack, sustain, and release tail playlists. The game engine or middleware tool would then randomly choose a sound within each playlist to trigger at the right moment, thus creating unpredictable combinations each time the event is triggered. On top of that, the three containers can be programmed with slight randomizations on pitch and volume as well. This takes our random possibilities and exponentially increases their variability (see Figure 8.1).

Programming randomization in middleware programs like Wwise and FMOD is a cinch! Each tool has its own method, but these programs are specifically built for these types of randomization and variation. Later in the chapter we'll take you to the

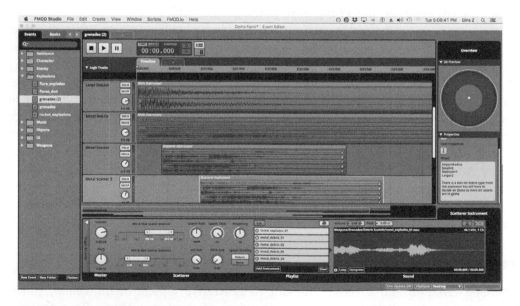

FIGURE 8.1 Screenshot of multi-instruments in FMOD Studio.

Sound Lab for some tutorials and practice, but you can also check the documentation pages to dive right in.

Spencer

Loops

In Chapter 2 on the companion site we discussed creating and preparing looped assets for use in game. Here we will focus on implementing them. By default, implemented audio events will play once from beginning to end. Depending on the method of implementation, it will be necessary to set instructions for the audio engine to loop an asset instead of triggering it just once. This process is usually relatively easy, regardless of the implementation medium, and loops can be set to either loop once or a predetermined number of times.

Audio designers make use of loop functions within the audio engine to get more mileage out of audio assets, which helps keep the memory footprint minimal. Loops also compensate for the differences in timing of each playthrough. Imagine if the soundtrack needed to be as long as the game. Sixty hours of music would certainly take up a large amount of memory and be CPU intensive to stream, not to mention the time it would take to compose. Implementing loops instead of longer linear cues makes the soundtrack more efficient and adaptable. Looping events aren't just for music either; they can also be used for continuity of background ambience and continuous triggered sound effects.

Looping events is a great solution for keeping memory and CPU usage in check, but the designer must be mindful of avoiding repetition. The length of the loop is something to consider based on how long a player might spend in a specific area where the loop is triggered. We will discuss these benefits and some challenges of looping later in the chapter.

Dynamic Mixing

Mixing for linear media is a process implemented during the post-production stage of film or TV. At this stage designers ensure that signal level, dynamics, frequency content, and positioning of audio are all polished and ready for distribution. However, because games are nonlinear we are essentially working with a moving target when mixing. Dynamic mixing is a multi-faceted process which requires the same attention to signal level, dynamics, frequency content, and positioning, but also requires that the mix be flexible enough to change along with the game state. To do this designers often use snapshots, which allow audio engines to trigger mix changes in synchronization with visual events. Later in this chapter we will discuss in greater detail dynamic mix systems and resource management.

Implementation Preparation

Before we dive into the micro details of audio engines and asset implementation, let's discuss preparing assets. We will start with the process of mapping out the implementation system, and then we will move into exporting assets ensuring proper volume level, format, and file naming.

Sound Spotting

In the film and television world spotting is the process of determining how the musical score and sound effects will be tied to visuals. During this process the sound designer and/or composer will sit down with the director to review scenes and determine the **hit points,** or moments that require audio emphasis. In games there is a similar process, but it is often less formal. A game design document (or a GDD) may be provided as a guide for narrative and gameplay. It may contain a synopsis of the story, concept art, and gameplay mechanics. A style of music or a list of musical references may be included to guide the sonic direction. Early gameplay in the form of a **build** (a pre-release playable version of the game) may be provided to the audio team to prepare concept audio for review. Regardless of how formal or informal the process might be, there is still a need for this pre-production phase.

If audio references from other games are provided, the audio designer should have a good idea of the direction to take. Direction is an important part of the process as visuals can often work well with different sonic styles. Choosing a direction during the spotting process will clarify exactly what style and mood the audio needs to be in to best support the game. For example, a sci-fi shooter as a descriptive can be interpreted in a few ways. *Metal Gear Solid* and *Dead Space* are both in the sci-fi genre, but the former is militaristic and technology-driven while the latter is dark and at times horrific. These two games would each require their own particular audio direction to be chosen during the spotting stage.

The spotting session might simply consist of the audio designer playing a build and creating an asset list to review with the developer. Sometimes the game developer will already have a list of assets in mind and present these to the audio designer. Should you receive an asset list as an audio designer it is wise to review the list with the build to ensure sound events have not been overlooked. These asset lists should be kept in a spreadsheet and shared with the rest of the team. A good asset list will allow for new content to be added and the status of assets to be tracked and shared over a cloud-based app like Google Docs for real-time updates. In Chapter 2 we directed you to the Sound Lab (companion site) for a review of a typical asset list.

Building the Sonic Maze

After spotting the game you should have a clearer understanding of the types of assets required. The next step is planning how those assets can be implemented so they sell the scene and support the narrative. We call this *building the sonic maze*. Essentially you are building the walls of the maze through which the player will move. Your game will consist of a number of assets including sound effects, music, and dialogue. As the audio designer it's your job to ensure all audio events are implemented with purpose, and are responsive to the player's input. You must identify the placement of these sound events in the game world by determining where in the scene a player should hear the various sounds as they move about the environment. Taking into account the full player experience and how it impacts the way sounds are triggered is key to effective audio implementation.

To understand how to map out the sounds we must break down the game and its core mechanic(s), just as we did with the animations in Chapter 3. To better understand this, let's explore a hypothetical sci-fi game. The game will be a third-person shooter, which consists of a player character armed with an automatic weapon and four types of robot as the non-player characters (NPCs). The game is set in a high-tech lab containing five rooms per level with a total of three levels. Inside each room there is various machinery and technology such as LCD monitors (with visual static on the screen) and air vents.

The target platform is mobile and we will be using middleware along with our game engine. Since we know our game is being developed for a mobile platform we will need to be resourceful with memory and CPU usage. Later in the chapter we will discuss ways to manage resources per platform, but for now we just need to understand that there are processing and memory limitations on mobile platforms which are not as much of an issue on desktop or console platforms.

Now that we have an idea of our game, let's start with mapping out the ambient soundscape of our lab. We can take a macro look at all three game levels and decide what the ambience will be in each of the five rooms. We should first break down room size based on how much technology is in each room. This will help us define our background ambient loop, which will act as the glue to all of our other in-game sounds.

Next, we can take a more micro look at the elements that will emit sound from localized spaces. Let's imagine one of the rooms contains an air vent, two LCD monitors, and a large mainframe computer system. We would plan to hook sound **emitters** on those game objects to add some variation to the scene. They will blend with our background ambience, but will work as detailed focal points that bring the ambience to life. Just as sounds in the real world originate from various sources, we need individual sounds in the game world to add depth to the ambience. Once we have a working version of this game (vertical slice pre-alpha build) we can experiment with more complex adaptivity in the environment. We can decide how close or far away a player needs to be to hear certain sounds, and whether or not the acoustics of the location will require reverb or echo effects.

Next we can plan player character sounds, which include weapon fire and footsteps. Starting with footsteps, we would look to the gameplay and visuals to see how to create variation. In our game scene let's say there are two different floor types, one being solid metal and the other being a thin, wooden platform. With this information we can adapt the footsteps so the sounds change with each floor type. We can continue to think about adding layers of cloth movement and using randomization within our middleware to ensure our footsteps transition and vary smoothly per terrain.

At this point we have taken a very broad look at the sonic maze we would create for this game. From here we would continue with the weapon and robot enemy sounds. These are in a sense simpler because the animation and gameplay will give us specifics in how to create and implement the proper assets. These sounds would be the focal point of the audio because the gameplay is based around combat. We would then employ our numerous techniques from Chapter 3 to create complex and detailed assets for implementation. Although this was a quick overview of the process of implementation, it has served the purpose of introducing a general workflow which we will dive into in the following sections. For now, let's continue by discussing the preparation of assets for implementation.

Preparing Assets

Delivering quality audio assets begins at the very start of your recording and editing process. A poorly recorded sound that is then trimmed and normalized is not going to sound as polished as a sound that was recorded at a healthy volume to begin with. Working in games is no different than working in a music or post-production studio in this respect. You will want to optimize your workflow and the path the audio takes from beginning to end as much as possible. When preparing assets for implementation there are a few things that should be considered to make the workflow a smooth process. Let's explore those here.

Mastering Assets

Mastering is a process you may be familiar with in regards to preparing a final soundtrack mix to prepare for distribution. The process ensures consistency across all the music tracks in the album. Typically a mastering engineer will work with a full stereo mix to enhance the sonic character and correct any balance issues with EQ, compression, and limiting. This level of polish is what you are used to hearing when you download an album.

With game audio the assets aren't in a linear mix that can be processed in a static stereo or surround file. Game audio is dynamic, and assets are cut up and implemented individually into the engine. In this sense, mastering game audio *does not* produce audio in its final format, the *game engine* does. Because of this, the mastering process is really meant to prepare the individual assets for interactive mixing. The goal at this stage is then to ensure a cohesive and balanced soundscape once implemented. EQ, compression, and limiting are common tools used to maintain continuity in terms of frequency and volume between all assets. Reverb is sometimes applied to assets to add a sense of "space" to each sound.

> The most important point to remember about "mastering" audio for a game is that the engine is an important part of the "mastering process." The final mix is whatever the game engine spits out in real time as the player interacts with the game. Try to think of it as if you were using EQ, compression, and a bit of reverb on a vocal track before sending off to the main mix bus. It will sound clean and polished on its own, but it is still only one aspect of the final mix.
>
> Gina

During this phase of asset preparation, the audio designer must think about the full mix including music, dialogue, sound effects, Foley, ambiences, UI sounds, etc., and get a sense of the frequency spectrum the assets will occupy. Mid-range, around 1.5 kHz, is the frequency range that our ears are most sensitive to and usually holds the core elements of a soundscape. You may have been introduced to the **Fletcher-Munson** curve when you first started working in audio. This curve is a good representation of where our ears are most and least sensitive. The midrange is also the part of the frequency spectrum that is most *consistent* across different speaker types. It's important to ensure that the assimilation of all assets into the soundscape won't leave any holes (or buildups) within the 20 Hz to 20 kHz range of human hearing.

The mastering process can also prepare the assets for the lossy compression we mentioned above. This process of encoding attempts to remove frequency content undetectable to the human ear through psychoacoustic modeling. If certain groups of sounds need to be compressed (i.e. low-frequency effects), you can improve the audio quality of the compressed sound by applying high or low pass filters to remove portions of the frequency spectrum preemptively. This then allows the lossy encoding process to dedicate more space and bandwidth to the frequency ranges you, the audio designer, have already identified as important. Different lossy compression encoders, schemes, and formats will provide very different results. It can be helpful to experiment with compression formats beforehand, and evaluate their effect on various categories of sound. A certain amount of trial and error should be expected, and (as always) listen critically to your work to assess whether it suits the needs of the game.

It's important to note that the mastering stage often carries into the dynamic mixing stage as you test and adjust the audio systems integrated with the game engine. We will cover dynamic mixing in more depth later in this chapter.

Exporting Assets

An important part of bouncing down the assets and preparing them for implementation is ensuring the audio is loud enough so it will not need to have gain applied in the audio engine. Our DAWs are built to provide good gain-staging and signal flow, so doing this work in our specialized audio tool (the DAW) will result in the best-sounding audio in game. Working with a range of reference volumes for each asset type is a helpful way to ensure you have enough audio resolution to work with in the final mix. If you must boost the volume in the audio engine beyond 0 dB, you risk introducing artifacts and system noise. There is only so much gain that can be boosted on an asset that was bounced out of a DAW at a very low volume.

When delivering assets to be integrated by the developer (asset cannon approach), you may want to control the loudness by exporting assets at specific levels for different categories of sound. Here are some rough reference volumes to use as a starting point when you are delivering assets to be integrated by the developer:

- Dialog -6 dBFS
- SFX -6/-9 dBFS
- Music and ambience -20 dBFS
- Foley -18 dBFS

If the audio designer is in charge of implementing natively into the engine or middleware, typically all of the assets can be exported from the DAW around -3 to -5 dBFS. If the project has an audio director, she may offer her preference on the matter. Some might require setting up a limiter with -1 tp (**True Peak**). The idea here is to ensure you have plenty of audio resolution for fine-tuning volume levels in the audio engine.

Use these signal references as a starting point and adapt them to each project's unique needs.

Effects Processing

Next we can explore effects processing. The decision to bake in effects processing or to apply real-time effects in the audio engine will influence how the final assets are rendered for implementation. Generally speaking, effects processing in engine can be very CPU intensive. Therefore, baking effects into the audio asset is helpful when you are working with memory- and CPU-limited platforms like mobile. Baking in effects also works for the asset cannon implementation method since you may not have control over defining the effects settings in engine. The negative side of this method is, of course, the effects not adapting to the player's location. If you apply reverb on footsteps they will sound with the same reverb indoors or outdoors, which doesn't really make sense. Later in this chapter we will discuss in more detail resource management and optimization, which will allow you as the audio designer to make more educated decisions in regard to planning effects processing.

File Format and Compression

Now that we have a signal level and effects processing covered, let's talk about file format.

Audio implementation takes into consideration not only what sound events will trigger and when, but also looks at the complete audio storage footprint to ensure that the sounds will fit in their allotted memory. In the past, the limitation might have been the size of the delivery format (CD/DVD/Blu-ray). In today's world we must also consider the size of the install package, or the DLC download target size. This forces us to use **file compression** (not to be confused with a compressor, the audio processing tool), which reduces the file size of our audio assets. When the target platform requires us to compress our audio we try to intelligently target different compression approaches for sound categories. For instance it may be acceptable to compress a folder of footstep sounds far more than the background soundtrack.

> As mentioned previously, when working with looped assets like music and ambiences, we try to stay away from the mp3 compression scheme as it fails to loop correctly in most cases. With these types of sound we are better off using the widely seen Ogg Vorbis compression scheme. Not only do Ogg files loop the same as previewed in your DAW or two-track editor but they are also less CPU intensive when being decoded for playback during run-time.
>
> Gina

Most modern game engines can utilize a variety of audio file types and formats. In many instances a decision will need to be made about whether the application itself can handle all audio at its highest resolution. Different hardware platforms (console vs. PC vs. mobile, etc.) may run natively in a particular sample rate and bit depth, or even with an expected audio file format.

Let's explore the different ways we might handle file formats for two different delivery methods. If you are working within the "asset cannon" scenario, you may want to handle file compression ahead of time. If this is the case, ask that the developer do no further compression of the audio in the game engine. The file format would then be determined ahead of time by the programmer based on the target platform and memory allocated for audio. If you are also in charge of implementation, the files should be exported from your DAW at the highest possible quality as the audio engine will handle conversion settings which can be defined per target platform. We recommend 48 kHz, 24 bit .wav files.

File Naming Standards

File naming is an important part of the asset preparation process. Naming conventions set rules for the expected character sequence of audio file names. This is important for a few reasons, the first of which has to do with being organized when working with team members. The second, and more critically important, has to do with how the assets are embedded into and called from game engine components or scripts. Operating systems, programming languages, and programmers have their own rules and best practices for file naming. If you aren't well versed in these it's a good idea to discuss any file-naming specifics with the programmer prior to delivery. This will save you and your development team a lot of back and forth to communicate what is what and where it should be placed when delivering a folder of audio assets.

In the Sound Lab (companion site) we discuss file-naming practices in more detail. Be sure to check it out when you can. For now, we leave you with some of Damian Kastbauer's thoughts on file-naming standards.

Visiting Artist: Damian Kastbauer, Technical Sound Designer

Thoughts on File-Naming Standards

The opportunity to understand the direct connection of an event-based audio system and game engine begins by leveraging a consistent naming standard. Drawing this line between the input and output is the clearest way to understand their relationship. Building systems with a naming standard in place allows for the flowing of text-based strings from the game to be received by the audio engine. These audio events then arrive loaded with context about what system it came from, what it might represent, and where to find the place it originated from.

This can be as simple as adding a prefix to define the system of origin, for example:

play_vfx_ = the name of the gameplay system originating the audio event (prefix)

explosion_magic_barrel_fire_17* = the name of the object requesting the audio event

Audio event: **play_vfx_explosion_barrel_fire_17**

This can be further extended to include available actions specific to a system (in this case, as a suffix):

play_vfx_explosion_barrel_fire_17_**start**

(Continued)

....

play_vfx_explosion_barrel_fire_17_**stop**

*Leveraging other disciplines' naming standards can also help lead you to the asset where an audio event has been authored. If you have the name of the object built into your audio event name, it can easily be used to search within the game engine.

Whether you choose to concatenate (build) different aspects of the event name dynamically or establish the rules up front for use by programmers, the effort of establishing a naming standard is rewarded with a clearer understanding of a sound's origin during debug. When you find yourself digging through an audio event log trying to understand what, where, and why a sound is playing, the ability to parse the event name and begin your search with a map leading right to the sound is a tremendous way to navigate the project and solve problems.

In the world of abstractions at the core of game development, anything that can be done to build clarity into a pipeline, which can be easily understood, is of value to the process.

Asset Delivery Methods

Delivery for the asset cannon method is often via a secured shared drive on a cloud-based storage site. An asset list in the form of a spreadsheet should be used to keep track of the status of assets, provide notes for integration, and feedback from the developer.

When importing assets natively into the game engine (or when using audio middleware) the process usually involves **source control** as a delivery method. We will discuss source control in more detail later in this chapter, but for now let's define it as a server or cloud-based system that hosts source files. In our case the files would consist of game assets and project sessions shared by everyone on the team. It offers control over versioning, like a backup for all of the changes made to the project. It's important to note that sound banks generated from middleware can be shared across cloud-based storage as an alternative for smaller teams that do not wish to put source control in place. Whether through source control or a drive share, the process allows for sharing sound banks without the programmer needing to have a copy of the middleware.

Source Control

Source control (also known as revision or version control) is an important part of the game development workflow, but is not often discussed. Working with teams means that multiple people need a way to collaborate on a project. This method will need to allow for merging new assets, making updates, and making changes to a shared **repository** – a central location where data is stored.

Whether you are working on an indie or AAA game, you may be given source control credentials and left wondering what that means. Each development team might have their own preference for source control, so you should familiarize yourself with software like Git, SVN, Mercurial, and Perforce. Git is used quite often among indie teams and there are quite a few "Getting Started" guides and video tutorials on the internet.[3]

Whenever you have more than one person editing and interacting with assets, you will need a safeguard to avoid changes that might break the game. Source control offers merging of assets and rolling back or reverting to a previous version of an asset, since it keeps a version history. It sounds pretty straightforward, right? Well, it is and it isn't. For some it can be tricky checking out a repository and creating a local copy of it. The workflow can be a bit daunting as the local repository will consist of **trees**. The trees are simply objects that create a hierarchy of the files in the repository. You will need to get comfortable working with a directory, checking out files, as well as adding and committing them. We highly recommend starting with exploring Git and the corresponding tutorials.[4] Source control is something you have to jump in and use to become comfortable in the workflow.

QA before Asset Implementation/Delivery

Prior to delivering or implementing audio assets, we prefer to audition assets bounced from the DAW in a two-track editor (Adobe Audition, Audacity, etc.). Keep in mind this isn't a workflow everyone chooses, but we find it useful to ensure our assets are ready for implementation. While introducing another piece of software into the workflow does add an extra step, two-track editors are destructive, which allows any last-minute editing of the file to be easily saved to disk and quickly shared for delivery. The two-track editor outside of the DAW offers a quick look at the head and tail of the assets to ensure the fades are at the zero crossing. Loops can also be checked for seamless playback and levels can be reviewed. Sometimes listening to the asset outside of the DAW environment can reveal issues like tiny pops and clicks, or issues with reverb tails.

This is also a great time to close your eyes and assess the relative levels across similar groups of sounds. By playing all of your swoosh variations for example, you may start to hear the slight volume differences that will make one audio file stand out in game – and not in a good way! Audition lets you rebalance the volume of any sounds that are poking out more than the others. It also allows you to make the final determination of a sound that isn't quite working with the others, and should not be brought into the game at all. This can also be done visually using loudness meters as part of your mastering process.

Game Engines

Game engines manage numerous tasks to render and run the virtual world of every video game. At any given time while an engine is running it will be handling 2D or 3D graphics, animations, physics, shading, rendering, artificial intelligence, streaming, visual effects, interactive logic, memory management, and audio. The game engine tracks the location of media assets (including music, sound effects, and voice-over assets) as well as when (and when not) to call events. Even in a simple game, this can add up to a large amount of data that needs to be coordinated. In a more complex game the scope can extend into hundreds or even thousands of assets. To render the game the engine might also handle porting of the game to multiple platforms.

There are many game engines available for use and, by extension, there are many plugins and tools to use with them. This means you will find a good amount of variety in the workflow for each project. You might find yourself familiar with some of the more widely mentioned game engines like Unity or Unreal, but there are a variety of other engines like Corona, Game Maker Studio, Cryengine, and Amazon Lumberyard being used by game developers all over the world. Refer to the "List of Game Engines" Wikipedia site[5] for a more detailed list. Some developers decide to design and build their own proprietary engines, but third-party software provides programmers with the necessary tools to build games more efficiently.

These engines make use of various scripting languages like C#, C++, Python, Xml, Javascript, and Lua. As you can see it helps to be flexible and open to learning new tech when working in game audio. This doesn't mean you will need to learn all of these engines and languages, but keeping abreast of the latest trends in the industry and familiarizing yourself with the most utilized tools will help you position yourself as an artisan in the field.

In this chapter we will focus on the Unity engine and its nomenclature as we examine the process of implementing audio natively. At times we will mention similar features in Unreal to demonstrate adaptability between tools. We will also demonstrate the power

Table 8.1 Outline of the terminology differences between Unity and Unreal.

UNITY	UNREAL
Project Browser	Content Browser
Scene	View Port
Hierarchy	World View
Inspector	Details Panel
GameObject	Actor / Pawn

of audio middleware (specifically Wwise and FMOD) by showing a reduction in the time programmers need to spend on the scripting side, as well as a reflection of the middleware offering much more control over sonic events to the audio designer.

Regardless of the engine used for development, there are common techniques and workflows across the discipline. In Table 8.1 we will break down these common elements, which are an important part of game audio implementation and can be applied to projects of varied configurations.

The Engine Editor

The game engine is defined as the software application that provides the necessary tools and features to allow users to build games quickly and efficiently. The game engine editor is a creative hub visualized as a graphical interface with customizable layouts of windows and views. These layouts are used by designers and programmers to piece together elements of the game.

The Sound Lab

A typical Unity engine layout contains a Hierarchy, Project, Console, Scene, Inspector, Audio Mixer, and a Game window. They all work in conjunction with each other in this creative hub. In the Sound Lab we will break down how they are linked. There are also some important components integrated into the engine to help the user define and provide audio functionality within the game. We also discuss the views and components as well as provide a walkthrough of the Unity engine. When you are finished, come back and move onto 2D versus 3D events. If you skip past this companion site study you may not fully understand the information to come.

2D and 3D Events

The additional resources on the companion site have already covered the functions (such as Audio Source and Audio Listener) available in the audio engine. Here we will explore the difference between 2D and 3D events. If you didn't take the time to explore "The Engine Editor" material on the site please do so now or the keywords and terminology used going forward may seem confusing.

Spencer

In linear media the final mix contains volume attenuations that are "burned" in. This means a sound coming from an object or person far away from the camera will sound

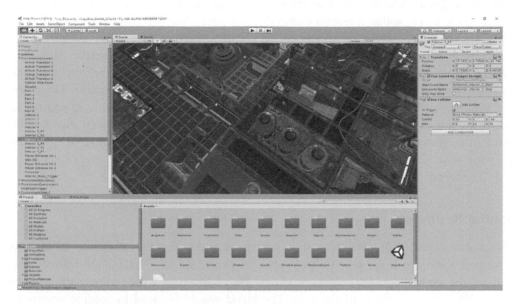

FIGURE 8.2 Screenshot of the Unity engine editor.

softer and more distant, and objects closer to the camera will sound louder and closer. This is typically accomplished during the post-production stage. In interactive media instead of "burning in" those automations, we make use of 2D and 3D events. These events define the attenuation and panning of sound in real time to approximate distance and spatial information relative to the **audio listener**. The audio listener is essentially a virtual surrogate for our ears in the game environment. It is almost always hooked into the game engine via the camera so that what the player is seeing matches what the player is hearing.

A 2D (or two-dimensional) event is "unaware" of the three-dimensional, 360-degree environment. 2D events don't have a specific location as a source, so they don't adapt spatially as the player moves about the game environment. In a sense, 2D events are non-local – the only way they can adapt is to pan left or right, not forward, backward, up, or down. Instead they play through the listener's headphones or speakers just as if they were playing out of a DAW. This makes them effective for audio that is intended to surround the player at all times. Common uses for 2D events are background music and environmental ambiences.

3D (three-dimensional) events *do* include spatial information. They set the volume attenuation and pan based on the source distance and direction from the audio listener. In other words, 3D events are **localized**. 3D events are great for specific sound effects where location is important to the gameplay. However, they can also be implemented with a wide spread so the sound source simulates a stereo image when the player or camera is within a certain radius of the emitter. We will explore uses for hybrid techniques in the "Ambient Zones and Sound Emitters" section later in the chapter.

A simple way to look at 2D and 3D events is by thinking of diegetic and non-diegetic sounds. Underscore is non-diegetic. Players can't interact with the sound source because there is no source! So 2D events work well for that scenario. However, if a sound is diegetic, then the player will likely be able to interact with its source. In this case we can guide the player to the location of an interactable object (like a radio that can be picked up and added to your inventory) by setting the event to 3D, thereby allowing it to adapt to distance and direction. There are many more ways to use 2D and 3D events, but these examples are the most typical.

Spatial Settings

The audio engine will need a spatial setting in order for the audio to behave as described above. In Unity the **Audio Source** (as explained in "The Engine Editor" section on the companion site) offers a Spatial Blend function which allows the user to define 2D, 3D, or a blend of the two. The blend would be used in specific situations where the spatialization of the sound needs to change based on the perspective. When a player character (PC) is driving a vehicle in game, music might be a 2D event to fill the space of the vehicle's interior. When the PC exits the vehicle the sound will behave as a 3D event so it can be spatially located, emanating from the vehicle. When a pure 3D spatial blend is selected there are a few options to control the behavior of the sound's playback. We will discuss those options a bit later in the chapter.

Doppler level is another important setting within a 2D or 3D event. The Doppler level will raise the pitch of an audio source as the audio listener approaches the object, and then lower the pitch as it moves away from the object. This emulates the way we hear sound coming from moving objects (or as we are moving) in the real world. Keep in mind that when the value is set to zero the effect is disabled. You may want the option to disable the effect when dealing with music as underscore is most often thought of as non-diegetic.

Typically 2D event assets will be imported as stereo or multi-channel files, while 3D event assets are imported as mono files. The mono assets will sound more consistent and localized when positioned in the 3D space. Multi-channel sounds and sounds that need to cover a wide stereo field can be utilized as a 3D asset, but you will have to use **Spread Control** (as explained in "The Engine Editor" section on the companion site) to separate the channels in virtual 3D space. If done properly, the perceived width of the sound will diminish as the source moves further away from the audio listener, just as it does in the real world. The Spread will allow you to mimic the effect on multi-channel sound.

The **Volume Rolloff** function controls the attenuation as the listener moves to and from the audio source. The volume curve offers a logarithmic, linear, and user-defined custom slope setting. A minimum and maximum distance setting offers further control over how loud a sound is over a given distance. A lower minimum will play the sound at its loudest volume setting when the listener is very close to the object. The values are in meters, so a value of one means when the audio listener is one meter from the object it will play at the full volume at which the file was imported and consider the volume and mixer level slider settings. A maximum distance value determines the distance at which the sound will *stop* attenuating. If the slope falls off to zero the sound will become inaudible.

To determine the best settings we need an understanding of how sound behaves in the real world. Lower frequencies travel further because higher frequencies lose their energy faster. In the real world it's difficult to discern the difference at shorter ranges unless there is high-density matter absorbing the sound waves. This behavior dictates an *inverse curve*, which is exactly what the logarithmic setting in Unity offers. Most often this will be the most natural sounding option. The linear roll-off attenuates a bit too quickly. Adding a low-pass filter curve to a logarithmic roll-off that attenuates (or reduces) the high-end frequencies as the volume fades out most closely simulates real-world sound. However, there will be times the natural-sounding curve will not work best in game. As always, the definitive answer for which curve to use will come from listening yourself and deciding on the one that best fits the game. If you aren't happy with the logarithmic results, the curve can be further defined and saved as a custom value.

As you can see, there is much more that goes into implementing a sound than simply choosing the asset and dropping it in the audio engine. Be sure to always test in game to ensure the 3D settings are accurate, and that they add immersion to the environment.

Audio Middleware

The GameSoundCon yearly survey[6] polls audio middleware usage between AAA, indie, and pro casual game developers. It's a great reference for keeping up with the latest implementation trends. It also includes custom and proprietary engines as well as native engine implementation in the poll for a well-rounded snapshot of how audio is being integrated into games. Currently, Wwise and Fmod are the two popular middleware options for both sound and music. Elias is focused on music integration, but can be easily integrated with Wwise or FMOD to combine a sound design and music solution. It isn't hard to find numerous games developed with these tools.

The Sound Lab

In the Sound Lab Table 8.2 compares middleware implementation tasks broken out into FMOD and Wwise terminology.

The choice of audio engine for a project is usually determined by the game developer or programmer, but a knowledgeable audio designer can help sway this choice. Larger developers with an audio director might rely on their recommendation, or they might decide a custom and proprietary engine is best for the project.

There are many reasons to use audio middleware as opposed to native engine integration. These reasons include the additional licensing costs for using middleware or a programmer's concern for handling bugs or errors they may be unfamiliar with. As an advocate for working with audio middleware, you should be familiar with licensing costs and the indie budget-friendly options. You should also be well versed in the software and able to assist the programmer with any issues that may arise. Middleware developers do offer support and there are plenty of forums on the internet to provide assistance.

The point of middleware is to offer the sound designer or composer (who may not be familiar with programming) more control over the final in-game mix. It also works to offload the detailed and time-consuming process of audio implementation to the audio designer as opposed to the programmer. Having this control over the implementation process will make your game sound better, period. Middleware allows you to bypass the asset cannon and native engine implementation approaches, and get right into the business of triggering and mixing adaptive audio.

Here are a few other advantages to using audio middleware.

- You have better control over the audio footprint with file compression.
- Dynamic music systems are much easier to configure and implement.
- Debugging and profiling allows for quick ways to uncover issues and resolve them.
- It allows for easier testing for revision and iteration.
- You have control over the final mix.
- You can easily add variety to sound events.
- Flexible licensing makes middleware accessible for smaller teams.

The Sound Lab

 In the Sound Lab Table 8.3 looks at some advantages of using middleware over the native audio function in most game engines.

Fmod and Wwise have similar functions, but their graphical user interface looks a bit different, and naming conventions for certain modules and functions vary. We will also have a look at some middleware tasks and how they are referred to in each application.

Integration Packages

As mentioned earlier, by default the game engine and the audio middleware won't automatically connect and speak to each other. The middleware software developers provide an integration package, which needs to be installed into the game engine. Because the integration process connects two different software packages, the packages are specific to individual software versions. If you update either the game engine software or the middleware version you should *reintegrate the package*. For this reason, it is best to hold off on upgrades to your tools until you are finished with the game if at all possible. Once the integration package is in place there are several components that can be used in the game engine without any additional scripting. The audio designer will have instant access to add an audio listener, load banks, trigger game state switches, and audio event emitters into a scene.

The integration package also provides a few **classes** or libraries of pre-programmed functionality, which can be used with minimal code. As you use these integration packages you will appreciate things like calling banks, events, triggers, and real-time control parameters all without needing any support from the programmer. An API (Application Programming Interface) offers the game's programmer deeper control, while the provided stand-alone app offers the audio designer a graphical hub in which to import sound assets, configure them for playback in game, and test the audio systems.

Middleware Events

The audio functions we discussed in the Sound Lab earlier in this chapter can be applied to middleware functions and modules. Events in middleware are similar to the Audio Source component in the game engine. The Audio Source acts as a container that holds information for the audio to be triggered in game. Middleware events act similarly in that they hold the audio assets and data which determines how and when it is triggered in game. This data is what determines how the event should behave in game. Additionally, middleware events contain more advanced functionality like **scatter sounds**, RTPC (real-time parameter control), and **nested events**, which allows the user to organize large and complex events into one or more master event. We will discuss this further on the companion site.

Firelight Technologies' FMOD allows the user to determine if an event will be 2D or 3D upon creation. When a 3D event is selected a spatializer planner will be available on the master track. In Wwise an audio object can be defined as 2D or 3D in the Position Editor. Once the user has a solid understanding of the functions of audio implementation that knowledge can be applied when acclimating into different tools.

Using middleware, audio designers can easily configure tempo sync transitions, set loop points, create randomization, delays, and and apply real-time effects to events. Programs like Wwise and FMOD accomplish this by taking **game data** (values that originate from the game code) and synchronizing it to various audio events. The audio events are made up of all kinds of audio assets and automation stored within **containers** (a broad term used to describe the instruments and modules that can be created with middleware) and triggered in real time. Each middleware program has a different method of synchronizing with game data. FMOD uses **parameters**, tabs with a pre-set value range that mirrors the game data. Wwise uses **game syncs**, which function similarly, but without the visual tab that FMOD offers. Native integration into a game engine requires programmer support to create these elements. Additionally, the mixer system allows for grouping events into buses that allow the use of sends and returns. Snapshots can provide pre-configured audio commands which will be applied to specific actions in game. These audio configurations will act as a dynamic mixing board, adapting the mix on the fly. Figure 8.3 demonstrates the flow of data from game engine to middleware.

Integrated Profiling

A **profiler** is a tool that provides detailed information about how a game (or in our case a game's audio) is performing. This tool can uncover various problems or resource management issues and provide some information on how to fix them. Later in the chapter in the "Testing, Debugging and QA" section we discuss profiling as a quality-control and bug-fixing tool. Here, we will explore using profilers

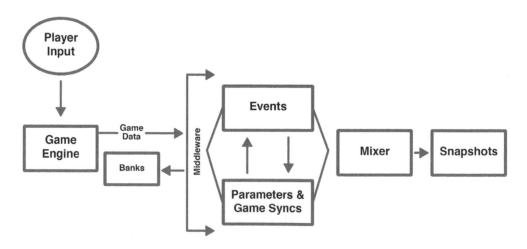

FIGURE 8.3 The flow of data from the game engine into middleware and back into the game engine.

for resource management and other tools to audition a project's events and snapshots in a game-like context prior to connecting to a running instance of the game.

In additional to a profiler, the **Wwise Soundcaster** and **FMOD Sandbox** features provide exactly this kind of functionality. These tools provide the audio designer with a way to configure various game state changes without the need to connect to the game. By being certain that the state changes and triggers are working correctly before updating a project into the main development branch, it is far easier for the audio designer to ensure the cohesion of the mix without interrupting workflow for the rest of the team.

Once the designer has created and configured events to her liking, sound banks can be generated based on the target platform. This will create a package from which the game engine can interpret all the necessary audio functions and trigger them appropriately in game.

Software development is never simple. During these tests, issues with implementation often occur. Any small discrepancy such as a mislabellled event can cause an expected sound to improperly trigger. This is where the profiler comes in. A profiler offers a connection to the game through the game engine editor or stand-alone build of the game. A profiler lets the audio designer track the communications between the game and the middleware to verify what data is being sent in between the two programs. This is a crucial step toward optimizing your game's audio and uncovering bugs or issues with implementation. Later in the chapter we will further explore resource management and optimization.

It's not unusual to have UI sounds and recorded dialogue implemented into the game engine's native audio system while sound effects and music are implemented through middleware. Different developers have their own preferred workflow so audio designers should be willing to adapt on a per project basis.

Gina

The Sound Lab

 Before moving on, head over to the Sound Lab for tutorials on Wwise and FMOD. When you are finished there come back here to explore individual components and features of middleware that help improve the final in-game mix. If you skip this companion site study, you will not fully understand the information to come.

Achieving Variation within Limits

Since we have limited resources per development platform, it isn't possible to have 24 hours of streaming music and infinitely varied sound effects for every possible action in the game.[7] This means we have to be creative in how we generate audio content.

To account for longer play times, we can create continuous loops of music and ambience. We can also string together shorter clips, triggered randomly, that fit together to make a longer more generative loop. This offers a bit more variety. Adding short linear sounds (typically referred to as **one-shot sounds**) on top of our continuous loops can help further improve the variation. This can be done via random playlists, which we discussed at the beginning of the chapter. For example, a playlist of gunshot sounds may contain four to six one-shot assets. If the same gunshot sound were to be triggered ten times in a row, it would sound synthetic, and the player would lose her sense of immersion. The random playlist instead chooses an arbitrary gunshot sound from the playlist. Every time the player chooses to fire the weapon in game a different sound triggers, adding variety and maintaining a sense of immersion. Designers can also set containers to randomize the pitch, volume, pan, and delay time for the sound to trigger, and multiply the output so that it sounds varied and scattered around 3D space (like simulating a forest of birds chirping). Random settings on pitch and volume should not be overly exaggerated. Similar sounding assets will produce an organic experience, but sound that is too heavily randomized will feel confusing to the player. A small 2 dB fluctuation in volume and a few cents in pitch are enough to provide the subtle randomization necessary to avoid listener fatigue and add generative detail.

Randomization can and should be applied to a variety of sounds in game, but some thought should be put into what *kinds of sound* need randomization in addition to *how much* randomization should be applied. Some things will sound consistent with volume and pitch randomizations, others will sound chaotic. For example, UI button presses in game with random pitch or volume applied won't sound consistent to the player and therefore will not relay the intended effect. It will actually be confusing to the player. Ambient bird chirps however are great candidates for subtle pitch and volume randomization.

In the Sound Lab (companion site) we will provide a video tutorial on randomization. You can visit the site now or come back to it later in the chapter. We will be sure to remind you.

Ambient Zones and Sound Emitters

To create an immersive soundscape in a game scene we start with an **ambient zone** (see Figure 8.4). Ambient zones are a defined area within a game scene to which designated components such as audio event emitters can be added to trigger background sounds to match the environment. Instantiating this will create a looping background event which triggers over a defined area. To bring the space to life we can then introduce positional sounds by placing sound emitters throughout the scene. These emitters add detail and specificity. When combined with the looping ambience, these events define the sonic setting for a given scene.

Let's take a look at a scenario in the likes of a FPS (First Person Shooter). We will need to start with an ambient zone. We'll add a simple ambient loop of a deserted city. However, our battlefield won't sound very realistic with just a static ambience looping in the background. By adding positional sound emitters scattered around the environment we can provide a heightened sense of realism. For example, our battlefield might need some randomized debris to crumble now and then, or the shouts and cries of unseen NPCs still fleeing the city. By placing emitters on game objects we can add those sonic details to the scene. When the player moves through the environment and into the radius of the emitters she will hear the soundscape change and adapt.

One important function of emitters is to approximate sounds relative to distance. When the source of a sound is far away, as the audio listener (if you can recall, the audio listener is almost always attached to the camera, which we are controlling in a FPS) moves closer to the emitter the mix responds appropriately. As we previously mentioned, as the source moves closer, the volume of the sound event must increase to mimic the way we hear in real life. Additionally, a filter can to be applied to re-introduce the sound's higher frequencies as the listener approaches the source, which adds more detail and realism. Back in our FPS example, the player will see an NPC shouting in the distance and, as she moves closer, she will hear the shouts louder and the HPF will open up to include higher frequencies. On top of that, the left–right spatialization (and in some cases even up–down) will automatically adapt to

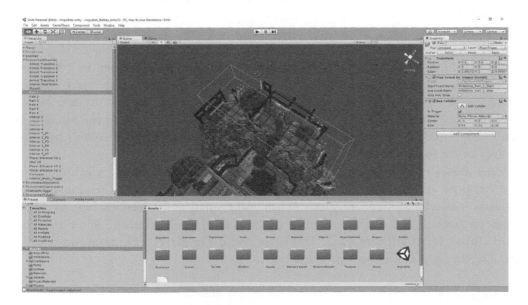

FIGURE 8.4 An ambient zone defined by a spherical game object in a Unity scene set to trigger an audio event when the player enters the area.

the angle of the audio listener. She will perceive all of this as cohesive and realistic details in the soundscape.

It's important to note that sound emitters don't necessarily need to be hooked directly to a visual object in the game scene. The audio designer can place sound event emitters in a scene to create positional *random sound triggers* within an ambient zone. This approach is great when a game object is not visible to the player, but exists as an element regardless (i.e. wind, bird chirps, crickets). These randomized triggers add detail to the soundscape without cluttering the visuals.

The ambient zone/emitter system can be implemented into the game in a few ways. One way is to tie the ambience and emitters to a *single event*. One middleware event can handle multiple audio objects, so we can set an ambient zone in the game engine that triggers a single event in the middleware program. In the event we might combine one or more looped assets and layer them with multiple one-shot sounds. Spatial settings can then be applied to each asset as well. The static loop would be a 2D event, covering the full area of the ambient zone, while the one-shot sounds would be set to scatter around 3D space. The second method is to place one or more looped assets into one event and create a *separate event* for the sound emitters. Just like the first method we can define the spatial settings so the static loop is 2D and the emitters trigger 3D events. The difference here is that the ambient zone will trigger the static loop event over a defined region, while the positional emitters will be *independently placed* and separately triggered around the game scene. This method is a bit more work, but it offers more flexibility with the emitters.

A third option for a responsive soundscape involves a more dynamic ambient system in which multiple loops and emitters trigger sounds according to the game state. This kind of dynamic system can adapt to game context, and can be controlled by real-time parameters (see the section below). A good example of this is Infinity Ward's *Call of Duty (COD4)*. The team used a dynamic ambient system which streams four-channel ambiences and switches between them based on the amount of action happening in the scene.[8] This systems helps to create a detailed soundscape that evolves as the battle intensifies or vice versa.

Some games take advantage of quad ambiences for ambient zones, which make up two or more sets of stereo environments of the same recordings. This can provide a denser sonic background in game. When using quad ambiences it's best to utilize captured audio that doesn't have much in the way of easily identified sonic elements (bird chirps, tonal artifacts, etc.). This helps avoid the player picking up on a small detail in the loop as it repeats over and over.

Reverb Zones

Reverb can make a notable difference in how realistic a scene sounds and how well the scene conveys a sense of space. Reverb zones, just like ambient zones, are defined regions for which reverb settings can be applied. This offers the audio designer a way to create specific reverb effects on sounds triggered within this area.

As we mentioned previously, smaller mobile game projects may require the audio designer to bake the reverb into the sound effect before importing into the engine. The problem with this is lack of adaptability to the player's location. When resources are available the best way to create space in a location is by setting up reverb regions that will trigger different reverb presets in predetermined locations with predefined effect settings.

Reverb utilization in games has evolved and is still evolving as limitations in CPU and memory are improved. Now, more games can benefit from realistic-sounding ambient spaces. Game are already using **ray tracing** and convolution reverb for simulating truer reflections in a space. As we push forward we can look at more advanced ways to create believable aural spaces. **Auralization**[9] is the process of simulating **sound propagation** through the use of physics and graphical data. This is another example of more advanced techniques for implementing reverb zones.

To conclude our thoughts on reverb zones, let's discuss applying reverb to the zones. Presets are a great starting point, but reverb settings should be tweaked to fit the scene. Pre-delay and reverb tail can make a huge difference in how the scene sounds. Pre-delay is the amount of time between the dry sound and the audible early reflections. Adjusting the pre-delay can add more clarity to the mix by opening up space around the initial sound. Adjusting the pre-delay can change the room size without changing the decay time. This will help avoid a washed-out mix from too much reverb versus dry signal.

Looped Events

Middleware offers the audio designer an easy way to define the start and end points of a looped asset. The ability to define a **loop region** allows the designer to define the start and end points of the container. When the event is triggered, the loop will then continue to play back that region until the event is stopped. The audio designer can specify a particular number times to complete the loop, or set it to infinite for continuous looping.

Loops are an important part of game audio for assets like music and ambiences, but there are a variety of other sounds that benefit from looping. An automatic weapon fire sound would usually consist of a start, a loop, and a stop. These **modules** (sometimes called **sound containers**) could be single files, playlists, or separate events entirely. The start sound would contain the initial fire and the stop sound would contain the last bullet fire and tail. The loop would contain a stream of the weapon's consecutive bursts, and a loop region would wrap around the entire duration. All three modules triggered in synchrony will allow the player to fire the weapon (thus triggering the start module into the loop) and hear a continuous barrage of bullets until the player ceases firing and the stop module is triggered, ending the event.

Variation and contrast in a looped event can help avoid listener fatigue. Avoiding specific tonal elements can also prevent the listener from picking out the loop point. Using a playlist of one shot-sounds in lieu of a static loop can add even more variation. In this case the user can define the random playlist so it doesn't play the same asset twice in a row. If we move back to our automatic weapon example, the loop module would then be replaced by a playlist of short single-fire sounds. The loop region would remain an essential element of the event.

Looped events can be stopped and started based on ambient zone triggers. This is useful in managing resources because looped sounds will not continue to play in the background if the audio listener is not inside the defined ambient zone. Imagine that our PC is approaching a forest. As the audio listener moves closer to the forest and into the ambient zone, the forest ambience event will trigger and loop. The problem with this scenario is that the event sound starts from the beginning of the loop each time. If the player decides to run to and from the forest six to ten times in a row that loop start might become aggravating. By using a **random seek variable**, a different point in the loop will randomly be chosen to initiate the playback every time the event is re-triggered. This can also be a great solution for when two positional sound emitters in close proximity are triggering the same looped event. Phasing issues can be avoided if the files start their playback from different points in the loop.

> Much of the information provided here on looped events may seem simple or obvious, but loops are the foundation of more complex events. Make sure to get comfortable with all their variations.
>
> Spencer

Real-Time Parameter Control

Earlier in this chapter we briefly defined real-time parameter control (RTPC) as the ability to control specific properties of audio events based on real-time game data. Here we will take a microscopic look into some examples of RTPC usage. To put it concisely, RTPC is a method of taking values from the game engine, and using them to automate volume, panning, effects, or really any other sonic properties found within a middleware event. The only difference between automation in a DAW and automation in middleware is that your DAW is locked into a fixed timeline – the SMPTE code. The playhead *only* moves forward as time passes. In middleware we can use any in-game value that we want to move the playhead forward, and we can have multiple playheads that correspond to multiple parameters, each independent of one another. In the following sections we'll dig into what these values are and where they come from, and then we'll share some examples of RTPC usage.

Game engines manage a large amount of information when the game is in runtime. At any time an engine can pass values that define the game state. These could include information on player location, time of day, velocity of an object, player health, vehicle RPM, and more. Pretty much any action or change in the game state can provide data which can be used to adapt the audio around. For example, when 3D events attenuate and position sound, what is actually happening is the game engine is tracking the distance (typically in meters) from the audio listener to all objects in the scene. These distance values are then passed from the engine to the banks generated by your middleware program. These values can then be linked to parameters in FMOD (or game syncs in Wwise) in your middleware program, allowing the volume and panning to adapt in real time, as needed. Game data isn't limited to distance. Information can include number of enemies in range, time of day, or mission completion.

Another of the many applications of RTPC is to adapt footstep playlists to different types of terrain. The programmer may pass the values for the terrain types to you, or you can do it yourself using the profiler. Either way, these terrain values must be linked to a parameter (let's call it "terrain," although you can technically name it whatever you like) via a short scripting process. Usually this part is done by the audio programmer, but it's important to communicate your ideas first so you can synchronize event and parameter names and ensure appropriate values are being called. Once this is accomplished you can use the parameter values to trigger the corresponding footstep sounds in an event. Random adjustments to the parameters for pitch and volume can also be applied to the footstep playlists. Additionally you can add layers of cloth or armor movement to the event and position the modules with a slight delay. The delay will avoid the footfall and cloth from playing at the same time and sounding too robotic. The delay setting can also be randomized so it plays back with a slightly different delay time for every trigger.

You can also use RTPC to adapt music and sound design based on the player character's health. Lower health values on a parameter can be used to automate a low pass filter, which can then affect all sounds in game. When the health parameter decreases into a predetermined threshold, the LPF will kick in. This will give the player a feeling of losing control. It will also force the player to focus on the game mechanics as opposed to more superficial sound effects. Alternatively, low health parameters can also be used to *add layers*. In this case, instead of using a parameter to automate the cutoff of a LPF *down*, try using it to automate *up* a layer of music when health falls below a certain value. This can increase the drama of a battle scene or boss fight.

The automations we've mentioned are drawn in with curves on the parameter window. Just like volume curves in a DAW, the curve can be defined so it rolls off smoothly. Sometimes the speed at which the parameter value falls in game can sometimes be quick or uneven. This could cause the audio to follow along abruptly causing the transitions to feel rough. In FMOD, a **seek speed** module will allow you to set the speed at which the value is chased as it moves across the timeline. Small details such as this will help the audio adapt smoothly to game states.

The real fun with RTPCs begins when you begin working with multiple parameters independent of one another. Let's imagine a hypothetical rainforest scene that adapts to the time of day. Ideally we would have two or three different ambient loops on the parameter tab: one for morning, one for afternoon, and one for night. First, we need to create a parameter – let's call it "Time of Day" so it corresponds with the time of day values being passed from the engine. In Wwise a **blend container** can be used to seamlessly crossfade these loops. FMOD similarly allows us to add these loops to a single parameter timeline with a crossfade between them. As the time of day data changes in game, we are left with a single ambient event that evolves over time using real-time parameter control. This will provide a more realistic soundscape since the bird chirps of the morning can slowly fade away into cricket chirps as daylight hours move into evening and night.

But that's not all! We can also add a *second* parameter to our event. Let's call it "Weather." Using this second parameter we can add a layer of sound to a new audio track. This layer will consist of three looping ambiences: low, medium, and high intensities of rain. To keep it simple these loops will again be added to the parameter tab with crossfades (and a *fade in* on the low-intensity loop). The only problem here is that rainforests don't always have rain! Let's compensate for that by moving the three loops over so that lower parameter values contain no sound at all. Now we have two parameters working *independently* to adapt to time of day and weather values from the game.

The cherry on top of this wonderful rainforest ambience cake will be the addition of scatterer sounds (FMOD) or random containers (Wwise). These audio objects are sound containers which can be used to sprinkle randomized sounds throughout the 3D space. This allows you to input a playlist of bird chirp sounds and output a pretty convincing army of birds. Let's add a few scatterers with playlists of birds chirping and some other animal sounds to fill out the sonic space. These scatterers can be placed on the main timeline itself and set with a loop region. At last, we are left with an ambient system that adapts smoothly to two game parameters, and generates a forest full of animals using only a few containers.

It's important to note that while parameters and RTPC controls can be assigned to all audio objects, busses, effects, and attenuation instances, you should still use them selectively as they can consume a significant amount of the platform's memory and

CPU. We will discuss this in further detail later in this chapter in the section titled "Resource Management and Performance Optimization."

Nested Events

Just like other forms of software development, the parent/child hierarchy is an important structure which allows for manipulating game objects and events. A nested event is a category of event that is referenced from its parent event. Both Wwise and FMOD offer use of nested events for extending the flexibility of a single event. This opens a variety of possibilities for delaying playback of an event, sequential playback of events, and creating templates across multiple events.

For example, a complex event can be created and nested into a simpler master event. You can then combine *other* nested events within the master event, and control all the events via parameters/game syncs in the master (like a Russian nesting doll, hence the name). It allows for more cohesion and a level of depth that would otherwise require actually scripting in the game engine. This kind of system is great for complex music systems or vehicle engines that require macro and micro parameter control of individual audio assets. For example, a parent music event will be configured with parameters defined by game states. This macro control could be exploration or combat. The child events that are nested into the parent will have micro control via parameters which could vary the soundtrack in either exploration or combat states.

Game States and Snapshots

In theory a game consists of a sequence of states, which consist of a combination of visual and aural cues. The game states are made up of menu, exploration, combat, boss battle, pause, game over, and scripted logic controls the flow of the game states according to player input. These states are used to trigger the ambient zones we described earlier in the chapter. Each ambient zone can have its own events and mixes applied to increase or create sonic diversity in the game. We do this using snapshots. Snapshots store and recall a range of sound properties, mixer settings, and effects settings. Essentially they are mixer presets for your game that can be triggered as players interact with the game. Snapshots are extremely useful as they can be used to create alternate mix settings and transition between them in real time according to game state changes. They are commonly used to dynamically reduce the volume of music and sound effects to leave room for voice-over. If you've ever played a game where the mix changes as your PC dives underwater, that is mostly likely the work of snapshots. In this case the default ambience has been faded out, and a new ambience has been faded in, likely with the addition of an LPF to simulate the physics of sound in water. In effect, what you're hearing is the *game state* changing from land to water, which triggers two different snapshots.

For instance, you can duck sound effects and music when the player presses pause using a snapshot. Simply save a snapshot of the mixer with SFX and music up, and store it with the default game state. Then save a second snapshot of the mixer with the music and sounds ducked. When the pause state is triggered snapshot 2 will activate, and the music and SFX will duck out of the way for the player to experience a brief respite from the in-game action.

There are other creative uses of snapshots as well. By creating a snapshot that combines a LPF with volume automation it is possible to dramatically impact the player's focus. In a battle arena environment, this snapshot can be triggered when a player is firing a powerful weapon (or standing next to it). The snapshot will effectively filter and attenuate the volume of all sounds *except* for the weapon, which ensures that other sounds don't mask it. The result is that for a brief moment the player will hear almost nothing but the massive weapon, emphasizing its power and engendering laser-like focus on the act of firing. This can be a very satisfying and fun effect for players if it is not overused. We encourage you to think of your own snapshots that influence player focus on different aspects of gameplay. Try it!

Transitions and Switches

Middlewares provide audio designers with ways to smoothly transition from one audio event to another in real time. This control defines the conditions of tempo/beat synchronization, fades, and the entry and exit points of music cues. We can also delay the transition after it is triggered by the game event, which allows for more control and a smoother change. In both FMOD and Wwise, game data can be used to drive these transition changes. For example, the velocity value of an object in game could define which impact sound is triggered. Below is a comparison of the two transition methods.

In FMOD transitions are triggered by game state values which allow an event playhead to jump horizontally to **transition markers**. A quantization value can be set to define the delay in either milliseconds or in beats per minute (bpm). This is useful particularly for music, so that transitions can occur in a range of timings, from immediate to any number of beats or bars. FMOD also offers a "transition to" flag, which will allow the transition to happen when the playhead crosses paths with the flag. A transition region is another option; it can span the length of an instrument module, and it continuously checks the parameter conditions. When the predetermined conditions are met, the transition occurs.

In **Wwise** a transition matrix can be set up to define music switches and playlist container transitions. For sound effects, these switches are used to jump from one sound to another. For example, a **switch group** that defines the terrains available in game will allow switches to be triggered by a game sync. In this case, the game engine sends its data to Wwise, informing the audio engine which material the player character is

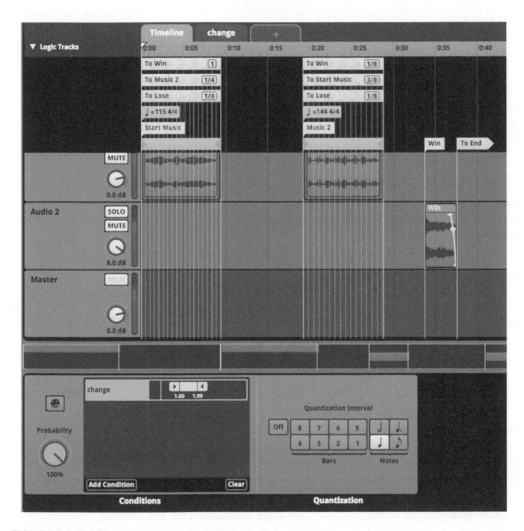

FIGURE 8.5 FMOD transitions.

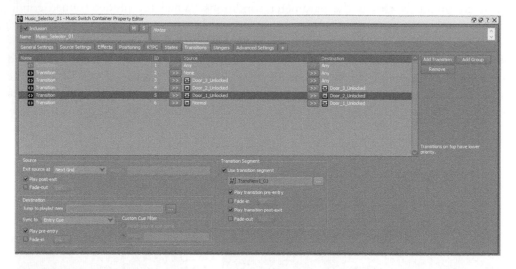

FIGURE 8.6 Wwise transition matrix.

colliding with in game. A switch group would then transition between footstep play-lists corresponding to the terrain. When the player character steps on solid metal the solid metal footstep event will trigger. When the player moves to dirt the dirt footstep event will trigger and so on.

Obstruction and Occlusion

Humans are good at detecting the location of a sound's source. Our brains are highly sensitive to the change in panning and frequency spectrum, which tells us not only *where* the sound sits in relation to our ears, but also whether or not the source is located behind an obstruction. It's an important part of navigating the real world. It makes sense then to mimic the psychoacoustics of a space in a game scene.

Objects in a game that block sound are called **obstructions**. These obstructions can be made of any material and come in varying shapes and densities, which change the sound. Sound has the ability to travel around obstructions, but it can still bounce off surrounding surfaces. For this reason, to mimic the physics of sound through an obstruction a low pass filter could be applied to the sound source, but the reverb send should not be affected.

Occlusion occurs when the path between a source and the listener is *completely obstructed*. A good example of this is a wall between the source and the listener. Sound cannot travel around this wall, so it must travel *through* it. In this case a volume attenuation *and* a low pass filter should be applied to the source. The reverb send will also need to be reduced since no reflections from around the wall can occur. In all cases of obstruction and occlusion the programmer is responsible for defining the obstruction and occlusion values in the engine. The audio designer is responsible for defining the volume attenuation and LPF (low pass filter) to reflect how the audio is affected.

A common technique used to emulate real-world sound propagation is ray tracing (or ray casting). Ray tracing is a function utilized by graphics in game to determine the visibility of an object's surface. It does this by tracing "rays of light" from the player's line of sight to the object in the scene.[10] This technique can be used to define lighting and reflections on objects with which to create realistic visuals. To use this technique for sound we can cast a ray from the audio listener to various objects in the scene to determine how the sound should interact with those objects. For example, in game you might have a loud sound emitting from an object inside a room. The player could leave the room and still hear the sound if there is no door closing off the previous room from the next. The "rays of light" would detect the walls between the rooms, but take into account the "rays of light" that continue *into* the room via the open doorway. In this way, ray tracing yields game syncs that we can use to trigger obstruction and occlusion automation as detailed above.

Ideally we would want to hear proper sound propagation in every game. Unfortunately each game has its own priorities and budgetary limitations, therefore some games will need to do without an obstruction and occlusion system. Depending on the target development platform, implementing ray tracing could be too costly on the CPU. In this case occlusion and obstruction may have to be "faked" by putting distance filters on the sound events.

Gina

Simplicity vs. Complexity

Simplicity vs complexity is both an aesthetic choice and a technical one for some games. Just because you have the ability to craft a complex event does not mean it's the best choice for the game. The old saying "just because you can, doesn't mean you should" is something to remind ourselves as game resources like bandwidth and memory are integrally intertwined with the decision on how sound can be leveraged and implemented in a game. The end product is what matters, and if a simple solution works better than a complex one, so be it! A focus on the theory of game audio, experimentation, and reiteration, along with a solid understanding of what the game needs, will lead you on the right path.

DIALOGUE SYSTEMS

Earlier we discussed the use of recorded speech to provide information to the player and support the narrative. Here we will discuss some other ways voice-over can be integrated into the game so it adapts to gameplay. To start off, imagine you are immersed in a role-playing game (RPG), exploring a village. You come upon a blacksmith who tells you a tale of a sword that will help you through your next mission. But wait! You already completed that mission. Without proper logic in place to handle dialogue in game, the intended information could easily become misinformation.

A great example of this kind of dialogue logic is *God of War* (2018), by Sony Interactive Entertainment. In this game, time spent walking or boating to the next location triggers dialogue. These instances offer players a chance to hear the story progress, and help them prepare for what lies ahead. For example, when boating to a location, Kratos tells stories. These snippets of dialogue set the scene and provide the backstory. Since the game is interactive, and these travel scenes are not pre-rendered, the player can choose to dock and leave the boat at any time. However, the developers ensured that conversation on the boat would naturally end, regardless of timing. In one scene, Mimir is reciting a tale to Atreus while traveling by boat. When the player docks the conversation will naturally end with a line from Kratos saying "Enough. No stories … not while on foot." Mimir then replies with "Completely understand, I'll finish later, lad." The story picks back up when the player heads back to the boat. Atreus says "Mimir, you were in the middle of a story before…" This elegant dialogue system is essential to pushing the narrative of *God of War* forward without hindering gameplay. The developer's attention to detail keeps the player tuned in to the story without losing the element of choice and interactivity.

One important method of employing dialogue logic that you can use to make voice-over more interactive is **stitching**. Stitching is a process in which the dialogue is tied together in game. Some games have full phrases or sentences recorded as individual lines (this is the case in the above *God of War* example). These lines are then are mapped out to follow along with gameplay. Other games like EA Sports *Madden NFL* require dialogue to be recorded in smaller bits, and broken down into phrases, numbers, and team names. For example, the voice artists playing the role of the announcers might record the same phrase for every single football team. A more practical and memory-saving approach would be to have both the line and the team names recorded separately. Then they could be strung together by the audio engine. This could be achieved in middleware by creating two sound containers: one would contain a playlist with every possible team, and the other would contain the phrase "… win!" In effect the player would just hear a seamless rendition of "The Giants win!" With this approach it would be crucial for the voice artist to match the inflection correctly with each take.

Typically a programmer will create the logic which triggers voice-overs in game, but you will be responsible for the actual implementation of the assets themselves. Here we will discuss some techniques for implementation of those assets into the game engine.

Native audio in Engine: Regardless of which audio middleware option is being used for the game's audio integration, voice-overs may be handled directly by the game engine's native audio system. You would then deliver assets for the programmer to import and control via a script. This operation is simple, but it doesn't offer much control over the assets in game.

FMOD: Games with a large amount of spoken dialogue content typically trigger assets from a database using a method which FMOD calls a **Programmer Sound**. This technique requires a script to trigger the Programmer Sound Module and determines which line(s) of dialogue to trigger in game. The audio engine API documentation usually covers how to work with these.

With a small number of assets, each spoken dialogue file may be integrated into its own events. To add variety to the voice-overs a **parent event** can be set up with several **child events** referenced within it (this is referred to as a nested event). The parent event would control how often a child event would be triggered, and the child events will hold a playlist of random takes on the same line. Play percentages and probability settings can then be added to the events to avoid the same line of dialogue triggering one after the other in sequence.

Wwise: A specific function called a Dialogue Event in Wwise will allow you to create various conditions and scenarios for triggering voice-over in game. Game state and switch values can then be used to create a matrix of conditions from which dialogue is stitched together seamlessly.

Scripting

If you've never had to work with **scripting**, now might be the time to start. It can seem overwhelming, but with patience and practice it will become a second language. In the long run your audio workflow will be greatly improved with an understanding of the scripted logic behind your games. Not all audio designers will have a need for these technical skills, however. You can probably get by designing sounds and delivering them to the developer for implementation without ever touching the game engine. Just know that sound designers with scripting skills have an edge over those that don't. While the complexities of scripting are outside the scope of this text, we have dedicated this section to the benefits of diving into this field as a game sound designer, as well introducing some of the basics.

Scripting is a form of programming, and the syntax for integration changes depending on the game engine. **Scripts** contain game logic, which defines the behavior for *all* game objects, including audio sources. Scripts are typically written by the programmer but can be edited or updated by non-programmers as well. A "technical sound designer" with coding experience, or an audio programmer, is likely to handle all of the scripting for the game's audio, but sound designers may need to edit existing scripts.

Developers are often wary of committing to audio middleware solutions due to a fear of licensing costs and possible issues with integration. For this reason it can be useful to be familiar with some basic scripting because it will help you navigate the native audio engine. Being familiar with middleware scripting can also be an asset when arguing in favor of middleware use. Knowing how to script will allow you to effectively demonstrate how much work will be taken off the programmer's plate by using a tool like FMOD or Wwise. Being able to call sounds and declare events and parameters without the programmer will perhaps be the deciding factor!

```csharp
// Declare your FMOD Sounds in this section

public float speed;

Rigidbody rigidBody;                              //rigid body component

[FMODUnity.EventRef]                              //Declare FMOD Sound Event for Unity
public string music = "event:/Music";             //Public string is the display property and type of variable we are declaring.
//'music' is the name of the variable, and we also assign the path to the FMOD Event.
FMOD.Studio.EventInstance musicEv;                //cube event music
FMOD.Studio.ParameterInstance musicChangeParam;   //end param object for transitioning to the end of music)


void Start ()
{
    rigidBody = GetComponent<Rigidbody>();

                                                  // Create FMOD event instances and get parameters in this section
    musicEv = FMODUnity.RuntimeManager.CreateInstance(music);
    musicEv.getParameter("change", out musicChangeParam);

    //We use the event name that we specified up in the Public class section. Change is how the parameter is named in FMOD Studio.

}

void OnTriggerEnter(Collider other)
{

    if (other.gameObject.CompareTag ("Playcube")) {
        // When collision with the Playcube is detected, and if not already playing music event, play the music

        FMOD.Studio.PLAYBACK_STATE play_state;
        musicEv.getPlaybackState(out play_state);
        if (play_state != FMOD.Studio.PLAYBACK_STATE.PLAYING)
        {
            musicChangeParam.setValue(0);
            musicEv.start();
        }
    }
    if (other.gameObject.CompareTag ("Playcube"))
    {
        // When collision with the Stopcube is detected, set end param to 1, which transitions to the end of the event music

        musicChangeParam.setValue(0);

        // musicEv.release (); if you do not intend on playing this sound again
    }
```

FIGURE 8.7 Declaration of parameter in script and Void Update section of script to show how the parameter is attached.

Learning scripting is not straightforward and errors are an inevitable part of the process. Internet searches, software documentation, and software developer forums are great places to help with troubleshooting. YouTube can even be a valuable resource. It's a good idea to read through manuals first to get comfortable with each software package used in your workflow. This is the quickest way to learn all the features the application developers have worked so hard to include.

Programming Languages for the Audio Designer

The question of which language to learn comes up often when speaking to those looking to jump into a game audio career. The answer is, "It depends on what you want to accomplish." It may seem vague but there are many types of programming languages and each can lead you toward a different goal.

C# and C++ are the two most likely languages you'll need to be familiar with if you are interested in digging into the scripts in your game. If earning a computer science degree does not fit into your schedule, you might consider online scripting courses or dedicated books on game engines. *Game Audio Development with Unity 5.X*[11] and *Unreal Engine 4 Scripting with C++*[12] are helpful starting points. The website asoundeffect.com[13] also offers a great introduction to audio scripting in Unity, and there are also many good online courses offered by sites like Udemy[14] with an introduction to programming in Unreal and Unity. In Unity most of the programming you do would be C# based while Unreal offers Blueprints (node-based visual scripting specific to the unreal engine), or C++. Blueprints might be a bit easier than C++ if you're a complete newbie.

There are plenty of other languages to choose from as well. Perhaps you are interested in creating custom actions in Reaper. In this case Python and Lua would good places to start. Python is also a great language to start with if you have an interest in adapting sound through machine learning. Javascript also has some great audio libraries you can build on for web applications while JUCE offers the ability to create your own audio plugins.

XML is another tool to know, but it isn't a programming language per se. It it a *format* used by programming languages to represent data. For example, a sound designer might find themselves tasked with editing an XML document to add voice-over file names to a database.

Regardless of your end goal, a great way to get started is checking out some of the tutorials and books we mentioned above. It doesn't really matter where you start, just pick an engine and read through some of the documentation or check out video tutorials on YouTube. It can seem daunting, but there are varying degrees in which you can learn and use the knowledge in the real world. Just having a basic understanding of what can be done with game audio programming can build a better bridge between you, as the sound designer, and the programmer. At the very least it will give you a better idea of what might be possible in a game engine. You will feel more confident as you map out the audio system for the game. Try it and you may realize you like it!

Programming Languages for the Audio Programmer

Developers creating games with larger budgets will often have an audio programmer role. In "Game Development Roles Defined" in Chapter 1 we touched upon this role. Each studio has different requirements for the technical sound designer and audio programmer roles, but audio programmer roles most often require well-rounded programming knowledge along with DSP (digital signal processing) expertise. Experience with audio middleware like Wwise and FMOD will also be important. *A Digital Signal Processing Primer* by R. Steiglitz[15] and *The Audio Programming Book* edited by Richard Boulanger and Victor Lazzarini[16] are great resources to get you started.

Visiting Artist: Brian Schmidt, Audio Designer, Founder and Executive Director at GameSoundCon

On the Note of Programming

I'm often asked "how technical do I need to be to be a game composer or sound designer? Do I need to learn to program? How important is the tech stuff?" I usually reply as follows: Games run on code and tech; they are a game's DNA—it's lifeblood. Having a basic understanding of programming concepts will help you understand how the game is put together, and how the different pieces of the puzzle work as well as provide you with an insight into what issues the game programmer may have to address when you request that your sounds be played back in a particular way. Taking a simple "Introduction to Programming" at your local community college can be of great benefit, even if you never write a line of code professionally.

Taking the next step and obtaining a working facility in programming can enable you to directly implement the game's audio *precisely* how you, as the composer or sound designer, want it to be implemented and in doing so, make you a more indispensable member of the game team. If all you do as a game composer/sound designer is upload cues to a Dropbox as .wav files for someone else to implement, you become one of the most easily replaceable members of the team.

DYNAMIC MIX SYSTEMS

Great game audio requires an equal blend of asset creation and implementation. Producing a well-balanced and polished dynamic mix is essential to the implementation stage of development.

What Is Mixing?

Mixing as a process consists of bringing all the audio assets that exist in the game together. All sounds must have their place on the frequency spectrum. Additionally, the volume and spatialization from sound to sound should be consistent. Even as the mix changes with gameplay, the mix needs to sound natural and cohesive.

In Chapter 3, we discussed adding clarity to sounds so they fare better in the mix. The technical elements of the full mix should avoid too many sounds sharing the same sonic characteristics because it makes the aural experience cluttered. Creatively, an effective mix should be dynamic and focused. The mix as a whole needs to have elements that set the mood and inform the player of relevant aspects of the game.

Sometimes mixing is left to the final stages of the development cycle, after all content is already implemented. However, it's good practice to set up the necessary technical elements for the mix and keep house as you go. In film, the final mix is printed and synced to picture. In games, mixing happens at run-time as the player plays the game, so technically the mix is prepared *prior* to the game shipping. Below we'll discuss some techniques that result in an effective mix.

Mix Groups

Grouping, a very basic mixing concept adopted from film, is a technique where the audio designer assigns individual assets to a parent group. The idea is to route similar assets into submixes so adjustments to the group are applied to all the sounds within it. Similar to the mixer in a DAW, game audio engines offer a master bus, which can be branched out into the necessary parent/child busses.

In a film mix it might be enough to keep it simple and have only a few groups such as music, Foley, dialogue, and sfx. This can work just fine for a game with a smaller number of assets and a less complex adaptive system, but larger games with more complex interactive requirements need to go deeper. In these cases groups can be further broken down into additional sub-categories such as SFX>Weapons>Pistols. It doesn't make much sense to add a new bus for each individual sound in the game, but the mixer should be set up in a way that is flexible. These more complex parent/child bus structures offer control over the mix on both large and small scales (see Figure 8.8).

Auxiliary Channels

Film and games share the use of auxiliary (aux) channels for routing effects. In run-time the game engine will send values to the audio engine. These can be used to adapt reverb and other effect levels when the player triggers events in game. For example, if the player enters a tunnel from an outside location, a trigger can

FIGURE 8.8 Screenshot of Wwise group buses.

automate the aux reverb channel up to simulate the reflections one would expect to hear in the closed space.

Side-Chaining (Ducking)

Earlier in the chapter we covered snapshots, which allow designers to change mixer settings based on game states. In addition to snapshots, the audio designer can adapt the mix to singular triggered events using **side-chaining** (ducking). Side-chaining is a common technique where the level of one signal is used to manipulate another. In music production this is used to reduce the bass level when it receives an input signal from the bass drum, which adds clarity in the low. Build-up in the lower frequencies is avoided here because the bass and kick will automatically alternate instead of playing on top of one another.

In games, side-chaining can be used to control player focus and prioritize events in the mix. The audio designer can set up a priority system which will duck the volume of one group of sounds based on input from another. Wwise has a feature called Auto-Ducking which makes it easy to keep critical dialogue up on top of the mix. It does this by routing the dialogue submix signal to other sub-mixes and assigning a value (in dB) to attenuate the volume of the non-dialogue channels. You can then add a fade in and out to smoothly transition the ducking. It's important to use your ears to ensure the transition doesn't stand out too much. The process should be inaudible to the player. Side-chaining can also be set up using parameters or game syncs, which allow the user to set an attenuation curve.

In games like a MOBA (multiplayer online battle arena), shooters can have a lot of sonic action happening during the heat of battle. In such instances the side-chaining is crucial to duck the volume of NPC weapons so the player's weapon sounds clear in the mix. This technique can also be used to shift the player's focus. Some games duck in-game audio when UI text or dialogue instructions are guiding the player through a tutorial. This shifts the focus from the gameplay to the UI instructions. This is especially common with important or first-time UI instructions.

Snapshots vs. Side-Chaining

Earlier we discussed snapshots as they apply to ambient zones. Game voice-over is another great example of how a sound designer could make use of snapshots and side-chaining. In-game dialogue can be broken down into high (critical) and low (non-critical) priority. Any dialogue that drives the story or provides necessary information to the player can be considered critical and should always be made audible and intelligible over all other sounds in game via ducking or snapshots. Non-critical dialogue such as barks or walla are typically used to fill the scene with background murmuring, which is not critical to gameplay. It is therefore not necessary to hear it over louder sounds like explosions and weapon fire.

When using a snapshot in the example of a MOBA, we would first set our dialogue to run through Channel "A" and the explosions and weapon fire to route through to Channel "B." Snapshot 1 might have the two channels equal in volume, but snapshot 2 would attenuate the volume of Channel "B" (the non-dialogue channel), and possibly even use EQ to carve out frequencies around 1.7 kHz–3 kHz to increase the speech intelligibility for the player. We can then choose when to transition to each snapshot, and how smooth (or crossfaded) the transition should sound.

The process of side-chaining is similar to snapshots, but it is a bit more specific. Side-chaining itself is just the process of telling an audio plugin (usually a compressor, or something capable of attenuating volume) what to "listen to" (i.e. what audio signal we want to *trigger* the mix change). In our example, we want to be *listening to* the dialogue so that we can *duck* the explosions and weapon fire. This can be done in various ways depending on the plugin, but it is usually a matter of selecting channels from a dropdown menu. This is a bit simpler than a snapshot because the interaction is automatic once the side-chain is set up. However some combination of these two approaches will likely yield the best results.

We don't know when important dialogue may trigger, so snapshots and ducking methods are crucial in allowing the engine to control clarity in the mix. These techniques are also discussed in later chapters as part of more advanced techniques for controlling dynamic mixes.

Dynamic Processing

Control over loudness is a key element in a great mix. Dynamic processing such as compression and limiting can be used to add further mix control. Dynamic processors on subgroups can help soften (or thicken up) transients like explosions, weapons, or impacts. These final finesses can make a game sound much more polished. A compressor and/or limiter can also be placed on the master bus to control the dynamic range of the full mix. Doing so can help increase the intelligibility of recorded dialogue in game and add more punch to transients. As always, dynamic processing should be used carefully to avoid squashing all of the dynamics out of the full mix. Use your ears to guide you, and try to always be listening from a player's perspective.

Mobile puzzle games with simpler mixes work well with dynamics processors on the master bus from the start of the project. This acts as a dynamics "safety net." More sonically complex games are better off having several mix buses that have compression/limiting suited to each group. Louder levels in a weapons and explosions subgroup would have a different dynamics setting than a footsteps subgroup, or a character sounds subgroup.

The practice of using dynamics on the master bus is a subjective choice and you will find that audio designers have varying opinions on the practice. Some will agree it's wise to decide on using master bus processing early on as all other mix decisions will be affected. Others will say it's better working out the mix through attenuations and levels. Once the mix sounds good, bus compression can be applied to glue it all together. In either case it is essential to compare and contrast your dynamic processing by **bypassing** (or A/B) as you mix. This will ensure that you are aware of all changes you make to the dynamics, and can accurately evaluate whether those changes sound better or worse.

As we have been saying throughout the book, there is no strict set of rules to follow. If a process works for you then by all means use it!

Visiting Artist: Jason Kanter, Audio Director, Sound Designer

Thoughts on Bus Compression

Imagine you have a 10,000-piece jigsaw puzzle and your plan is to preserve it with puzzle glue once it's assembled. With this goal in mind, would you pour the glue out on the table and assemble the pieces into it?

(Continued)

Of course not! It would be a sticky mess and as the glue spread on the pieces it would blur their detail, making it harder to fit them together. So initially it might seem like the pieces fit and the glue would certainly do its job of holding them together, but in the end you'd just have a jumbled blob that wouldn't make much sense.

Attempting to balance your mix through the dynamic control of a bus compressor is like assembling your puzzle in a bed of glue. Balancing a mix into a bus compressor may work fine if you only have a few sounds to contend with at any given moment, but if you have a big game world with hundreds of sounds being triggered concurrently, it can lead to a cacophonous mess. The more elements you add, the more dependent on the compressor you'll be to hold it all together. As the mix takes shape, it will seem to be well balanced but removing the compressor often reveals an unbalanced heap of pieces that don't truly fit together, making it nearly impossible to make any significant changes without breaking the entire mix down and starting over.

A bus compressor can be an incredibly useful tool to help control the dynamics of your game but only after the mix is balanced. Establishing a well-balanced mix and then applying a compressor will give your mix the control you need while still allowing you to make some adjustments up until the game has shipped.

High Dynamic Range (HDR)

High Dynamic Range (or HDR) is a dynamic mix system that prioritizes the loudest playing sound over all others in the mix. In other words it's a dynamic system that operates during the run-time system and turns off softer sounds when important louder sounds are triggered. This is another way to add focus and clarity to the mix as well as help reduce the overall voice count, which reduces strain on the CPU. Developers like DICE make use of HDR for many of their games.

HDR action can be complicated in FPS games. Instead of assigning importance based on volume, Blizzard's *Overwatch* designers used "threat level."[17] Enemy volume is based on their threat level to the player. Sounds that are more threatening have a higher priority than sounds that are less threatening.

AI and machine learning may eventually aid or replace these event-based mixes, which will allow the system to be "aware" of the audio output and make mix decisions based on that information.

Gina

Loudness

There are various development platforms (such as desktop computers, consoles, mobile, and handhelds) that host games from a variety of publishers and developers. Without certain standards for development and delivery, the user experience may be negatively affected. As technology advances and we continue to have access to various media types within a single platform, it's important to focus on a consistent experience.

There are standards for display area settings and file-size restrictions, but at the time of writing this book the game audio industry has not yet adopted loudness standards for video games. Publishers like Sony are working to define and encourage standardization, and it makes sense to look to the broadcast world for guidance as well.

Mix Considerations

When going through your final mix pass, consider your audience and how it might be listening to your game's audio. Everyone loves a stellar sound system, but not all development platforms can support one. Mobile games can be heard via the onboard speakers or headphones. These platforms also provide the user with the ability to move about in various locations, so you will want to test your audio in a few different real-world environments. *Pay close attention to how the external noise affects the listening experience as you test.* The user might be on a noisy train or bus, or at home and connecting via airplay to TV speakers. Test out the game with different consumer equipment. Be sure to check your dynamics. Are the quiet moments too quiet? Are the loud moments too loud? Adjust the mix accordingly. If you deliver your mix with the expectation that the player will have a superb surround system with the ability to handle the heavy low-frequency content, you are missing an opportunity to deliver an appropriate aural experience.

Most of your time will be spent listening in a treated room on studio reference monitors while you design, but you want to be sure to take time to listen on a **grot box** (industry slang for a small, low-quality studio monitor) as well to verify the integrity of your mix. Your grot box will simulate playback on consumer devices such as mobile phones, televisions, and radios. Speakers like the Avantone MixCube, and plugins such as Audre.io[18] are other great tools for preparing your mix for consumer products.

Listening volume when mixing is important. Everyone seems to like mixing loud, but consider listening to your mix at lower volumes so you aren't missing out on a typical consumer experience. Not everyone has the ability to blast the audio they are listening to. In addition, your own ears will get less fatigued and you won't suffer from hearing loss.

If you find your UI sounds are masked by the music and SFX in game you could try to widen their stereo field to give them their own space in the mix.

Gina

Reference in Mono

While you will spend most of your time mixing in stereo, you want to be sure to spend time referencing in mono as it can uncover various issues with your mix. It's especially helpful in revealing any phasing issues you might have. It also works well to determine balance for low-frequency elements in your mix.

Final Thoughts On Dynamic Mixing

With dynamic mixing systems in place, the audio will now change based on player input *without* losing balance or consistency. It's important at this point to test for all variations on gameplay. Running through a level without creating any mayhem will be a very different aural experience than running and gunning through a level and interacting with all the objects in the world. Players may want to get away from the action sometimes, and just stand on a beach or in the middle of a forest and take in the ambience. For this reason you should take the time to bring the ambience alive with even the smallest of details.

The techniques we've mentioned so far are extremely important for the audio quality in every game, but resource management and optimization are also critical. In the sections below we will discuss managing resources and achieving the best game performance possible.

RESOURCE MANAGEMENT AND PERFORMANCE OPTIMIZATION

Nonlinear audio uses resources such as CPU, GPU, memory, and bandwidth which need to be shared with other aspects of the game development pipeline. The percentage of these shared resources that are carved out for audio is dependent on the game and development platform. You should work closely with your programmers to determine the audio resource budget for each of your projects. This includes an estimate of shared resources that can be utilized by audio, the amount of audio assets each game requires, the level of interactivity in the audio systems, the required real-time effects needed, and the target platforms. If the audio budget is not looked after, the result could create serious lag in game.

With this information a proper plan for resource management can be mapped out prior to implementing audio. Although we are covering this here in Chapter 8, it doesn't mean this process should be held off until the last step before implementation. The pre-production stage we discussed in Chapter 2 is an ideal time to lay out the resource management plan. Start by determining the target platform and discussing the estimated audio memory and CPU budget.

To begin, we'll break down the process of how games make use of their resources. When a game is downloaded from an online asset store or physical disc it is stored on the system hard drive (or flash drive on mobile). When the game is started, assets are loaded into memory (or RAM). How much of the game's assets are loaded into RAM at any particular time is based on the overall footprint of the game. Larger games, like open-world AAA titles, will have too many assets to load all at once. Asset and bank management are crucial to offset this issue, and we will explore these in a bit.

To continue with our resource pipeline, RAM then works with the CPU, which processes the game logic and player input. At any given point, the CPU might be charged with handling a whole list of tasks such as collision detection, audio events, AI, animations, or scoring systems to name a few. The CPU also works with the GPU to process and render graphical assets, which are stored in **VRAM**. It's important to note that some games have made use of additional GPU resources as a co-process for audio DSP (digital signal processing) to help balance the CPU load.

Now let's take a look at how audio events in particular make use of these resources. When optimizing audio in a game project the focus should be on the processor and memory. All the information from the audio engine will need to be processed, and some of it will be stored either in memory or on the hard disc. Middleware programs back all of the audio engine information into multiple sound banks. Data such as audio sources, events, dynamic mix systems, and asset streaming can be found in these sound banks. Specific sound banks can be generated for each target platform, which allows for resource management across multiple systems. In short, sound banks are extraordinarily powerful tools for organization and resource management.

When a 3D audio event is attached to an object in the scene, pre-planned game logic will determine when and how the event is triggered during run-time. This sound object holds onto its slot in the virtual **voice map**, whether it is within audible range to the listener or not. The audio event will use up valuable CPU resources despite the fact that the player cannot even hear it! When the game is loaded, sound banks are loaded into the target platform's RAM so audio will be ready to play. It then falls onto us as audio designers to prioritize these events to minimize the CPU hit.

Hierarchy and Inheritance

Game engines and audio middleware operate under a parent/child hierarchical structure. A child object will have a parent object, which may have its own parent, and so on. The idea is to have a "container" with defined settings that offers organization and group control when applied to multiple audio objects. This structure is also utilized for visuals and logic, but here we will explore the hierarchical relationships as they pertain to audio only.

The parent/child structure in software like Wwise can go deep and become complex quickly. Child audio objects can inherit settings from the parent which may offset or override any child-specific setting. Because of this, it's important for audio designers to fully understand how the structure is configured for each project, and how to control it to produce a desirable result.

Sharing properties from a parent audio object across a large group of child objects isn't just for organizational purposes. It's a CPU and memory saver as well. Imagine that in your DAW you have five tracks, each with a different sound-effect layer. If you apply a reverb plugin as an insert on each of the five tracks with all the same settings, you are using five instances of a plugin, which will be taxing on the CPU. Alternatively, you can apply the plugin on the master bus or a group bus that all five tracks are routed to. This reduces the number of plugins to a single instance, thus drastically reducing the CPU load. Now, getting back to audio engines, each time we override a parent setting on a child it uses more memory and CPU at run-time, just as each instance of a reverb plugin does. By grouping child objects to a parent object with the same settings, we can apply the necessary settings while optimizing our use of resources.

Bank Management

We briefly mentioned banks in the introduction to this chapter. Here we will further discuss their use and how to manage them. Games with smaller amounts of audio assets and more available free memory would be fine using only one sound bank containing all events and assets. In this case, the programmer can load the single bank when the game is initialized. This will keep things simple as it avoids loading and unloading banks and tracking which sounds are available or not. The downside is the inefficient use of system memory as all events, media, and data structures are loaded at once.

Larger game productions with more assets and less available memory benefit from multiple bank solutions. A plan for distributing the assets across the banks is necessary to load and unload banks as the levels, environments, or game states change. Scripting will be required to initialize the sound engine and handle the load and unload process. With enough memory available, multiple banks can be loaded at one time to avoid delayed transitions between scenes.

Games can vary from simple audio event triggers to very complex triggers. In the latter case, banks can be micromanaged with events in one bank and their associated assets/media in another. This allows the banks with *events* to be loaded into memory while the banks with *media files* wait to be called. The media banks can then be loaded based on proximity to objects in the game. This is an efficient use of memory as the banks holding instructions are far less of a resource hog than the actual media files. This requires a lot more communication between the audio team and programmers to integrate, but it does offer a cost-effective solution for managing memory.

There are other ways to ensure media and information from banks are ready and available when needed. Events can be analyzed by the engine to check if its associated media is loaded. If it is not, selecting a streaming option allows the media to be pulled directly from the disc. This method will also allow you to decide which banks will be loaded based on specific game data. Streaming can be an essential function on platforms with minimal RAM, but with larger projects for console and desktop the choice between memory savings and potential problems should be weighed. With games that have a boat load of audio assets, streaming approaches can have problems surface at later stages of development once all or most of the assets and logic are in place. While streaming can be great for saving memory and CPU with longer duration sound assets, lack of bandwidth can introduce small delays during loading which isn't ideal for sound triggers that rely on precise timing. This can impact music and ambiences in particular as the cost of time to load a longer asset results in the game feeling "laggy" and unresponsive.

Due to the benefits, some audio engines default to streaming for imported files that are longer than a set number of seconds. Even though this is the default, it's best to work with the programmer to define the best solution for handling assets with longer durations as there is a variety of options. Chopping up longer files into shorter bits to play back in a playlist can help avoid having to stream from disc, but it isn't a good solution for all lengthy audio content. Seek times can also cause delays as storage capacity expansion seems to outpace transfer speed. Audio engines do however offer some setting options for storing part of the file in memory to compensate for the delay during load time. No matter what the approach, you will need to be active in testing and observing how the solution is handling starts and stops of audio, and everything in between.

> We should point out that while the majority of projects still utilize banks as a way to categorize events for loading, some audio designers have issues with banks in that they're either going to be big and bulky or lean and plentiful. Banks cost processing time to load and unload, so in order to minimize this banks can be loaded once and ready to go. The issue with this approach is that larger banks cost more as they are held in memory. There is always a struggle to find a balance between processing and memory.
>
> Gina

One exciting recent development in optimization is that Audiokinetic has offered a new solution in Wwise which allows the split of information across multiple banks. For example, if all the music in game has only one event, in the traditional method that event would be in one bank. This new Wwise functionality allows management of memory by splitting the banks so you can control what loads and when it loads. Essentially, you could prioritize bank contents to load less critical information into a separate bank, which will then load only when there is memory available.

Visiting Artist: Alexander Brandon, Composer

The Realities of Bank Management

Banks are a way to manage loading of assets, with the ideal scenario being that assets are only taking memory space when they are needed, and kept in storage when they are not. However, organizing assets is becoming increasingly difficult, as it is uncertain when sounds will be needed at any given time. Most often audio designers become frustrated by assigning many sounds to individual levels or scenes in a game only to discover they're needed in other locations. The sounds then are needed as global sounds, or permanently loaded. Since early game development, the concept of "on-demand" loading was introduced, and streaming audio became possible in the early to mid-1990s either from hard drives or discs.

Streaming does come at a cost, however, in that a minimum amount of memory must be allocated to "buffer" at least metadata (or information about the audio) to allow the game engine to call the sound at the right time. But this buffer is typically far less in memory than the full sound. The flip side to this is that it does add to the delay of playback.

These challenges are becoming less and less of a concern in modern development. In the game *Horizon: Zero Dawn* by Guerrilla Games, almost zero load time is experienced by the player, and sounds along with art and everything else are indeed loaded on demand and streamed ahead of player locations and encounters. This makes the necessity of banks less critical; rather memory management is kept to real-time scenarios where an audio implementer is required to ensure that not too many sounds are loaded in situations with the most music, sound, and voice happening simultaneously.

In the end, what matters most is being able to iterate your audio until it works ideally for each situation. Bank management is one of the biggest roadblocks to iteration as sounds and events need to be created and tested, but loading in game is not something that Wwise or FMOD can simulate in their toolsets. As such, a significant amount of time is spent making sure banks and events are loaded, and at the right time during a game. It will not be long before banks are entirely unnecessary, and all that will be needed is simply organizing audio according to design-related needs rather than technical considerations.

Event Management

During development the game is a work in progress, so it can be common to run into performance issues. Controlling the number of sounds playing simultaneously is part of resource management, and is critical to minimizing the load on CPU. When you have several game objects within a small radius emitting sound, event management will be necessary to stay within the allotted audio resource budget. An audio resource budget is concerned not only with how much audio is in the game as measured by storage space, but also how much bandwidth the audio engine is consuming as it plays back multiple sounds concurrently. Added to this is the audio that is sitting resident in memory at any given moment in time. Events and audio sources can have **priorities** set to control what the player hears based on the context of the game. It will also ensure that players do not hear sounds out of range or sounds that are inaudible.

Limiting playback (or managing polyphony) also helps reduce clutter in the mix and maintain clarity. Let us imagine we are playing a third-person open-world game. There are six gas tanks next to a building and each has an explosion event attached. If we fire at one of the tanks the explosions will chain, and six explosions will simultaneously trigger. This not only gobbles up virtual voices, but it may also cause the overall volume to clip, and possibly introduce phasing issues. At the very least it will sound muddy or distorted. By limiting the number of explosion instances per object, the volume will remain moderate, reducing the load on both CPU and virtual voices. It will also safeguard against one category of sound inadvertently overwhelming the mix. Polyphony management is accomplished in a variety of ways depending on the implementation method, but the simplest way is to set **priorities** using middleware (see the "Probability and Priorities" section below).

Probability and Priorities

Games with a larger number of audio assets can benefit from the ability to control the probability and priority of sounds in the soundscape. This is another way to optimize our games audio and prioritize our resources. Random sounds triggered in game help add variety, but it could become tiresome if the sounds were to constantly trigger. Control over the probability (or weight) comes in handy to define how often a sound will trigger. By adding weight to sounds so that some trigger more than others will allow for a varied soundscape without overuse. For example, in a forest we might have an event triggering bird chirps. If the same bird sound triggers constantly the sound would stand out to the player. By reducing the probability that the bird chirp plays we are left with a more realistic experience.

When you have limits on the number of channels that can play audio at any given time, a control system is necessary to limit the number of sounds triggering to avoid bottleneck. When the channel limit is reached, a process called **stealing** can be used to allow new voices to play while stopping older ones. Without defining this system, sounds might feel as if they randomly drop in and out, as if it were a bug in the audio system. Priority systems allows the designer to set a high, medium, or low property to an event. Footsteps triggering in a very busy battle arena will be given lower priority than sounds like weapons and explosions, thus dropping the sounds without the player noticing.

Distance from the listener can also be used to change the priority of an event. Events that trigger further from the listener will automatically be lower in priority, which allows closer events to be heard. "Height" as a game parameter can be used in games with multiple vertical levels to prioritize sounds. It would sound weird if an NPC was positioned directly one level up from the player, but their sound played back as if they were in the same space. Using height to prioritize audio makes sure that only the most relevant audio is heard by the player.

Best Practices for Real-Time Effects

There are a variety of options for real-time effects processing in audio engines. Being aware of the effects of real-time processing on the CPU when implementing can save resources and provide a better mix.

There is no way around the CPU cost of effects processing but some techniques will offer more efficiency. Applying effects on the mix bus as opposed to individual events or objects can save resources. This is equivalent to grouping a reverb effect using an auxiliary track. Determining ahead of time if effects will be more beneficial when used in real time or baked into the sound will also help manage resources. Make use of profiling tools to determine the relative cost of the effect in question. Generally speaking, if the effect is aesthetic, baking it in is a good idea. If the effect needs to be dynamic and adapt to gameplay, it probably needs to be real time. It is often worthwhile to make real-time events reusable. It might take time to set up a particular system, but being able to reuse this logic across other audio events in game will be worth the time and resources.

> As a side note, applying time-based effects processing directly on music events or objects may interfere with beat-synced transitions or other time-based properties. It's best to apply these effects at the mixer level to avoid this interference.
>
> Gina

Remember that **codecs** (encoding format), number of channels (stereo vs. mono), events, banks, real-time effects processing, and basically all nonlinear audio demand resources from the CPU. These resource limitations are a challenge, but by setting priorities and limiting playback audio designers can save CPU resources and produce a clear and dynamic mix. Keep in mind that these are not steps that need to be taken in the process of sound design for linear media. These methods are *only* employed when dealing with nonlinear and dynamic sound.

Platforms and Delivery

Understanding target delivery platforms will ensure a smooth delivery process. Most platforms have their own dev kits and requirements for audio (file format, size, etc.). The dev kits are provided to developers after signing an NDA (non-disclosure agreement). These agreements require the teams to keep certain information regarding the kits confidential, so it can be difficult to easily find specs for each platform online. Regardless, platform specifications are a major factor in defining your optimization plan. Consoles and desktop gaming platforms may have up to 80 channels available while handhelds and mobile may only have between 40 and 60 available channels. These figures are estimates, but can be a good starting point for defining the audio budget.

The list below covers major delivery platforms, in no particular order.

• Nintendo Switch, Wii, 3Ds
• Microsoft Xbox One
• Sony Playstation 4/PSVR/PSP/Vita
• Mobile (Apple and Android)
• Apple iWatch
• Web/HTML5
• PC
• VR Head Mount Displays
• Coin-operated games (includes arcade and casino)

While PC users can upgrade their CPU and graphics card, there is only so much room to improve due to motherboard limitations. Console, handheld, and mobile users however don't have the luxury of updating components. It's best to plan for the lowest common denominator either way.

Game audio engines have built-in audio compression schemes that allow you to choose between different formats for each platform. Researching the platform will reveal a bit of information about the playback system the player will be listening to during gameplay. To be safe, it's best to prepare assets to sound good on a variety of listening devices. Since there are many ways to manage the overall audio footprint, it's important to understand all your options. Developers will appreciate you as a sound designer if you are fluent in the variety of file delivery options.

As we discuss the various format options for the different platforms, keep in mind you will want to record, source, and design your sounds in the highest quality format. Working with 48 kHz/24-bit sessions is standard, but some designers choose to work with 96 kHz/24-bit. Once you bounce the file you can convert it to any file type and compression in the audio engine or prior to implementing the sounds.

Mobile and Handheld

A game developed for mobile or handheld means the player will most likely be listening through a small mono speaker built into the device or a pair of stereo earbuds. Neither will handle extreme ends of the frequency spectrum very well, so use EQ to manage your mix.

Since mobile and handheld devices have memory constraints, you will also need to consider file format and file size. All assets have to share the limited memory available, and audio will have just a small percentage of it allocated. Some mobile projects have an audio footprint limit between 2–5 MB which means you'll have to pay close attention to file format and compression settings. This will require careful planning regarding the number of assets.

The most common format options for mobile and handheld are Ogg files or raw PCM. While mp3 files are supported on most devices, they are not ideal for looping assets, as we discussed in Chapter 2. You may have played some mobile games and experienced this gap in the music. Some formats are platform specific so it's a good idea to research the platform you are developing for during the pre-production stage. Since creating immersive audio for games is 50 percent implementation and 50 percent design, it's important to ensure the assets play back seamlessly in game.

Compressing files reduces the dynamics in the audio and some assets will sound better at lower compression settings than others. Limiting the number of assets you deliver on mobile and handheld devices can help you avoid over-compression while staying within the allotted audio file size. If possible, reuse the same assets in multiple events to maintain efficiency.

Mobile games likely won't need 20 different footstep variations per terrain for each character for example. You can plan to have between six and eight variations of footsteps per terrain for the player character, and non-player characters (NPCs) can share footsteps based on group type. Similarly, delivering a few smaller ambience loops instead of three- to four-minute files can help add variety while keeping the file size in check.

Considering stereo vs. mono files is another way to reduce file size. If you plan on implementing 3D events into your game you will need to deliver the files in mono regardless. You may have to consider converting 2D events into mono as well to save file space. When designing assets that will eventually be converted to mono, be mindful of overpanning, which will not translate well in the mono file.

During the implementation process you will have to consider which files can be **streamed** vs. **loaded into memory**. Typically, looped files and longer assets are set to stream. Don't stream too many files at once or the quality of playback may suffer.

Console and PC

For game builds deployed on console or PC, the player will typically be listening to playback over TV speakers. However, over-the-ear headphones, surround sound systems, and sound bars are all common as well. The frequency range of these playback systems is wider than for small mobile speakers. This allows for a much fuller sound with rich dynamics and spatialization.

While console and PC have far fewer limitations than mobile and handheld platforms, you still need to consider overall file size and CPU usage. Some platforms use a proprietary audio decoder, which will save processing power. More information on these file types can be found in the dev kit documentation or as part of the engine or middleware audio scheme. For example, viewing the build preferences in FMOD Studio will reveal that Playstation 4's proprietary decoder is AT9 while Microsoft Xbox One's is XMA. Console and PC games can range from 20 to 50 GB, and even though audio will only be allotted a portion of this space you'll have much more to work with than the mobile limitations.

On console and PC the main audio concerns are the number of available virtual channels (managed by setting priorities for sounds) and CPU usage (managed by limiting real-time effects and choosing audio formats judiciously). Earlier in this chapter we covered virtual channel allocation and CPU usage. To review, audio engines and middleware allow you to limit the number of sounds playing at one time. You can also choose when to use real-time effects versus baking them into the sound before implementation.

Web and HTML5

While you may think of the abovementioned platforms as having far more game titles than the web, consider that Facebook has a large games platform with thousands of games. There are also many social casinos popping up online. Plenty of companies host web games for advertising their products or promoting their brand. There are also a number of stand-alone sites that host games, such as Pogo, MiniClip, Kongregate, Addicting Games, and Big Fish Games. These sites host hundreds of games and add new content regularly. To disregard web and HTML5 games is to disregard a large chunk of the market.

Bandwidth and file size are the biggest limitations in web-based games. You need to consider the game's overall file size. Web-based games often have a limit of 5 GB for all assets, which means audio might be limited to 200–500 MB. A bit of audio trickery is required to achieve this slim file size, namely, music and ambience may have to be baked into one asset to save file space. You may also have to consider using mono files for streaming assets. When you have to use heavy compression it's best to consider the dynamics in your assets since heavy compression may squash the sound.

Considering the development platform as you design the assets can help you stay in line with the requirements imposed by each device or system. In the Sound Lab you'll find a practical exercise to get you more familiar with file types and compression. Visit the site now, or come back to it after further reading.

Optimizing for Platforms

The native audio engines will make it manageable to publish assets per platform. Audio middleware goes a step further by allowing designers to work in a single project, and generate separate sound banks for each platform. Selecting the right codec for each platform can be helpful in managing CPU and memory usage. Audio designers take great pride in their work, and often cringe when faced with compressing assets. But high-quality WAV files take up a lot of storage space and can be a hit on CPU during playback. It's a good idea to come to terms with compressing audio as it will offer you more control over the final product.

A good way to get started understanding different codecs is by getting familiar with the conversion settings in your DAW. Grab a freeware version of Audacity or another audio converter, and test out how each compression type and amount affects a single sound. Try this with a few categories of sounds like underscore music, explosions, and UI. It is easier to hide artifacts that can be introduced during compression in some files than it is in others. It all depends on the type of sound, and its frequency content.

Some platforms (like Playstation and Microsoft consoles) have hardware decoders, which have some advantages as well as disadvantages over software codecs like Ogg Vorbis and mp3 (keep in mind the gap that the mp3 encoding process adds to files – refer to "Looping," Chapter 2).

Most codecs (whether hardware or software) have a processing cost, so it's best to read through the information that comes with the dev kits to get more familiar with each system. Hardware codecs may have limits on asset duration and loop points, but may work better natively than a software codec.

Having an understanding of the cost differences between codecs can help define which conversion settings to use in the audio engine. For example, Ogg files require more CPU but less memory than PCM files. Additionally, mobile platforms like Android and Apple have specific preferences for audio codecs. Some developers may ask for Ogg files for Android builds and AAC for iOS. As an audio designer you should be knowledgeable with the process of compressing sounds and file types, just as a graphic artist should have a strong understanding of file formats for various delivery spec.

TESTING, DEBUGGING, AND QA

Testing the game is an important part of the implementation phase. Although you'll be testing as you go (as we describe earlier in the chapter, see the "Implementation Cycle"), **playtesting** often occurs toward the end of the development process. Playtesting is a fancy word for playing the game and looking for glitches, errors, and bugs. As a sound designer you are the expert in how the game should sound. Therefore you are the best person to evaluate whether or not the sound is triggering properly and the mix is clear and effective. However, this is typically a collaborative process and the game's budget will determine the amount of testing that can be done.

Playtesting

Testing is a process that occurs during all phases of development. For instance, prior to importing assets into the audio engine, some audio designers mock up the audio against video capture of gameplay in their DAW. To take things a step further, middleware offers a space to mock up and audition sound systems for testing, prototyping, and mixing.

In Wwise this feature is called Soundcaster, which allows triggering of events and all of the effects and controls that can affect or evolve them. In FMOD, the Sandbox offers similar control. Both will allow you to apply parameter and game sync control over the sounds to simulate in-game events without having to go into the game engine itself. While these simulation environments allow you to test event triggers, they are also really useful for adjusting distance attenuation and other settings that might need some dialing in. In the Sound Lab (companion site) we have created a video tutorial for using the FMOD Sandbox and the Wwise Soundcaster. Be sure to check it out before moving on to later chapters.

Playtesting is equally important for composers as it is for sound designers. You might have composed the greatest adaptive system in history, but the only way to test if it works for *your* game is to play it with the music implemented. This will give you valuable insights on whether or not the music hits the mood properly, what transitions are awkward, and whether or not you should even have music in every area. Believe it or not, it is very common to realize you have simply implemented too much music. Wall-to-wall music is not always the best choice, and in many cases you will find yourself either recomposing to make certain cues more ambient, less complex/dense, or taking them out entirely.

Regardless, playtesting is what separates mediocre game scores from great ones.

Debugging

Profiling is both a game engine and middleware feature for examining activity such as memory, streaming, effects, voices, and game syncs for **debugging** and resource management. Both Wwise and FMOD have profiler features where you can profile locally or while connected remotely to the game engine. In this section we will focus on profiling as it pertains to middleware specifically. It's important to note that while we have chosen to discuss profiling in the "Testing, Debugging, and QA" section of this book, it can certainly be used earlier on in the implementation process.

Performance profiling allows the user to check for issues in their resource and asset management plan and make quick fixes by accessing the events that are overloading the system. Object profiling focuses on individual audio events triggered by player and non-player characters, ambient objects, weapons, and more. The profiler will generate a preview to monitor which sounds are being triggered in game. Sound-bank memory size, CPU, memory, and bandwidth can also be monitored. A spike in any of those resources can be further investigated in the captured profile session.

Earlier we discussed limiting event sounds based on distance attenuation. If the audio designer sets a voice kill at a max distance variable and you can't hear the sound in game, it doesn't mean it isn't still triggering. With a busy soundscape it is easy for sounds to get masked by other sounds. The profiler lets you look at voices triggered and their volume levels. Additionally, profiling sessions can be saved and recalled at a later time.

Quality Control

The QA, or quality assurance, process is often perceived as "just playing games," but there is much more to the process than that. A QA specialist needs to define how various players might play the game. Many games can be "broken" if the player's style falls outside the defined norm. In the context of general game development, QA testers work to push the game to its limits to unveil issues in all areas of the game. Here we will discuss QA as it pertains to audio. There are many organizational aspects related to the job of QA, and an audio designer will be well served to connect with the QA department to build a relationship with it as early as possible. Bug reporting, testing schedules, and protocols are the responsibilities of the QA team. Properly communicating how the team should listen and test audio can be a huge benefit, and an important step toward knowing that your audio is working as expected.

The size of the team and project budget will determine whether or not there is a dedicated QA team. Larger teams will have someone in QA specifically testing audio issues. This person might be someone who is an audio enthusiast, or even someone looking to make the jump to the audio team. Regardless of whether there is a dedicated audio QA tester, the audio designer should be testing, simulating, and profiling as well to ensure audio is behaving as it should. It's important to test both in the game engine editor and in local builds. Sound issues may not be present in the editor but will present in the build, so check both. Testing on dev kits for the target platform, or various quality computers and mobile devices will offer a view into how the audio might perform across various devices.

In the case of the audio designer tasked with only delivering assets to be implemented by the programmer, testing will be an extremely important part of the process. Perhaps the developer compressed the files too far or a sound isn't triggering where it was intended. To help resolve issues with incorrect in-game triggering, video mockups from a DAW can be used to demonstrate how sounds should trigger in game. Either way, you will have to insist on leaving room in the development timeline to test and make adjustments.

Contract or freelance audio designers should have a variety of testing devices when working remotely. For mobile games, it's good to have at the very least an Android device for testing but it can help to have an Apple device as well. Generally speaking, when the game is being developed for both mobile platforms **build pushes** are usually faster to Android. If you only have an Apple device you may have to wait for the approval process before the build is available. Working with consoles and purchasing dev kits can be expensive so it's not usual for a remote designer to be given one by the developer. The kits are only available to registered developers so it can be difficult to obtain one on your own. However, VR, AR, and MR **head mount displays (HMD)** can be purchased by an individual for testing purposes. If you plan on working on these mixed-reality projects it's a good idea to have a HMD.

Understanding how to test is an important part of the process. Go out of your way to "break" things in game. Thinking outside the box can help with this. Try going *against* the obvious flow of gameplay as much as possible. Since you are familiar with the audio system you should be prepared to test the game with certain expectations of audio behavior. To be thorough you will have to play through in unexpected ways as well. For example, transitions should be tested for smooth movement between assets, regardless of play style. Material collision sounds might sound good when the player kicks a box once, but if the player decides to continue to kick the box in short sequence it may trigger too many sounds. This kind of play style would then suggest that you may have to set a priority limit on the event. Again, the overall goal is to try to apply various play styles to find rough spots in the audio. Avoiding being cautious and try your hardest to trigger awkward combinations of sounds. It's better to find the issues during development than for a player to find them in game.

The Sound Lab

 Sound and music creation is only half the battle in game audio. Implementation takes technical knowledge, problem-solving abilities, and out-of-the-box ideas to make an immersive sonic experience for the player. Now that you have a better understanding of implementing assets and resource management, stop here and head over to the Sound Lab where we wrap up what you have learned in this chapter.

NOTES

1   Wikipedia. "List of Game Engines."
2   R. Usher, "How Does In-Game Audio Affect Players?"
3   https://guides.github.com/activities/hello-world/
4   R. Dudler, "Git – The Simple Guide."
5   Wikipedia. "List of Game Engines."
6   www.gamesoundcon.com/survey
7   Of course, limitations are less of a concern with next-gen consoles, but a majority of games are being developed for mobile or handheld systems which do have restrictions.
8   M. Henein, "Answering the Call of Duty."
9   https://acoustics.byu.edu/research/real-time-convolution-auralization
10 J. Peddie, "What's the Difference between Ray Tracing, Ray Casting, and Ray Charles?"
11 M. Lanham, *Game Audio Development with Unity 5.X.*
12 J. Doran, *Unreal Engine 4 Scripting with* C++.
13 A.-S. Mongeau, "An Introduction to Game Audio Scripting in Unity."
14 www.udemy.com/unitycourse/
15 R. Steiglitz, *A Digital Signal Processing Primer.*
16 R. Boulanger, R. and V. Lazzarini, *The Audio Programming Book.*
17 www.reddit.com/r/Overwatch/comments/50swfk/ama_request_the_overwatch_sound_design_team/
18 https://audre.io/

BIBLIOGRAPHY

Boulanger, R. and Lazzarini, V. (eds.) (2010). *The Audio Programming Book*, Har/DVD edn. Cambridge, MA: MIT Press

Doran, J. (2019). *Unreal Engine 4 Scripting with* C++. Birmingham, UK: Packt Publishing.

Duder, R. (n.d.). "Git – The Simple Guide." Retrieved from http://rogerdudler.github.io/git-guide/

Henein, M. (2008). "Answering the Call of Duty." Retrieved from ww.mixonline.com/sfp/answering-call-duty-369344

Lanham, M. (2017). *Game Audio Development with Unity 5.X.* Birmingham, UK: Packt Publishing.

Mongeau, A.-S. (n.d.). "An Introduction to Game Audio Scripting in Unity." Retrieved from www.asoundeffect.com/game-audio-scripting/

Peddie, J. (2016). "What's the Difference between Ray Tracing, Ray Casting, and Ray Charles?" Retrieved from www.electronicdesign.com/displays/what-s-difference-between-ray-tracing-ray-casting-and-ray-charles

Raybould, D. (2016). *Game Audio Implementation: A Practical Guide Using the Unreal Engine.* Burlington, MA: Focal Press.

Steiglitz, R. (1996). *A Digital Signal Processing Primer: With Applications to Digital Audio and Computer Music.* Menlo Park, CA: Addison-Wesley.

Usher, R. (2012). "How Does In-Game Audio Affect Players?" Retrieved from www.gamasutra.com/view/feature/168731/how_does_ingame_audio_affect_.php

Wikipedia. "List of Game Engines." Retrieved from https://en.wikipedia.org/wiki/List_of_game_engines

Music
Implementation

In this chapter we will build on what we learned of audio implementation in Chapter 8 and discuss some advanced methods of music implementation. These methods will be ingrained with music composition, at times overlapping quite a bit with orchestration and arrangement as well. We will then go into middleware integration and cover some more experimental music implementation methods. We will end with some helpful tips on experimenting with game music and implementation.

APPROACHING IMPLEMENTATION

There are many approaches to musical implementation, but composers tend to divide themselves into one of two camps. Some composers think of an implementation style before ever writing a single note. Others work diligently to perfect the musical style before even considering how it will work in game. There is nothing wrong with either of these approaches, but there are positives and negatives to each.

For composers that begin with a clear idea of implementation *before* writing the music, the ability to hone in on the perfect emotion and style to fit the game is severely limited. It's roughly akin to making a frame before you know anything about the picture that will go inside. Starting with an implementation idea works if you are very confident in the intended musical style already, and are looking to make integration either very specific or very smooth. It also might be necessary if the game requires music that adapts to the mechanics in a very particular way. This may need prototyping of the implementation system *first*, and a focus on stylistic considerations afterward.

This method *does not* support a very creative or experimental compositional process because the implementation method dictates the type of music you will be creating, almost like painting by number. In our experience, this approach usually leads to smoothly integrated adaptive systems, but the music itself can sometimes be generic and uninteresting. All that being said, some composers lead with implementation because it inspires them! There are certainly games where the technical aspects are equally as important as the music itself. If using implementation as a starting point is a source of this inspiration, and it suits the needs of the game and your workflow with the developer, then by all means use it to your advantage. Just be sure to leave adequate time to fine-tune the composition of the music itself as it is of equal importance and is much more "visible" to the player.

The other way to approach implementation is, interestingly enough, to completely ignore it. Guy Whitmore has coined this as the "play it, don't say it"[1] method, and it works very well for digging into the music composition itself before worrying about the technical aspects. This is also a very compelling way to show your developer what you'd like to ideally do with the game from a musical standpoint. Although it doesn't address issues of implementation right away, it often leads to highly creative, envelope-pushing implementation techniques later on in the process.

To approach this method you would first need a gameplay capture. The next step would be to score the capture as you would a linear film. Do not hold back due to implementation concerns. Force yourself to write music that fits as tightly as possible, as if it were a film. Then show it to your client and assess how well it works. When everyone is satisfied with the music, *then* implementation can be addressed. It won't be easy because you will likely have written some very idiosyncratic transitions. Sometimes adjustments need to be made, but by thoroughly thinking through the transitions in the music you will be able to cut viable chunks and lay them into a workable adaptive system.

The draw here is twofold: for one it is *very* difficult to think about creating the most effective music possible while focused on implementation challenges. This method allows you to focus on one aspect at a time. The other positive is that it will give you more time to brainstorm and iterate different musical ideas in a short timespan. It takes virtually no time at all to sketch out five to ten ideas onto a gameplay capture and select the best direction. By contrast, if implementation were considered during that process the time would be tripled or quadrupled. It just takes too long to prototype adaptive systems and the composition itself simultaneously.

ARRANGING AND ORCHESTRATION AS IMPLEMENTATION

Many of the orchestration and arranging techniques in the previous chapter are applicable to all kinds of music. However, being an interactive medium, arranging comes into play in a very important way when writing for games. As a composer, the way you stack your chords and layer your parts (regardless of instrumentation) will have a significant impact on the effectiveness of a game scene. It's important at this stage of the process to start to think about the level of interactivity that is appropriate for your game, and by what means you will extract the desired mood from your music.

We've approached this topic a few times already, but what we are essentially concerned with now is the **modularity** of our music. Modularity can mean many different things in this context, but the basic idea is that our music has to be able to *change* as well as *remain static* so that it is adaptable to the game state. There are a number of ways to visualize this process. The end result is that our musical arrangements will always be dissected into smaller, moveable elements (sometimes we will call them modules).

There are two main categories that most types of adaptive scoring fall under (regardless of how implementation is approached), both of which we have touched on in Chapter 6. Together these categories constitute a broad framework of interdependent composition and implementation techniques. **Vertical layering** (sometimes called additive layering, mapped layers, vertical remixing, layering, or vertical scoring) and **horizontal resequencing** (sometimes called interior pathing or horizontal scoring) are commonly used and immensely powerful tools for implementing nonlinear music. These are loose categories used to *conceptualize* nonlinear musical systems, and each contain numerous specific techniques within them. There are many ways to utilize these vertical and horizontal methods both separately and in combination. Below we will give an overview of each category and then explore various techniques in depth.

There are many ways of "visualizing" modularity in game music. The vertical/horizontal framework is a two-dimensional approach which we use to simplify things. Think of vertical layering and horizontal resequencing as two sides of the same coin. The terminology is meant to elicit a "bottom-up" thought process, meaning that writing nonlinear music starts at the note-by-note level of composition, and can be expanded from there. This is in contrast to a "top-down" approach wherein the specifics of implementation are decided before any notes are written. There are many other approaches to game music however; some are three-dimensional (we'll explore this later on using nested events), others are what's called "nodal," and still others are actually generative in themselves. We will touch on these to a degree, but we recommend playing games and researching some of the processes that went into constructing these adaptive systems. Then you can find your own approach or combine a number of approaches that suit you best.

Spencer

ADVANCED VERTICAL TECHNIQUES

As you recall from Chapter 6, vertical layering is a method of implementation where *layers* of music are stacked on top of each other and activated by the game state. We can use these layers to control the mood and intensity of a cue. These layers can be comprised of any instruments, and can serve any number of emotional changes to a game scene. While exploring vertical layering systems, keep in mind that vertical layers *always* act as a means of adding *density* to a score. For this reason vertical scoring is commonly used to take a game scene from neutral to intense and back.

The strength in this technique is that it is actually quite easy to achieve. As composers we naturally think of adding layers of instruments to change the mood of a cue, so vertical scoring is usually intuitive to us. Another huge strength with this approach is that the layers can easily be tied to any game data, so the transitions sound smooth in game. For example, the layers could be set up to amplify intensity in a horror game. We might create a parameter or game sync called "distance to monster" that tracks the actual distance between our PC and an enemy. In this case the

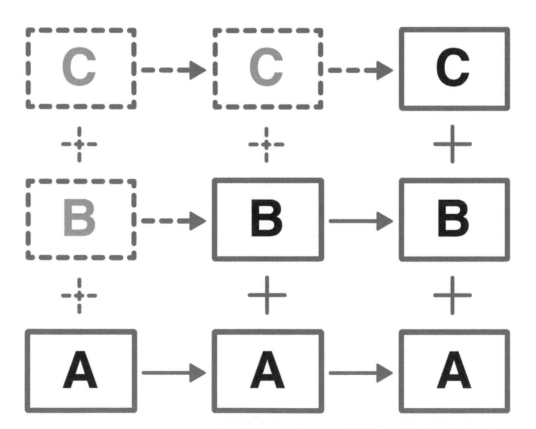

FIGURE 9.1 Adaptive music making use of vertical layering. Each block represents a loopable chunk of music. The clear blocks are triggered simultaneously with the opaque blocks, but are not yet active. Most if not all vertical techniques will have a similar structure.

closer the player is to the monster hiding in the shadows, the more layers are added to the cue. This type of system can be used for any number of scenarios, so it is a very flexible way to implement game music.

The limitation of this technique is that it also puts us in a box where the only way to change the mood of a cue is to *add or subtract* instruments. This can actually result in some very static and even non-musical cues. Consider that there are many ways to change the mood of a piece including modulation, chord changes, melody changes, and so on. In general, you are stuck with whatever chord changes, melodies, and key changes are present in your base layer, so use vertical scoring with caution.

Let's take a look at a few ways to use vertical scoring effectively.

Vertical Layering by Instrument Group

The natural starting point for vertical scoring is to split your layers up based on instrument section. For example, after we have written a looping cue, we could split the cue into four layers for implementation: 1) strings, 2) winds, 3) brass, 4) percussion. These layers could then be brought into middleware and triggered based on "intensity."

An important point to remember is that with vertical layering the ordering of layers should generally be from *least* to *most* dense. There is some ambiguity here depending on the parts you write, but if you start with a very dense layer of percussion and then add a sparse layer of woodwinds it is unlikely to make a significant impact on the mood of the cue. A far greater impact is made if you start with winds and *then* add the percussion on top, because the percussion will make a noticeable impact. You will be limiting the flexibility of your system if you choose to bring in your densest layers too early. This goes for just about every type of vertical layering system.

Vertical Layering by Timbral Grouping

A very close cousin to layering through instrument sections is layering through the timbral groupings we studied at the end of Chapter 8. Here, instead of splitting based on strings, winds, brass, and percussion, we would split layers based on wind group I, wind group, II, brass I, brass II, etc. This can be slightly more effective if the texture is chordal because it will help with achieving clear, balanced voicings.

Vertical Layering by Mood and Intent

This method is a combination of the first two with the added stipulation of *intent*. This means that you will split up layers not just based on balance in the orchestra, but on the intended *emotional* outcome for your player. Here we might make use of one layer that contains low strings and low woodwinds together, achieving a dark and ominous effect. We might make use of high strings and woodwinds as another layer, achieving a soaring and bright effect. The low strings alone are not as dark without the low woodwinds, and the high strings alone are not as bright as they are combined with high woodwinds. For this reason, these pairings are not split up by instruments or timbres, but are nonetheless extremely successful at evoking emotion.

Vertical Layering by Interlocking Motifs

Vertical scoring by interlocking motifs is a common and highly effective implementation technique. It can be heard in games ranging from the *Zelda* franchise all the way to *Nier Automata*, so it is important to have it in your toolkit. The idea behind this technique is not so much about the timbral grouping of each vertical layer; instead the emphasis is on *how the layers interlock*. With interlocking vertical layers, every group of instruments is meant to "lock into place," as if they were each a piece of the puzzle. Each layer then has a particular space in the overall mix, and every instrument should be audible and clear even when all the layers are triggered. In fact this is an important criteria for *all* vertical layering systems, but it is taken to the next level with the interlocking method.

In an orchestral setting vertical layers have a tendency to blend together, which is often the intention. As a consequence this method of vertical layering works extremely well with non-traditional instruments as well as orchestral instruments. Synthesizers, samples, processed effects, and orchestral instruments are all fair game with this method so don't hold back if you're thinking about using it. Hybrid templates sound fantastic when interlocking in a game soundtrack. It can be compared somewhat to a fugue, or to a heavily polyphonic piece of concert music. Every melody, countermelody, and rhythm contrasts so prominently that their entrances are noticeable. But when all layers are triggered each motif becomes a dot in a pointillist painting.

Adaptive Mixing in Vertical Layers

It is possible to adjust your music mix in real time using vertical layers. This is helpful for settings in which the composition is fixed, but the mood needs to adapt. In other words, you can use this method to change the timbre or instrumentation of your layers, but *not* the basic elements of the music itself.

One example of this is splitting a cue up into melody, chords, percussion, and background instruments. In a way this can be thought of as vertical layering by *function*. Here we might have two or three alternate versions of each layer, which differ in instrumentation. For the melody layer we may have a flute layer, a cello layer, and a pizzicato string layer. These alternates contain the exact same melody, but by switching the instruments we can change the mood or tone of the music as a whole. The flute might be bright and cheery, while the cello would be darker and more somber.

Pizzicato strings as a melody could be considered quirky or ironic. Based on the actions of the player, your vertical system will fade in and out of the corresponding layers.

Vertical Layering Limitations

An important consideration to make here concerns the emotional flexibility of all vertical scoring methods. Vertical scoring is powerfully adaptive and easy to control, but it has some limitations regarding mood. Because every layer has to work together in various combinations, there is a limit to the textural contrast of each layer. It is not likely that a heavily polyphonic layer with a wild harmonic progression will fit smoothly with three other layers that stay in the same key and operate homophonically. This is where we begin to see a need for horizontal scoring, which absolutely can accommodate this contrast.

What vertical scoring *can* do well is provide almost infinite possibilities for adaptive layering and mixing within a basic range of emotion. This range can stretch with a bit of creativity, but there is always a limit. It is important to understand vertical scoring as a means of adjusting *intensity* and *density* in an adaptive way. If vertical layers are meant to provide a spectrum of emotion from neutral to dark, then the *intensity* of that darkness is what we are changing by adding or subtracting layers. Equally important to note is that regardless of mood, adding vertical layers *always increases the density of the music*. Subtracting vertical layers *always decreases density*. From a compositional perspective this is incredibly important in games because there are many circumstances where increased density is not necessary or even beneficial.

One helpful example of this is with infinite runners where an increase in distance also increases the speed. It is tempting to pick a base layer and then add more layers on top as the player reaches distance milestones. However this isn't always the best approach. For many infinite runners the key emotion is *urgency*. If you start a vertical system with a layer at a lower tempo you may never reach the level of urgency that the game calls for. You will however be increasing the *density*, but as we have learned that is a completely different quality from *urgency* and the *intensity* of that urgency. In short, know what vertical scoring is capable of and what it's not, and then use it to your advantage where appropriate. In the case of our infinite runner, a horizontal system that transitions tempo might be a much better fit.

In the endless runner *ZombiED: Run Zombie Run* by zinWorks Studios we utilized a vertical system which introduces new music layers as the player passes predefined distance markers. For additional adaptability, we ducked the main in-game music layers to introduce a new musical theme when the player picks up the flying power-up. In additional to the new musical theme, which makes the player feel like they are soaring above the scene, we trigger an ambient wind loop to accompany the new music. Once the flying power-up runs out the music transitions back into the main in-game music under a stinger.

Gina

Visiting Artist: John Robert Matz, Composer

Orchestration, Instrumentation, and Interactivity

Frequently, when designing reactive scores, a composer will use orchestration elements to more effectively tie the music to the player's actions on screen. This can manifest in a multitude of ways.

A simple example might be to use unpitched percussion elements, tonally ambiguous synth sweeps, string aleatory, etc. to facilitate a change in music, a pivot point to introduce a new cue, a new feel, etc. In the acoustic/orchestral realm of my score for *Fossil Echo*, tam-tam and cymbal scrapes, chimes, shakers, and deep taiko and bass drum builds and hits were especially effective at providing simple, effective transitions from one cue to the next. The amount of the spectrum they used had an effect of "wiping away" the old key and feel, and setting the stage for the new one, without breaking the musical immersion by fading out awkwardly.

Another possible example involves layering. Say you need to compose an adaptive piece of music for a linear chase sequence, with a separate "danger" layer to be introduced should the player start to fall behind. You might tailor the orchestration for the primary layer of the piece to leave a space for a layer of driving drums, percussion, and high, dissonant brass that would be controlled by real-time parameter control (RTPC; see Chapter 8).

The biggest thing to remember is that you only have so much room in the audio spectrum at any given time, and, as such, you need to make sure that your orchestration choices mesh with each other well in any possible scenario. You don't want to step on your own musical toes.

Vertical Applications in Middleware

Most vertical techniques are accomplished in middleware through use of RTPCs and **automation** (see "Audio Middleware," Chapter 8, page 265). In short, RTPCs allow middleware to track values in the game engine and automate aspects of our musical system. Each program has a different method of application for RTPCs. Game parameters in FMOD and game syncs in Wwise (see Figure 9.3) track game data, at which point we can set volume automation to respond. These automation ramps will determine which layers the player is listening to as the game changes. The techniques described above are different ways of splitting musical cues into layers, so that they respond appropriately to the game data (see Figure 8.3, page 268).

An example of how this could be applied to middleware would be a vertical system tied to the number of enemies in a combat scene. First you would create a parameter or game sync (depending on whether you're using FMOD or Wwise) and call it "number of enemies." You would then split up a combat track into three layers based on their intensity level (low, medium, and high) and import them into your middleware program as separate assets. Finally, you would draw in volume automation ramps that trigger one layer at a time as the "number of enemies" value increases. We are left with a vertical layering system that increases in intensity as the number of enemies in the scene increases.

Note that the number of enemies in a scene is a value that the game is already likely to be tracking because it is necessary to the gameplay. Values likely to be tracked by the game are great starting points in developing your musical systems because they don't require much programmer support to create parameters/game syncs for them.

Spencer

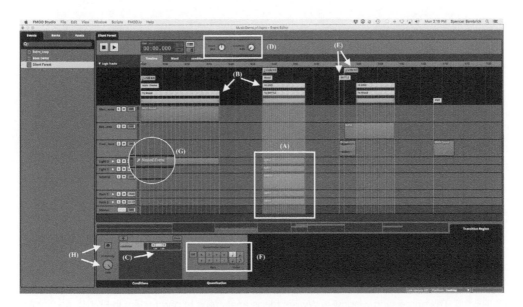

FIGURE 9.2 A screenshot from an FMOD session: this session employs both vertical and horizontal techniques.

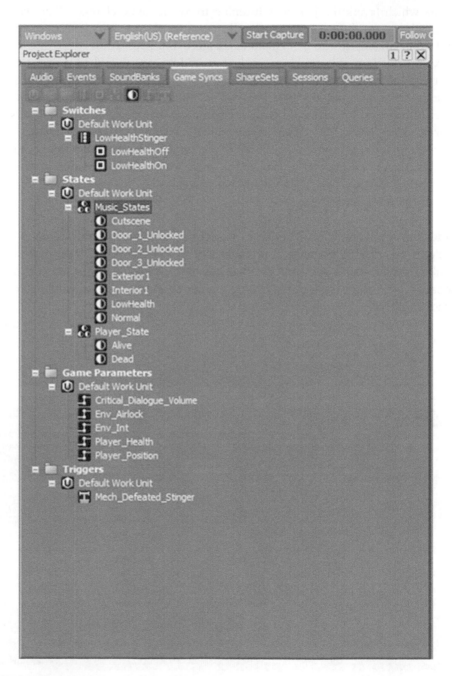

FIGURE 9.3 The game sync window in Wwise: this is analagous to parameters in FMOD, and it allows users to "catch" incoming data from the game engine and adapt the music according to those values.

Vertical Systems

• •

Figure 9.2 shows an example of a vertical system (among other things, which we will get to shortly) in FMOD. Here, the vertical layers (A) are equipped with volume automation and ready to ramp up according to the "mood" parameter (D). Note that the mood parameter begins in the middle and can be dialed both clockwise and counterclockwise. This determines whether or not the exploration music evokes a "light" or "dark" mood. It also controls the *intensity* of the two moods. A value at the center, which is where the parameter is initialized, results in a completely neutral, percussion-only layer. See the Sound Lab (companion site) for a full analysis, as well as a link to download it and play around with it yourself.

ADVANCED HORIZONTAL TECHNIQUES

Horizontal resequencing is our second category of adaptive music composition and implementation. What vertical scoring lacks – the ability to incorporate tempo changes and changes to the overall texture of a cue – horizontal scoring accomplishes naturally. Instead of using layers of sound like a vertical setup, a horizontal setup is much closer to a linear musical transition. In this way fully composed modules of music can be triggered, transitioned, and shuffled in whatever way fits the game state.

The advantage of this type of scoring is that it can accomodate abrupt shifts that drastically change the mood of a game. Most games that feature combat music will incorporate some kind of horizontal resequencing. Combat music is often at a faster tempo than exploration music, and these shifts necessarily occur immediately as a combat sequence is initiated. Horizontal resequencing is needed to trigger and transition these shifts.

As always, there are limitations with this method. The main limitation is the mechanism for transitioning. If a horizontal module is to be triggered immediately, the transition can often sound abrupt and non-musical. In these scenarios, it's important to consider your priorities to the game and to the player. Sometimes an abrupt transition is more useful to the player because alerting them to a potential threat is necessary for the player to succeed in combat. Scenarios where transitions have more flexibility are much easier to smooth over and polish because they can be faded over a number of bars to land on a strong beat. In addition, horizontal scores can easily sound mottled and dissimilar if careful attention isn't paid to the instrumentation and mixing of each cue. Meticulous attention must be paid to horizontal scores in order to ensure cohesion. In any case, using volume envelopes with increased release time can be a huge help when transitioning disparate segments. Well-placed stingers can also help cover up awkward or abrupt transitions.

Let's look at some effective ways to utilize horizontal resequencing.

Stingers and Transitions

Although we mentioned using stingers (short, linear chunks of music) earlier, we have not yet covered how useful and far-reaching their applications can be. Stingers are usually thought of as brief, catchy melodies that help players identify treasures or valuable pick-ups. However their true potential can be realized when used in conjunction with other loops. In this way stingers can act as intros and outros, adding purpose and development to the musical structure. They can also fire over the top of a loop, behaving almost like a sound effect to alert players to some item being used, or a special attack being triggered.

Lastly, and most importantly, stingers can be used as transition pieces to connect larger modules of music. This is an important function because it allows composer to write complex and diverse music in disparate styles. These modules can then be attached like Legos using stinger transitions. The best part of this method is that these stinger transitions are pre-composed and linear, so they will sound extremely musical and cohesive moving into or out of another musical cue. If care is taken to write effective transitions, it's easy to approach the tailored sound of film scores with this technique.

Horizontal Resequencing to Define Musical Structure

In traditional linear music we often take for granted its ability to take us on a journey. But how do we accomplish the same when composing for games if you can't predict player actions? The answer is by using horizontal scoring methods. To feel satisfying, musical cues need to have a clear structure (see "Form and Structure in Game Music," Chapter 6, page 184) that corresponds to game states. This means that any cue should have a compelling intro, a number of varying and/or contrasting sections (A, B, C, etc.), and a clear ending. It is possible to accomplish all of this using only horizontal scoring.

Our starting point will be the introduction. This is essentially a stinger which will trigger and move on immediately to the next section. The length will depend on the context of the game, but generally it will be a short (one- to five-second) introduction to a cue. Often these stingers are attention grabbing and fun. The intro needs to successfully alert the player to whatever new game context is about to occur (immediate combat, a potential threat, a mini-game, etc.), *and* it needs to lead into the main loop in a musically coherent way. Remember that as the introduction, this stinger is really asking a question that the player must answer with her actions. The stinger shouldn't feel like a musical resolution; instead it has to feel somewhat open ended so that the player feels like there is more to come.

Next comes the loop. This loop serves the sole purpose of setting the mood for this new game context. It's important that this module (we'll call it the "A" section) loops because we won't know how long the player will take to complete her task. On the one hand, if it's a combat task it could be as quick as a couple seconds. On the other hand, a puzzle could take minutes to solve. The priority here is to compose a loop that exactly fits the mood of the task.

Finally, we need an ending to our horizontal system. After the player completes her task the music needs to quickly arrive at a satisfying conclusion. This is accomplished by composing another stinger. This time the stinger *should* feel resolved. Our player has defeated her enemies in combat so this outro is her reward. An open-ended ending stinger would feel quite unsatisfying here. In this sense we are now aiming for a period, or an exclamation point, rather than a question mark.

Horizontal Staging

The above type of setup can be made more adaptable in any number of ways. The "beginning–middle–end" structure is just one possibility. Adding other tiers of horizontal staging is a great way to add complexity and adaptivity to a game scene. Take a boss level for example. Bosses usually have different forms that emerge when a certain amount of health points are lost. This usually means the combat will be split into multiple stages. For each stage the musical cue should change correspondingly. Perhaps Stage A is meant to feel easy, but Stages B and C are meant to be insanely difficult and chaotic. In this scenario the horizontal structure may now look like this: Intro–Stage A–Stage B–Stage C–End. In this way each stage can accommodate tempo changes, instrument changes, or basically anything you can imagine that will up the ante. This is noticeably different from vertical methods because we are actually transitioning to an entirely new piece of music, rather than simply increasing density in the same piece of music.

To smooth out the cue as a whole it is also possible to insert transitions that help the piece build organically: Intro–Stage A–Transition–Stage B –Transition–Stage C–End.

It's possible that each stage shift may be accompanied by a cutscene, in which each transition will be replaced by a linear score to fit the cutscene. In this case each stage might act as its own mini-horizontal system. By this we mean that after each introduction stinger, the stage in itself is a loop, followed by an ending stinger that transitions into another linear cue that synchronizes with a cutscene. A three-stage battle would then look like this:

Intro–Stage A–End–Cutscene
Intro–Stage B–End–Cutscene
Intro–Stage C–End–Cutscene

In these more complex examples it can dramatically help with transitions to **quantize** horizontal triggers (not to be mistaken with quantizing in a DAW). This means that transitions will only occur 1) after a parameter/game sync meets a predefined value (e.g. Boss reaches < 100 hp), and 2) after a pre-defined number of bars or beats (e.g. after the Boss reaches < 100 hp and the music reaches the next downbeat). This ensures that there are no abrupt cuts or unwanted shifts in rhythm. There are times when immediate shifts may be desirable (as mentioned above), but that is left to the composer's discretion. Volume envelopes can again smooth over these myriad transitions, as it did in our examples above.

> There are numerous ways to accomplish these horizontal transitions depending on how you are implementing the music. FMOD has **transition regions** that automatically shift the playhead to predefined modules after a parameter hits the threshold. Wwise has a similar method using game syncs instead of parameters. Both are relatively simple to learn, but complex to master. In many cases the best tool you will have to smooth out these transitions and make them musical is your ear. By listening critically you can adjust transitions at the composition level, and smooth them out using any number of middleware tools.
>
> Gina

Refer back to Chapter 8 for some specifics on transitions, and be sure to check out the Sound Lab (companion site) at the end of the Chapter 9 for some video tutorials and examples that go into detail on how to set up these systems.

Horizontal Markers

At times it will be necessary to take a longer cue and make it adaptive. For example, a developer might love a particular theme you wrote and want it used in a battle scene. In these scenarios it can be helpful to use **markers** to define sections of the track. It will be necessary to listen carefully to the track to find points that will loop seamlessly, and designate them with markers. Since the track is linear, it will likely already have a structure to it, but only parts of it will sound cohesive when looping. Drop a marker on these sections and use horizontal transitions to trigger each section of the track as the gameplay dictates. Note that you may have to cut up some sections and move them around in order to ensure seamless looping.

The goal for this technique is to allow adequate flexibility of a track. It should function similarly to any other horizontal scoring method, but will allow you to adapt a linear track to fit the game. Usually the markers are set as *mood markers*. So Marker A will dictate a subtle exploratory mood, while Marker B might be designated for combat.

Horizontal Applications in Middleware

Most if not all horizontal techniques will be accomplished using sound containers and **transitions**. As covered in "Audio Middleware" in Chapter 8, sound containers are playlists of music. We can randomize them, weight them differently, loop individual cues within the playlist or the entire playlist itself, and basically perform any operation on them that we want. Transitions are functions in a middleware program that jump from one sound container (or marker) to another based on parameter or game sync values. We can define the threshold ahead of time, and when the value hits that threshold, the transition will occur.

For more resources (including a downloadable version of this session) see the Sound Lab (companion site).

ADDING RANDOMNESS AND VARIATION

Randomness and variation are essential components of game music. We discussed many ways to increase randomness and variation in Chapter 8, but now we will discuss them as they relate exclusively to music. Much of the terminology remains the same however, so feel free to refer back to Chapter 8 at any point if you are confused, or need to refresh your memory on some of these topics. Keep in mind that many of the techniques we discuss in these later chapters will not feel comfortable until you take a practical approach with them, so be sure to check out the Sound Lab (companion site) at the end of each chapter.

Playlists and Randomization

Complexity and variation can be added to the horizontal and vertical methods by using playlists (see Chapter 8). For example, instead of using one loop (A) in between intro and outro stingers we can instead randomly select from multiple loops (A, B, C, etc.). This will add some interest to the main module of a horizontal system and keep things fresh for the player. We can use playlists for stingers as well. This is especially effective when you want the player to feel excited and/or uncertain as the game state changes. Horror games in particular are great opportunities to employ playlists of stingers. Each stinger can be a new and horrifying way to alert the player of an incoming threat.

To make a horizontal system as random and varied as possible you can also use randomized playlists for the stingers *and* the loops. This adds some unpredictability to the scene, without changing the mood. This works well for open-world games where players will be logging hundreds of hours. It will get you the most mileage out of your system and keep players from muting their TVs.

In the case of a vertical system, playlists can be used to randomly (or intentionally) vary instrument timbre for a given melody. Rather than having one layer with the violins playing a melody, you can instead use a playlist for that melody layer. The playlist would then contain multiple loops of the same melody performed by violins, a flute, and a vocalist. The audio engine would then alternate between them either randomly or based on game data. The same process works just as well for chords, background parts, and variations for all of the above. In effect you could have an entirely randomized system where the compositional components (melody, harmony, rhythm) of a track remain the same, but the instruments that play them are constantly shifting!

Most middleware programs will allow you to set a **play percentage** or **weight** on each item in the playlist (again, see Chapter 8). This allows you to control the amount of randomness in your music system by *limiting* variability. Having a playlist on each module of music will vastly decrease the chances of your system becoming repetitive or stale upon retriggering.

Figure 9.2 (H) gives us an alternative look at some basic randomization settings for horizontal transitions. In this case, we can actually add a bit of chance to a horizontal trigger by altering the probability settings. For values below 100 percent, the horizontal trigger will sometimes fail to execute.

Variation for Musical Development

From a compositional perspective, it can be very effective to rearrange themes and motifs using different instruments and ensembles. This method can effectively create variation on a small scale by providing a fresh sound for frequently used themes. On a larger scale, it allows further development of your music. A theme can be used and reused many times if the arrangements are sufficiently different. A character theme for example can be arranged for a chamber ensemble during intimate moments, and then rearranged for full orchestra during a battle. In most cases the context of a theme has more of an impact on the player than the theme itself.

As you're using this method, keep in mind the story arc of the characters and the game itself and try to update your arrangements to match the story. Characters that are descending into villainy can have an arrangement that matches this descent. Likewise, characters that begin the story in a dark situation can also have arrangements which develop into something more hopeful or triumphant.

Lastly, in terms of variation and development, there is no replacement for well-designed linear cues. By recreating a theme with some new elements it is possible to make the original theme feel fresh when it triggers again later on. In many cases these new elements can be adding or changing instruments as mentioned earlier. It is also equally as effective to compose variations in harmony, melody, and any other compositional elements themselves to match the game context.

The Sound Lab

Before moving onto "Composing Complex Adaptive Systems Using Middleware," head over to the Sound Lab for a practical look at vertical and horizontal basics.

COMPOSING COMPLEX ADAPTIVE SYSTEMS USING MIDDLEWARE

Now that we have covered a few horizontal and vertical scoring techniques in depth, we can discuss how to combine these methods into complex, adaptive scoring systems. Refer back to Figure 7.5 (both the full score *and* the middleware session of the cue are available on the companion site under Chapter 7 and in the "Study Score" section). This polyphonic example employs an introductory stinger, and a loop that contains vertical layers. In this way it combines vertical and horizontal methods to achieve adaptability. Complex systems like this are used to emulate the style of **tight scoring** that can be heard in many films. Older cartoons often referred to this style of composing as "Mickey Mousing" because the music would follow along closely with Mickey Mouse's movements. Highly adaptive music systems are the video game analog of tight scoring, and they can be extremely effective in the right context.

There is an infinite number of ways in which vertical and horizontal systems can be combined, which makes it difficult to establish foundational tenets. Usually these systems employ multiple parameters/game syncs in creative ways to deepen the adaptivity. In general the most effective way to understand highly adaptive systems is to make them yourself. That being said, we have provided a few more advanced theoretical concepts below that can be achieved using virtually any middleware tool or proprietary audio engine. In the Sound Lab we detail more specific and practical methods using Wwise and FMOD. To reiterate, the technology of game audio changes constantly. What is important is honing your ability to plan and implement a system that will support gameplay in a flexible and coherent way.

Combining Vertical Layering and Horizontal Resequencing

A fantastic example of an extreme musical adaptivity is the game *Peggle 2*.[2] The composer, Guy Whitmore, used a combination of vertical and horizontal techniques to score the player's actions very tightly. *Peggle 2* is similar to pinball in that players shoot balls onto a two-dimensional map to hit pegs and progress through to the next level. Musically, this game uses many short, vertically scored loops in conjunction with horizontal stingers and transitions to create a sense of development and progression throughout each level.

The system as a whole can be thought of as a series of horizontal stages with added vertical layering. The vertical layers allow for subtle shifts in intensity while the horizontal elements produce more drastic shifts in urgency. These shifts include changes to the orchestration as well as modulations to closely related keys. Modulations are generally avoided in game music due to the immense difficulty of coordinating them, but with *Peggle 2* composers can no longer use that as an excuse. The horizontal elements also act as natural climaxes in the music that sync to the animations and provide some structure so that it feels like a tightly scored movie.

On top of all of this, the actual pegs themselves generate musical notes when hit. These notes are tied to the scale that Whitmore uses at each horizontal juncture in the system. The notes proceed in ascending scalar order to relay a sense of anticipation onto the player. The amount of thought and planning that goes into a highly adaptive system like this is monstrous, but the end result unequivocally pays off. *Peggle 2* wraps its musical system around minute player actions and gives players an experience as tightly scored as any Disney or Warner Brothers cartoon.

Visiting Artist: Dren McDonald, Composer

Inspiration and Interactive Music

Inspiration is that piece of magic that keeps us going when everyone else you know is asleep and you are solving musical/gameplay puzzles when your body wants to lurch you off to dreamland. So let's start with that!

Interactive music is quite unique, and I've found that for myself, the gameplay (and narrative, if that is a strong component) is usually responsible for sparking the initial musical inspiration. If you are lucky, that spark can carry you through a project from start to finish, like a pair of wings taking you through the fluffy clouds, which brings me to a story about a game that did that for me, *Gathering Sky* (hence, the wings metaphor … you'll see.)

I first experienced *Gathering Sky* (initially titled *Apsis*) when it was an entry at IndieCade and I was volunteering as a juror, reviewing games. A lot of the games were broken, builds weren't loading on devices like they were supposed to, many games felt like obvious "homages" to other games, and there were times that this volunteer gig wasn't exactly what I hoped it would be. Then I came across *Apsis* in my queue. The game actually opened and I could play it, so they had that going for

(Continued)

them. The game began with a singular bird and you would guide the bird through the sky ... until the bird finds another bird friend, and when the birds interact, the new bird friend will follow your first bird ... and then you continue to build up more bird friends as you fly through this mysterious sky of clouds, wind currents, and rock formations. Before you know it, you are guiding an entire flock through the sky, and you feel somewhat responsible for these pixel creatures in a way I can't explain. You'll just have to play the game to experience it.

I think it was this initial feeling that hooked me with this game, and really sparked my imagination. Why did I care so much about my flock? How did I become so emotionally engaged with this experience that did not include other humans or human forms or speech or an obvious storyline? I was emotionally invested in the experience and I couldn't stop thinking about it.

During that first playthrough (in which I played the entire game, for 45 minutes straight, no breaks), there was music in the game, but no sound design to speak of. Somehow the music was appearing to work with the game, however the songs were linear, and would just end, leaving silence. So something strange was happening.[3] I gave a detailed review of the game, lauding its virtues and then giving an incredibly detailed list of improvements that they should consider for the audio. I did not forget the game, but I didn't expect to hear from the developers about any of my ramblings.

Fast forward a few months, and I was at an indie game event in San Francisco. Devs were in a room just showing off their games, as is usually the case with these smaller events. And then I saw it ... the BIRD GAME. Holy cow, these developers are here! So I got to talk to them and tell them "Hi, yes, I was the one to give you all of that detailed audio feedback for your game, but don't take that personally, I loved the game; this is meant to be constructive feedback." After chatting with them for a while, we exchanged information and left it at "Well, I'm really busy with several projects at the moment, but I know a lot of other game audio folks who would probably love to work on this if you decide that you want help."

Long story short, all of those projects that I had lined up just disappeared or got rescheduled and I suddenly had time. Almost as if the universe made me available for this project.

So I began work on the game. It went on far longer than I anticipated that it would, but returning to the theme of "inspiration," even as the game development continued and started to feel like a lot of late nights in a row, I continued to raise the bar of possibility for this project. This was really only because

> *(Continued)*
>
> ....
>
> I believed so much in the vision of this "experience" (it's more of an "experience" than a "game"). I wanted to push myself to see what I could bring to this project to bring it to life. That inspiration can sometimes be rare, but when you find it and open yourself to it, it works like magic.

Nested Events: Music

Although *Peggle 2* is an example of adaptivity taken to the extreme, the possibilities are far from exhausted. A complex as vertical/horizontal combinations can be, they can be made even more complex through a process called nesting. Nesting events is a process by which entire musical systems can be made modular (see Figure 9.2(G) for an image of a nested event in FMOD). This allows adaptive systems to be reference *within* another system or **master event**. This can be quite useful at times, especially if the majority of music in a game scene is simple and one particular aspect of it is highly complex.

One example of this is if an exploration game includes a mini-game on a particular level. In this scenario most of the music may be a simple loop or vertical layers. When the player triggers the mini-game the nested event can spring into action at whatever level of complexity is required. In this case the mini-game music event can be as complicated as desired, but the master event will be clear and organized.

Another more complex example of nesting is what we like to call the 3D approach to implementation. If the Y-axis consists of vertical layers, and the X-axis consists of horizontal modules, then the Z-axis (the third dimension of our music system) could be a nested event. Let's say, for example, that all of our vertical layers are triggered, making the music as intense as possible. Our system is able to jump around to different horizontal sections as well at a moment's notice. This is a system we've seen in many examples thus far. But in this example we can nest the vertical/horizontal system and use a master event to add more to it. We can add an entirely new nested event, set up with the same functionality as our first vertical/horizontal system, but with entirely new sounds. We can then use a parameter/game sync to crossfade between these two sets (or *planes*) of events.

The power of nested events is their ability to organize large and complex events into one or more master event. These master events allow for simplified functionality, while the nested events maintain a high level of adaptivity. They also allow a complicated event to be used in more than one game scene, or as a template to be used with other music (as we saw in the 3D approach). In more involved scenarios master events can be comprised of many nested events. The only limitations are the needs of the game and your own imagination!

Branching Music Systems

The term **branching** refers to music systems where the outcomes split based on previous events. The musical outcomes, or branches, the player would hear would be different depending on how she played the game. With this logic, a linear system (as we have seen before) would look like this: if A then B, if B then C, and so on in a straight line. With a branching system we would get something like that shown in Figure 9.4.

Each of these letters represent possible outcomes, like a musical "choose your own adventure."

It's important to note that branching systems are theoretically technique-agnostic. This means that you could device a branching music system with only vertical methods, or only horizontal methods, or a combination of both. You could even devise a system where each cue is entirely linear, but the overarching logic is not. This makes it extremely adaptable to different game genres. In practice, these systems work great for decision-based games, visual novels, and text adventures. Exploration games

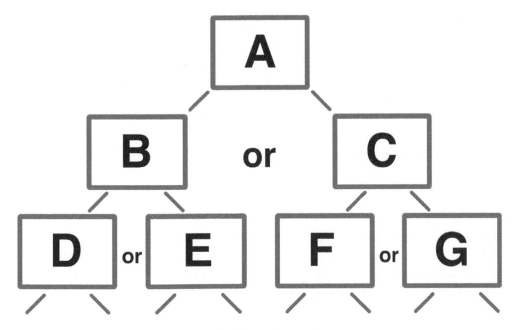

FIGURE 9.4 Example of music that branches into various outcomes.

are another great application for branching systems – you could device a system where the outcomes correlate to different pathways in a maze. In a loose sense, many adaptive scores are branching to a small extent. "Game Over" stingers are an example of one particular branch in a score. Regardless of the application, branching systems are incredible tools for telling a story through music. At the very least they can be helpful in conceptualizing entire trees of narrative logic. At most they can be the foundation of a highly complex and adaptive musical system.

AESTHETIC CREATIVITY AND COMPOSING OUTSIDE THE BOX

The issue of aesthetics in game music is extremely complicated. On one hand, as composers it is important for us to write music that speaks to the player on a personal level. It is equally important that the music in a game has a unique personality to it which helps set it apart from other games. On the other hand, the music needs to support the gameplay. Which of these is the priority? Looking at the big picture of game development, the latter has to be the priority. If the music is distracting or creates unnecessary confusion for the player, then the music is not a good fit for the game. However on smaller scales a little bit of outside-the-box creativity can go a long way toward making a game memorable. Both of these approaches to game music can coexist if you are able to keep the bigger picture in mind.

There are two areas of a game soundtrack where we as composers have full freedom to experiment and compose outside the box. The first is in the aesthetics of the music itself; the second is in the implementation. The aesthetics of a game soundtrack refer to the basic elements of the music itself; instrumentation, arrangement, harmony, melodic motifs, etc. The implementation refers to *how* the music is triggered in game. Both of these areas are rife with opportunities to experiment.

When (and When Not) to Experiment

Before we get into some techniques for composing outside the box, we must determine *when* to compose outside the box. The truth is that not every game needs hyper-complex adaptive music. And not every game needs the timbral density of a Mahler symphony either. Sometimes the best musical solution for a game is the simplest.

The best way to determine if this is the kind of project that calls for experimentation is to take a hard look at the game design pipeline. Ask yourself if there is room in the order of operations to insert an experiment and possibly fail. If the answer is no, then it is likely an unwanted risk. This is not always a bad thing. Some developers know exactly what they want and are therefore very clear about it. If your communications with a developer look similar to a task list with musical references, then that makes your job clearer and more organized. If you have an idea that will make the game better, float it by your client and be open to whatever response you receive. Also note that the idea should make the game better, not just your music. There is a clear difference there. Sometimes the results overlap, but not always. Be aware of what your goals for experimentation are, and are not.

If your developer has a much looser pipeline, and has given you some flexibility in terms of style, then you will likely have an opportunity to play around with some of your ideas. It is still important here to communicate before sinking exorbitant amounts of time into an experiment, so don't skip that step. Developers with this attitude are usually working on smaller games, possibly passion projects. They often look to you as an advisor if not audio director, deferring to your judgment on audio matters. This is a great space to try something new and exciting provided it fits the mechanics of the game and does not interfere in gameplay.

It is impossible to provide an exhaustive list of game characteristics which are better suited for this style of composing. The best way to determine if your musical experiments work with the game is to implement it and play, play, play! If you're having a hard time, ask your client if you can form a small sample group to playtest with the music. Sometimes it takes a non-musician to quickly determine if music helps or hurts the gameplay.

Experimenting with Aesthetics

The first area of experimentation with game music is the overall aesthetics of your soundtrack. The musical aesthetic usually comes down to your instrumentation and how you process those instruments – in other words (if you can remember back to Chapter 6), your sonic palette. Many composers instinctually adhere to traditional or well-trodden palettes. For example, a classically trained composer might write for string quartet or symphonic orchestra more often than not. Or a composer/guitarist might heavily utilize a rock band for most of her projects. This is perfectly fine most of the time, but it can limit the possible projects that you are qualified for. It also can be fun to experiment with your sonic palette by going beyond the norm in terms of instrumentation.

A simple way to create your own experimental ensemble is to combine elements from traditional ones. For example, how would it sound if a string quartet were combined with an electric guitar or other electronic elements? What kind of emotions would you feel when presented with this sonic palette? Is it capable of capturing the mood and emotional content required by the game? Thinking along these lines can yield some very interesting and powerful results.

Another way to experiment with the musical aesthetics of your project is to get creative with the way that you *process* your instruments. So in this case, instead of recording a string quartet in a concert hall and perfectly balancing the mix, try recording them each individually and adding distortion, delay, and reversed reverb. Or simply load up your string quartet into a granular synthesizer. You may come up with something totally new and eccentric. You can extend the string quartet far beyond the usual range of contexts with this method.

Experimenting with Implementation

Implementation can be a bit tricky to conceptualize as composers, but is an equally beneficial area to experiment in. In order to experiment with implementation we must go *beyond* the vertical and horizontal systems we've described so far. In these scenarios we are taking all possible player actions and selecting fragmented pieces of music to *react* to the player. This is true even in the most adaptive systems. To go beyond this we must design a system that acts *with* the player, and makes changes almost simultaneously, rather than waiting for a moment that makes musical sense. We're now entering into the territory of **generative music.**

The broad definition of generative music is music that emerges from a system, process, or framework.[4] In relation to games, generative music can be thought of as a means for music to be triggered at the level of individual notes (or even smaller grains). This can be an extremely powerful method of musical implementation simply because it has the potential to wrap so tightly around player action. Instead of using pre-composed modules of music that we see in the vertical and horizontal systems, the music is obeying a set of rules and being generated on the fly. It can make for a truly immersive player experience.

Generative music has a very rich history, one that pre-dates video games and goes well beyond the scope of this book. To put it simply, there are many, *many* ways to generate or automate musical systems.[5] But there are a few of particular interest to us as game composers. In a very loose sense, most video game music is in itself generative because the game acts as a framework and the player's actions then generate and influence the development of the music. However, when we talk about generative music within a video game we usually are referring to individual notes that are created or triggered by minute player actions, and a structure that is generated in real time. This means that basic musical elements like melody, harmony, counterpoint, tempo, and rhythm are all created the moment a player makes a choice. Even the most extremely adaptive scores are actually just reacting quickly to the player's actions, while generative music reacts *simultaneously*. In some ways games that incorporate generative music are really indirect methods of musical creation for the players, since the music that they end up with is not prewritten in any way. Rather it is the collective result of each and every choice that the player makes.

It's important to remember that games have always taken advantage of what developers call **procedural** generation. Procedural in the context of computer science means that the content in question has been created via an algorithm. For example, a procedural dungeon is one which is created on the fly as players are exploring, rather than being created piece by piece by artists and level designers. Often these include elements of randomness. So in this dungeon example, a new room may be created randomly, but with a set of rules or logic. In this case, dungeons will always contain the same or similar basic parameters such as room size, number of enemies, or items in order to maintain consistency and coherency in the gameplay. Each of these procedurally created rooms will also always have some kind of boundary marking the outer limits of where the player can move, otherwise they would not be rooms at all. The benefit of using procedural methods over traditional ones saves time and resources and makes processes more efficient. It also makes the experience a bit less predictable for the player. Generative music shares many of these benefits, and can be thought of in much the same way as procedural systems in games.

One iconic example of generative music is Laurie Spiegel's *Music Mouse*. Spiegel is one of the original composers who helped pioneer the field of generative music. *Music Mouse* is not a game per se, it is more of a musical playground. In it, the player's mouse movements are translated onto a two-dimensional piano grid. One axis triggers chords while the other triggers notes, all in the scale chosen by a player. This app is fun and can yield music that is quite beautiful without taking any time at all to master. Most importantly, it puts the power of musical creation into the hands of the player rather than the composer, which is a key distinction between traditional and generative music. Check out the Sound Lab (companion site) for a link to the game.

Another example of generative music is the game *Spore*, whose soundtrack was composed by Cliff Martinez, Saul Stokes, and Brian Eno. Eno is a long-time champion and innovator of generative music. This soundtrack is heavily synthesizer-based, and it has a lovely ambient mood throughout. The uniqueness of *Spore*, however, comes from the generative nature of the music. If a player enters the character-creation mode, she will hear a score that molds itself tightly around her actions. As more details are added to the character (color, texture, or body shape) the music will grow and evolve in timbre and harmony. This highlights a big difference between adaptive music and generative music in games. While adaptive music is capable of reacting to planned events in a musical and emotionally coherent way, generative music is often surprising and unpredictable because *players* are surprising and unpredictable. This is entirely necessary in the context of *Spore* because players are given free reign to create whatever character they want. This could be a seven-legged walking breathing lizard alien, and the music must reflect that! We as composers can plan for any event we expect a player to make, but by creating a generative system we allow for possibilities that we do not expect. This can add a great deal of depth and poignancy to a generative soundtrack.

Rich Vreeland's score for *Mini Metro* is a classic example of generative music in games. In this game players can choose to play in "creative mode," and are essentially given a sandbox to make a subway layout. Each choice the player makes yields musical results based on a set of rules. In this way players can create their own dynamic musical systems, and can listen as they change with the visuals. Exploratory games like *Mini Metro* are perfect for experimentation because they don't have specific tasks that players must tackle. They are meant to be experiential rather than goal-oriented, so players have the freedom to try new things and observe the results. In this case, the results are usually ambient and beautifully meditative.

Although the examples we've looked at so far are mostly synthetic and ambient that does not mean that themes cannot be made through generative methods as well. Daniel Brown (Intelligent Music Systems) created an artificial music intelligence system called *Mezzo* for his PhD dissertation at UC Santa Cruz (2012).[6] *Mezzo* composes game soundtracks in real time, including character themes. The themes, dynamics, and harmonies composed by Mezzo are adaptive, so they change based on the context of the game, just as a pre-composed soundtrack would. Although there are many ways to create generative music, the fact that *Mezzo* is algorithmic means that it falls squarely within the realm of procedural music.

With the advent of *Mezzo* and other tools like it some composers are worried that their work and creative input won't be of value to developers because an algorithm is capable of the same contribution. While we understand these concerns, we would urge composers to embrace these technological advancements rather than denouncing them. It is our firm belief that technological advancements like these can be used to aid the creative process, and will actually create opportunities for composers. Refusal to incorporate (or even to accept the existence of) new technologies and workflows is more likely to hinder our creative output than it is to sustain it. These new technologies may be daunting or threatening at first, but change is inevitable. By making ourselves more aware and flexible in our workflow we can learn to use new technology to create even more innovative soundtracks than might previously have been possible. Ultimately, tools like *Mezzo* will likely make our soundtracks more malleable than ever before. This can only serve to better support the player experience, which is, of course, the most important aspect of what we do as game composers.

TOOLS FOR EXPERIMENTATION

Experimentation is a very personal process, so it would be impossible to list all of the ways to experiment on a game soundtrack. We have instead made a list to introduce a few tools that go beyond the applications of middleware that we have mentioned already. These are meant to be starting points for you to think about possible methods of experimentation with sound creation and implementation for your game soundtracks.

PURE DATA

Pure Data (or Pd) is an open-source visual programming environment for audio and video. This means that composers can create custom audio and MIDI tools without writing lines of code. Instead, users create complex frameworks of objects and patch cords to achieve essentially any output they want. There are other programming environments like this (Cycling '74's Max is one example), but the fact that Pd is open source makes it invaluable. Pd integrates well with game engines like Unity and Unreal, and even middleware programs. It also runs on pretty much any device from personal computers to smartphones. Pd is often used on virtual-reality experiences by sound designers creating procedural sound effects. When working with Pd the sky's the limit. A simple Google search will yield crazy procedural music systems, Wii remote synthesizers, and interactive audio-visual sound installations.

Programming Languages

Programming languages aren't really tools, but they are absolutely effective ways to experiment with your game music. Knowing a programming language (especially one used commonly in the video game industry like C#) can open many doors with regards to implementation. You will find yourself better able to pinpoint music-related problems down the road *and* better able to propose creative solutions to them. Knowing how to program in some languages will also allow you to add custom tools to middleware, making the already streamlined implementation process flexible and powerful. You will also have fuller control over how your music sounds in its final form, which is an indispensable advantage. In Chapter 8 we offer additional information on scripting for games.

> Although working this way is fun and rewarding, the key to experimentation is to find the sound or implementation style that works best for the game and to develop that process into a controllable method. There are times when a more traditional approach is the best approach for the game, and this should be recognized. The ultimate goal is always to create music that best suits the needs of the game. Experiments that didn't quite make it into your projects are still useful though. Keep the creative process in mind throughout all of this experimentation. You never know when you'll need a more experimental sound or technique for another project.
>
> Spencer

Machine Learning

Machine learning is a vast and technical field. However, it's potential impact on game audio is equally vast and very exciting. Daniel Brown (Intelligent Music Systems) has essentially built a percussion "procedural-izer." This tool uses machine learning to take in pre-composed tracks, and outputs new tracks on the fly in the same style. It integrates to middleware programs like FMOD and Wwise, and has been used to great effect in *Rise of the Tomb Raider*. This is a huge first step toward what we hope will be a much more widely used technology in game audio.

Audio for New Realities: VR

New realities are upon us, and Virtual Reality (VR) is a great example of a tool that allows us to experiment with music and sound. Just as Stephan Schütze mentions in his book,[7] audio for VR is sometimes referred to as "The Wild West" because no one person at this point in time has all the answers, and there are few if any "best practices methods." But this doesn't mean we can't find our own way. Just as game audio took some workflows and techniques from film sound, we can take methods from game sound and bring them into the world of VR, along with some new tricks which we will inevitably come across as we experiment.

The Sound Lab

 In the Sound Lab (companion site) we present a very brief overview of audio for Virtual Reality (VR). For a more detailed look into the subject we recommend *New Realities in Audio: A Practical Guide for VR, AR, MR, and 360 Video* by Stephan Schütze with Anna Irwin-Schütze. This book goes into all the detail you would want to know about getting started in recording, designing, and implementing audio for these different types of realities.

Visiting Artist: Stephan Schütze, Sound Designer, Author

Thoughts on New Realities

VR and AR are developing so quickly that we have the freedom to stumble around and try something new and risky because the entire platform is still an unknown universe waiting to be explored. This is its biggest advantage and I encourage everyone to make the most of it.

I always understood that immersion would be a huge part of virtual spaces and I still think this is the case, but I have also discovered that scale is critically important and this is what I want to discuss here. Everything you need to know about VR audio you can learn from your car.

For games, movies, TV, or apps, we see representations of objects on screens. We are essentially viewing them through a window. Once we enter a virtual space those objects can take on a real sense of scale. A virtual car can be presented, right there in front of you where you can walk around it, climb into or on top of it, and get a real sense that there is a significant object that you are sharing the world with. For audio this is critical.

(Continued)

When creating for the traditional platforms, a car or truck could mostly be represented by a single sound source, maybe two if you wanted lots of detail. Essentially, the three-dimensional model of the car exists in the three-dimensional world and has sound emanating from it. This changes in VR.

When I am standing in front of a 1:1 scale vehicle in VR it means I can relate to it in the same way I could in the real world. The sound of the engine will be different at the front than what I hear from the exhaust. But more than that, in VR I can lean right down and place my ear close to the exhaust and really hear every aspect of the texture of the engine sound, how well tuned the engine is, the effect of the exhaust "wind" blowing out of the pipe and past my ear. The engine itself is a significant mixture of many different sound sources that all combine. Lean close to the fan and it will be more prominent than the alternator. Crawl underneath and the transmission sounds will be far more prominent.

In VR, scale plays such an important part of building the sonic world that you need to consider attention to detail to a far greater level than you might otherwise. Inside the car, if I lean closer to the internal loudspeakers, does the music from the radio get louder? It probably should. Instead of all the general "noises" that we might mix together to make the sound of a car driving over a bumpy road, is it worth dividing them into specific sound layers, each with their own location inside the car? So that bumping rattle actually comes from inside the glovebox where my CDs are stored, and the coins clanking are emanating from the center console, while my football boots thump around behind me in the trunk.

These kinds of decisions come down to exactly what effect you are trying to achieve, but the possibilities are there and the potential is there to create a sonic space that enhances the user experience by adding elements that emulate how we experience the real world. Sound is all around us, but as we scale things up in VR to simulate being in a "real" world, the level of audio detail should scale as well.

It is the scale and the ability to interact with the world in a more meaningful way that can really allow VR and AR to shine. You just need to consider what impact you want your audio to have on your audience and then craft your spaces appropriately. We can always add more sound; it is the choices of when and where we choose to add it that defines the experience we are creating for our audience. VR and AR provide us with some fun new challenges to play with.

SIMPLICITY VS. COMPLEXITY IN GAME MUSIC

We have had some real fun diving into the world of musical experimentation, but it's time to take a step back and look at game music a bit more broadly. As mentioned above, experimentation is only effective if it supports the needs of the game. It is very often the case that the simplest solution is the best solution. In this section we will discuss two areas where this comes into play: musical aesthetics and implementation workflow.

Musical aesthetics refers to the qualitative characteristics of your music. It basically boils down to the question, "How does your music *feel* to the player?" In some ways this is the most important question you can ask yourself as a game composer, and your job is not done until the answer is "It feels *right*."

We have exhaustively covered strategies for composing the basic elements of your music in earlier chapters. The outward level of complexity is equally important as it directly relates to the *feel* of the game. It is quite hard to solve a puzzle when the rhythmic complexity of a track is so jarring that the player is distracted. Likewise, a battle scene will in all likelihood lack a sense of urgency if the music is so simple that it becomes repetitive. It may even prove to be relaxing, completely undermining the aesthetic goals.

While there are no hard and fast rules dictating how complex a particular aspect of your music should be, it is important to understand that on an intuitive level, complexity – in any aspect of your musical aesthetic – draws attention toward your music. This effect can and should be exploited to manipulate the mood of the scene. When players need to focus on mechanics, keep complexity at a minimum to allow for full immersion into the puzzle. If attention needs to be drawn into the emotion of a scene, then writing quickly moving melodic ascent can effectively draw the attention from a more static gameplay environment. In other words, our brains focus on movement, whether it's visual or musical.

Further, virtually any aspect of your music can be considered "complex," not just the harmonic content. Yes, a melody can modulate into six different keys, but timbre can also be complex. Extremely novel or dense timbres tend to draw more attention. Imagine a pure sine wave compared with the sound of a distorted, mangled cello. Which is more complex? Which is more attention grabbing? Which better suits the intended *feel* of your game scene?

What matters most here is the overall perceived level of complexity. For example, a chord progression that is constantly wandering through keys is harmonically complex, but what effect does it have on the game? It depends on a number of factors. If the chords are played fortissimo by a sixteen-voice synthesizer in unison, then it will sound highly complex. However, if the chords are voiced by a harp at pianissimo then the overall *texture* may very well sound simple and serene if the voice leading is smooth. Blocky voice leading may yield a much more complex texture altogether. If you are incredulous, then think about a jazz ballad. Often these songs are incredibly harmonically complex, with major sevenths and minor seconds, yet they wash right by the audience like flowing water because the *texture* is simple and cohesive.

The takeaway here is not that either simplicity or complexity is a better choice over-all. The point is that taking the level of complexity into account is an important part of how a game feels to a player. Complexity can be hidden or emphasized depending on how you treat it in the arrangement, and this is something that should be taken into account. Depending on the particular gameplay mechanics and the desired emotional content of a game, complexity can be used to draw attention toward or away from the music.

Complexity in terms of implementation is a very similar story to complexity in aesthetics. The end result *must* be that the needs of the game are served. With regards to implementation however, complexity is much less noticeable by the player. In one extreme the complexity of implementation could be very jarring. For example an abrupt musical transition could occur every couple of seconds. This would certainly be noticeable, perhaps even in a fighting game or during a combat scene. On the other hand, if the transitions are smooth, and the voice leading is meticulous an extremely complex adaptive system can sound as simple as anything. We have seen this in games like *Peggle 2*, which champion complex implementation systems, yet yield extremely cohesive results without garnering attention away from the gameplay.

In some ways simple implementation methods can actually be the most striking for players. In a world of middleware and extreme adaptivity, a simple linear cue can be the most effective solution. Games that have many cutscenes are often in need of simple linear scoring. Additionally, open-world games that have vast expanses to explore can make very effective use of linearity. By triggering short musical stingers a composer can create emotional attachment to large swaths of land which otherwise might seem boring and lifeless. Using stingers in this very simple way, *The Legend of Zelda: Breath of the Wild* is able to maintain player autonomy while sustaining the sparse emotional content of the game's unbelievably large map. The music is always there, ready to add significance to moments of gameplay. Yet it never gets in the way of the experience.

The question of complexity in implementation is also important regarding implementation tools. The game industry is full of exciting and useful tools. Some feel that learning a particular tool is the same thing as learning "game audio." But there are many ways to approach implementation. Your ability to find the right approach for your project is the real trick to learning game audio. To put it another way, when it comes to workflow your tools should always simplify your process, increase the quality of your final product, or both. If your tools are doing neither then they are adding complexity to your workflow unnecessarily. At the end of the day the only thing that matters is how the game sounds, not the tools that you used.

* * *

PART III REVIEW

If you haven't taken the time to visit the Sound Lab (companion site) for this chapter, you must do so before moving on. The companion site for this chapter contains various examples and middleware sessions that are crucial to understanding the practical applications of the topics we have discussed. You will also find detailed information on various miscellaneous topics including adaptive recording, mixing, and part preparation for live sessions.

In *Chapter 8* we explored the theory and practice of audio implementation with an emphasis on audio engine capabilities and understanding resource management. We discussed the technical and creative skills that go hand-in-hand with shaping the aural experience of your game. We also broke down the fundamentals of implementing audio natively into game engines, and then covered the same processes using middleware. We have so far covered the entire implementation process from preparation of assets, to integration and optimization, and finally into testing and reiteration. Visiting artist Jeanine Cowen, sound designer and composer, discussed her thoughts on implementation and Damian Kastbauer, technical sound designer, discussed the importance of file-naming standards. Brian Schmidt, audio designer, founder and executive director at GameSoundCon, shared his thoughts on programming, Jason Kanter, audio director, gave us an analogy on bus compression, and Alexander Brandon, sound designer and composer, talked about the realities of bank management.

In *Chapter 9* we dug into advanced music implementation. Vertical and horizontal implementation techniques were covered in depth, followed by some other examples of complex adaptive systems. Later in Chapter 9 we covered experimental topics including aesthetic creativity, generative music, and VR with Stephan Schütze sharing his thoughts on audio for new realities. Dren Mcdonald also discussed inspiration for interactive music while John Robert Matz discussed his thoughts on orchestration, instrumentation, and interactivity.

* * *

NOTES

1 G. Whitmore, "Peggle 2: Live Orchestra Meets Highly Adaptive Score."
2 Ibid.
3 "It turns out that the devs had actually designed the levels to work with the music, so if there was a tempo increase in the music, they would blow the winds faster for the flock, etc. I think that might be a first!"
4 https://teropa.info/loop/#/title
5 G. Nierhaus, *Algorithmic Composition*.
6 D. Brown, "Expressing Narrative Function in Adaptive Computer-Composed Music."
7 S. Schütze with A. Irwin- Schütze, *New Realities in Audio*.

REFERENCES

Brown, D. (2012). "Expressing Narrative Function in Adaptive Computer-Composed Music." Retrieved from www.danielbrownmusic.com/uploads/1/3/2/3/13234393/final_dissertation_final_edit.pdf

Nierhaus, G. (2008). *Algorithmic Composition: Paradigms of Automated Music Generation*. New York: Springer.

Schütze, S. with Irwin- Schütze, A. (2018). *New Realities in Audio: A Practical Guide for VR, AR, MR, and 360 Video*. Boca Raton, FL: CRC Press.

Whitmore, G. (2014). "Peggle 2: Live Orchestra Meets Highly Adaptive Score." Retrieved from www.gdcvault.com/play/1020369/Peggle-2-Live-Orchestra-Meets

Part IV
BUSINESS AND NETWORKING

10 The Business of Games, Part I
What Do We Do Now?

Now that you have a clear understanding of the skills and process of creating and implementing audio for games, we will move on to an equally important topic: building your career in the game industry. In this chapter we will offer our philosophy on career development as a game audio professional and impart some helpful tips on sourcing and maintaining work. The information in this chapter should help you make a number of decisions regarding your career. The topics covered here – like choosing your career goals, finding and sustaining work, and building your network – are all essential elements of a healthy and sustainable career.

It's a good idea to keep abreast of the latest industry trends as they are always evolving. Head over to the Sound Lab for resources on the state of the industry.

CAREER DEVELOPMENT: OVERVIEW AND PHILOSOPHY

Students very commonly ask what it takes to be successful in the field of game audio. Our answer is usually something along the lines of "How do you measure 'success'?" It may seem trivial, but defining specifically what "success" means to you is a very important first step toward a career in game audio. Everybody wants to be "successful," but success means different things to different people. In effect, highly talented people can end up working hard to be successful without actually feeling like they have achieved much simply for the reason that "success" is subjective. Not only that, but its definition can (and will) change for you as you progress in your career.

A better way to frame this pursuit is to select tangible and specific goals, and then strive for *sustainability* rather than success. Goals take time to achieve, and if your career is sustainable you will have a better chance of accomplishing them. In our view the focus on sustainability yields better long-term career decisions, and a much happier, healthier work–life balance. Similar to the "spark of inspiration" cliché that we discussed in Chapter 5, the "starving artist" cliché is of little use when it comes to career development.

SELECTING TANGIBLE GOALS

One of the first things that should be considered when planning for your career is choosing tangible goals. By tangible we mean that these goals must be both specific, and grounded in reality. For example, a goal such as "compose the best game soundtrack" is *not* a tangible goal because it is not specific. How can you tell when you have composed the "best" game soundtrack? There is no way to measure such a goal, so it is essentially a lost cause. Likewise "getting paid one million dollars per minute of music" is also a poor choice of goals because it isn't exactly grounded in reality. There are thousands of games whose entire budget doesn't even reach one million dollars, so how can you reasonably expect to make that much for one minute of music? These are exaggerated examples, but they demonstrate why many of the goals we sometimes choose are not beneficial to us in the long run.

For a qualified, experienced game composer a goal like "getting paid $1,200 per minute of music" would actually be a tangible choice of goals. For someone with steady work coming in it is reasonable to take steps to increase your rates through the $1,200 per minute mark. Another tangible goal might be to compose music for "X" number of games in a given year because it is reasonable to assume that through vigilant **networking** and solicitations you can boost your yearly project output. For someone just starting out in the industry a great goal to set is something like "work with one or more live musicians on a project within the next year." This is a well-crafted goal because it will result in some great experience working with a musician, and it is a stepping stone toward larger goals. For example, the long-term goal might be to work with a full orchestra. This would likely be out of the question for a first soundtrack due to budget restrictions, but it is perfectly reasonable to convince a developer to pay for a single musician to record.

Don't take the idea of tangible goals to mean that you should not push yourself to achieve as much as you can. The point is to make sure you are being realistic about what is *within your control*. Hard work and effort are under your control, but often things like budget and market popularity aren't. The best goals you can set are ones you can consistently work toward and easily measure your progress in. Most people make the mistake of choosing lofty goals early on, and then becoming jaded and unsatisfied. Don't underestimate the power of achievable, tangible goals. When you reach them, be sure to celebrate the heck out of them! Then replace them with something new. This offers the added benefit of actually feeling a level of accomplishment each time you reach one of your goals. Using this framework to set goals will make your career sustainable and rewarding.

THE PYRAMID OF SUSTAINABILITY

The **Pyramid of Sustainability** is a learning tool that we like to use when students ask about career development. The purpose of the pyramid is to show the basic elements that go into building a career, and to help guide your focus. This pyramid is unique however, in that the higher you are on the pyramid the *less control* you have over that category.

As you can see in Figure 10.1, we have four components: personal health/happiness, technical kills, network, and luck. We'll discuss each in detail below.

Personal Health/Happiness

We have set the health/happiness category as the foundation of the pyramid for good reason: your career will not be sustainable unless you are physically and mentally healthy enough to deliver your best work. This may seem radical, but it is worth restating: your health and happiness is *a higher priority* than your career development. You can be happy and healthy without a career, but your career will be short-lived if you are unhappy and in poor health.

If we haven't made it clear enough, setting boundaries for a positive work–life balance is in our opinion the foundation of a sustainable career. The main things to consider here are 1) healthy sleep habits, 2) a nutritious diet and consistent exercise,

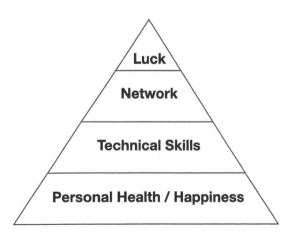

FIGURE 10.1 The Pyramid of Sustainability.

and 3) an unconditional feeling of self-worth. It can be tempting to pull all-nighters to get the job done, especially as a freelancer. However, in our experience it is crucial to *beware the culture of overworking*. By this we mean the glorification of the philosophy of "work till you drop." This philosophy seems noble in theory, but in practice it is demonstrably *not sustainable* and therefore detrimental to your career development.

The specifics of creating a routine that prioritizes physical and mental health are outside the scope of this book, but the basics are important for a sustainable career. Head over to the Sound Lab (companion site) where we've laid out a simple plan and included resources from the CDC to help guide you.

Technical Skills

The skills category is toward the bottom of the pyramid because we all have full control over the skills that we learn and practice. You might have a huge network of professionals and all the luck in the world, but without the hard skills to back it up your career will *not* be sustainable. Thus, a large amount of attention should be paid to this area of your career.

It's important to note here that the skills you should be learning are informed by your goals. If your goal is to get hired as a full-time sound designer for an AAA studio, then you should read every job posting you can and look at the requirements and responsibilities of each position. Be brutally honest about where your strengths and weaknesses lie. Then take the time to diligently work toward building your skills so that you check off each and every box for your dream job. If your goal is to make your living as a freelance composer or sound designer then you'll need to compare your work to the music and sound in the types of games you'd like to create for. Work on your music so that it is equally polished, and then build a reel that shows off these qualities.

When first starting out, a huge component of the skill-building category is working on your own projects. The importance of completing your own creative projects cannot be overstated. Any developer or audio director who wants to hire you *will* look at your past work. If you have none, it is highly unlikely that they will be interested in working with you due to the risk of inexperience. You need to have work that you can show potential clients whether it is previous professional work, school projects, or your own projects. Even if you have other work coming your way, it can still be immensely helpful to have your own skill-building side projects going on. Use these to improve on your weaknesses and hone your technical skills. They will very likely come in handy one day, and they may even give you the edge you need to secure a job or a project over some stiff competition.

One final note is that the importance of skill building does not go away, even after you have settled into the industry. Game audio is a field that requires hard technical skills as well as creativity. This necessitates a willingness to explore new skills, and a commitment to applying them in the real world.

Network

Networking is such a nebulous phrase that it can take on many meanings. For our purposes the term "network" simply refers to the people you know. This is surprisingly *not limited to people in the game industry*. It will be people in the game industry who eventually have the power to hire you, but a prospective job can come from *anyone*, and it is important to remember this. You never know who has a friend who happens to be an audio director, or who might secretly be a virtuoso cellist that could provide the exact sound you need for a project. To put it simply, networking skills are people skills.

We have placed "network" above "technical skills" on the pyramid because by their very nature the people you know are not under your control and therefore are not always predictable. For this reason "networking" is more about being yourself and building meaningful and trustworthy relationships than anything else. If you can earn someone's trust on a personal level (and trust is earned not given, as we will see in the following sections), it is very likely that they will trust you on a professional level as well. In a very basic sense, "networking" occurs naturally if you are committed to immersing yourself in your industry in a variety of ways.

Luck

The last category is luck. We placed it at the top of the pyramid because it is an almost completely uncontrollable factor. Despite this, it can play a large role in your career. Being in the right place at the right time, or making the right comment on a game forum are all examples of how luck plays into our career development. This factor is a significant reason why we stress the focus on sustainability over short-term success. We can't control how or when a lucky break will come our way, but the longer our career is in a stable place the more likely it is for luck to find us.

Because it is impossible to predict or control lucky breaks, there is little we can do about them. Yet it is still crucial to internalize the role that luck plays. It is a gentle reminder that every path is different, and that some factors are simply out of our hands. The most we can do to push our career forward is to be diligent in learning relevant skills and to make sure we take every opportunity to build meaningful relationships. Beyond that we simply have to trust that a lucky break will find us eventually.

CAREER PATHS IN GAME AUDIO

Your goals and your focus in terms of skills and networking will largely inform your choice of career path. This is a tricky topic, however because career paths are *not linear*. Like game music, your career will not proceed sequentially and predictably from point A to point B and then point C. Career trajectories can be surprising, and they are much more flexible than you might think. Often one role will offer experience and knowledge that leads to a totally new role, which in turn shifts your trajectory in a new direction and so on. This is not something to be avoided; it is something to be embraced. Follow your curiosity and try to cultivate a hunger for learning. This will ensure that every position or project you take on will broaden your skills and teach you more about game development and audio creation.

The Sound Lab

Many novice audio creators have a habit of romanticizing the work, and then feel disappointed and lost when the reality is quite different. This is not a sustainable state, so it helps to have a concrete idea of what shape your daily life will take before you make any big career moves. Head over to the Sound Lab (companion site) for more information on career paths such as in-house and freelance roles, and AAA vs. indie roles. As you digest and process the information, think carefully about what feels right for you.

11 The Business of Games, Part II

Gigs, Gigs, Gigs!

This chapter is all about networking and finding work. Here we will dig into networking in depth, as it is an important part of career development. We will begin by discussing how to network and create value. We will then show you a few ways to demonstrate that value. We will finish up by covering some methods of finding and landing work contracts.

NETWORKING AND CREATING VALUE

Some people think that "networking" is talking yourself up to as many people as possible. This is a logical way of thinking about networking, and meeting people is certainly an important aspect of it. But that definition is lacking quite a bit of nuance. Additionally, networking alone can only do so much if you aren't simultaneously creating value for yourself. This is a crucial step because effectively creating value will ensure that the people in your network will reach out to *you* when they need your particular skill set. That said, here is a more useful way to think about networking and creating value: Networking is the process of creating friendships; creating value is the process of building on those friendships in a way that offers something of value to your network. These are two sides of the same coin, and balancing them is the best way to create a sustainable career.

As a freelancer you will have to be extremely focused on your network and on creating value for yourself and your brand. This is something that many people excel at naturally, but some find difficult or uncomfortable. Make no mistake though – it is *essential* for your career development. Even for in-house professionals it is important to prioritize networking. It can give you the freedom to freelance if you decide to. It could also bring opportunities to move on to a better suited in-house position that otherwise wouldn't have been possible. The good news is that networking isn't magic, it is a *skill*. And it is a skill that can be fun and rewarding to learn. So here we have devoted a few sections to our philosophy on networking and some helpful tips to get you going.

Visiting Artist: Tom Salta, Composer

The first agent I ever had used to always say, "It's all about relationships," and how right he was. No matter what industry you want to be successful in, it's *all* about relationships. Everyone wants to work with their friends. Networking is simply a vehicle for making new relationships.

Starting Points

Now that we understand what networking is and that the goal is to create a network of professionals that value us and our work, let's explore some starting points. First, you can't create value for a network that doesn't exist. So where can you start meeting people?

Local Meetups

This is an often overlooked treasure trove of connections. Local meetups for game audio folks as well as game developers are usually free, and can at times be sparsely attended depending on where you live. For these reasons they are in fact the first place you should be looking to build your network. It will cost you nothing but time and energy. And the fewer people in attendance, the easier it is to spend quality time with them. This can lead to long-lasting connections. Check places like meetups.com, Facebook groups for game developers and audio creators, as well as local colleges. With social media it's easy to sniff out like-minded people and ask them out for a coffee or a drink. The more casual the meetup, the easier it is to make real friends instead of surface-level acquaintances.

Game Developer Lists

A very helpful place to start is with an online survey of the game studios in your area. Using a website like gamedevmap.com, devmap.gamesmith.com, or gamasutra.com can open your eyes to your local game community. Take a look at smaller developers especially. They may be in a position to work with a newer sound designer or composer, and you can learn a lot about the process of game development as a whole. Larger studios in your area may also have internship opportunities. Even if neither of these options seems plausible, look into the studios you can make a trip to and reach out to people. Be honest and curious, and offer to buy them a cup of coffee some day if they will let you ask a few questions about their work. You'd be surprised how often this can lead to a meaningful network connection.

Visiting Artist: Jason Kanter, Audio Director

Building a Community outside Industry Hubs

Game development outside of one of the world's few hubs can be a bit lonely at times. Austin, Seattle, Los Angeles, San Francisco, Boston, and the Raleigh-Durham area of North Carolina can all be considered game industry hubs, usually born out of the technical and computer science schools associated with those areas. And that's just within the United States – game industry hubs like these can be found throughout the world, each with a thriving game dev community surrounding it. Meetups, drink ups, networking events, and even educational talks can all be found within these communities, intended to educate, expand connections, and spark creative thought. But what do you do as a lone wolf composer or sound designer outside of one of these hubs? You start your own community!

New York City is considered an epicenter for many industries (theater, modeling, fashion, and finance to name just a few), but tech and gaming are not on that list. When I was hired to start the audio department at the then newly opened office of Avalanche Studios in New York I was a department of one. There were plenty of online communities I could turn to for advice and commiseration, but I was envious of the game and film audio events I would read about in those entertainment hubs far, far away. So we started our own community.

Way back in 2012 a couple of game audio folks and I would meet for lunch on a semi-regular basis. Members of this game audio lunch crew would bring their game audio friends and we'd invite new game audio folks who just moved into town. Over time this crew grew organically to over a dozen people and eventually someone suggested I start an online-based group to make it easier to connect and communicate. So I did.

New York City Game Audio (NYCGA) is a local online group for composers, sound designers, and fans of game audio. In order to join you have to be based in the NY Tristate area and have an appreciation for game sound. This requirement kept the group focused and prevented group collectors and bots from diluting the pool. We started by meeting up socially at pubs but as we matured we sought more creative and intellectual stimulation. We began hosting screening-type events at Avalanche, where we would watch a movie or play a game as a group and then all discuss the sound of it. Then we teamed up with some

(Continued)

....

members of Game Audio Network Guild (GANG) and began hosting presentations on various topics from folks within our own community or visiting guests.

Today the game audio community of New York is thriving. In the four years since that group of a dozen like-minded people last met for lunch, NYCGA has grown into a community of over 300 locally based audio developers teaching, networking, and sparking creative thought within the community. There have been multiple games whose sound has only been possible due to the collaborative efforts of game audio people who met through this network.

So while you may think you're stranded in a remote game audio wasteland, there are others like you just around the corner. Reach out to people. Start an online group for locals only. Encourage friends of friends to join you. It may be a lot of work on your part initially but the rewards are well worth the effort. And if you don't start a game audio community, who will? Special thanks to Damian Kastbauer for sparking yet another game audio blaze right here in NYC.

Advocacy Organizations

Organizations like the Game Audio Network Guild (GANG), the International Game Developers Association (IGDA), and Playcrafting are really beneficial to networking. These organizations have a slew of local meetups, which are great for all of the reasons mentioned above. Beyond that they often have educational seminars, networking mixers, and other events full of opportunities to meet people at similar places in their careers. It is also very likely that in attending these events you will meet industry veterans who may be able to impart some helpful advice.

Game Jams

Game jams are events where development teams make games in a short amount of time. These are incredibly fun and educational, and above all offer fantastic networking opportunities. You will actually walk away from a game jam with a product that you have worked on, so don't underestimate how helpful these can be in meeting people and creating value. Plenty of information regarding scheduled game jams can be found by doing a simple internet search. To find a team to jam with you can start your search on websites like crowdforge.io, where you can search for and join teams.

Conferences and Industry Events

Once you are in a place where you can afford to travel and put a bit of money into industry events, there are tons of conferences to attend. The Game Developers Conference (GDC), GameSoundCon (GSC), IndieCade, Penny Arcade Expo (PAX), and even MAGFest (The Music and Gaming Festival) are all great places to go to find like-minded people. These events have educational lectures as well as networking-specific events. A huge plus is that most of these events allow developers to share and promote their work, so it's a great chance for you to either share your work, or meet developers sharing theirs. Attending conferences can be a big expense. Planning and budgeting will help you manage which to attend in any given year. We recommend that newer game audio folks apply to the various GDC and GANG scholars programs to offset the cost. If you are already somewhat established in the industry it is always worth applying to be a speaker. This will allow you to share your work and if you are accepted it helps offset some of the costs related to attending the conference.

Social Media and Online Presence

This topic is starting to move toward creating value for yourself and finding work, but we've included it here because it is also a great way to make friends. When we say "social media presence" we aren't referring to cold-calling or solicitation (discussed later in the chapter). We are talking about taking an active role in the online community. Websites like IndieDB, TIGSource, Reddit, and game industry groups on Slack, Discord, and Facebook are all great places to post questions, feedback, or helpful comments aimed at offering something interesting to the community. This can lead to plenty of opportunities for work down the road.

Video Game Remixes and Collaborations

This is something many underestimate the power of, but video game music and remix communities like Materia Collective and OverClocked ReMix (OCR) are full of audio professionals having fun with video game music. You wouldn't believe how many industry professionals are part of these communities. So break out the synthesizers and recording gear, and get to remixing! Make friends, contribute to albums, and have fun. The worst-case scenario is that you end up having a blast remixing your favorite tunes and jamming with friends.

Ideally trying all of these approaches is a sure-fire way to start building a network. If it feels like too much, start by choosing one event to attend and choose one organization to join. Do it within the next two months. This alone will open up some new opportunities for you.

Strategies for Events

There is no substitute for making a face-to-face connection, so attending events in person is an essential aspect of networking success. When attending these events (especially GDC) try not to go in with the mindset that you will be leaving with a gig. This mindset tends to reek of desperation, and the classic "salesperson" attitude can be a major turn-off for potential clients. Instead focus on being as many places as you can, and enjoy absorbing the energy of the event. These events are meant to be fun. By being present and learning what you can, you will end up having meaningful conversations with tons of people. This is how you create and build friendships. Often these same people will end up being life-long friends, some of whom you may only see at these conferences.

A large part of the impact you have on people is how you act when you are having a conversation. A great read for this topic is Dale Carnegie's *How to Win Friends and Influence People*.[1] To summarize a very powerful book, the best way to make genuine connections is to *be* genuine. You should actually care about the people you are talking to and invest yourself in what they have to say. If you do this then you are 90 percent of the way toward making a professional connection. A common mistake people make is either overselling or underselling themselves. Oversellers may come off as cocky or overbearing, but undersellers can be just as self-destructive. Be confident in yourself, but leave plenty of room for others in the conversation. Making eye contact, using people's names, and actually retaining the information your peers are divulging all go a long way toward making a real connection.

If you are feeling overwhelmed or shy, a great trick is to simply walk around and play games. Every conference has an expo floor where various games and game prototypes will be available to play. So strike up a conversation. These devs have worked their butts off to make something for the public, and they are usually more than willing to tell you all about it. It is by far the easiest way to start a new friendship. This demonstrates an important networking concept: *real relationships are reciprocal*. Don't settle for talking about yourself and your achievements only – ask about their work and personal life. Asking questions and being genuinely interested in the answer shows that you are authentic, and that you are interested in them for more than just work opportunities. If you're more of an introvert, it also takes the focus off of you and makes it easier to dig into the conversation and find common ground.

Another tip is to have a basic "elevator pitch" for people asking about your work. Sometimes people are genuinely interested in what you do and you want to be able to effectively communicate what makes your work special. Keep it simple and to the point and try to avoid clichés. Most of all be honest. If you are interested in music for VR games, but have little experience then say exactly that. "I'm a composer and I'm interested in doing VR work." This frames it in a positive light without lying about your lack of experience. But it is also important to do your research beforehand so that you can effectively demonstrate your competence if a VR project does come your way.

One last tip for conferences is to avoid setting your sights on meeting *only* the well-known people in the industry. Don't cherry pick your connections. It is equally if not more important to build a network of people with similar career experience to you. These are the people you will maintain friendships with. One day you will realize that you are all at different studios, in positions of authority, and you will be reliant on each other to produce quality audio work. *That* is the goal. The goal is *not* to find the most famous person at the conference and talk her ear off. That person likely already has her own network which she made years ago in much the same way.

Creating Your Value

Once you have built a small but promising network of friends, what are the next steps? How do you make yourself valuable to them as a friend and professional? The bottom line here is that you need to ask yourself what your network needs and how you can be the person who fulfills that need. This doesn't mean constantly asking if they need extra music or sound design for a project they're working on. In fact, most likely that is exactly what *they* want from *their* network. It is much more likely that they need something that is not in their skill set. If they are composer and you are an instrumentalist then recording some parts for them would be a great start. Or perhaps they are under the gun on a project and need some help organizing a huge list of voice auditions. You could offer to set up a spreadsheet for them or be the go-between for the auditioners. People remember when you are helpful to them, and it will show your competence and commitment.

Another tip is to avoid letting everything be about work. Ask yourself if you're being a good *friend* or if you are just being a work acquaintance. Follow up with people you meet after conferences, even when you aren't looking for work. Offer to meet up if you're ever in the same city. Find some fun side projects to collaborate on. These are all great ways to show that you're in it for the long haul.

At the crux of the issue of offering value is the fact that we *all* want our network to provide work opportunities for us. Knowing that, a hugely successful way to create value is to provide work opportunities for your network. If you are in a position where you have budget to hire musicians, or voice artists, or sound designers, or composers, then pay it forward and offer an audition to people in your network. This puts you in an amazing situation where you get to collaborate with your friends, and come away looking like a hero. They will likely remember you when they are in a similar position.

Visiting Artist: Xiao'an Li, Composer, Partner at Li & Ortega, Founder of Business Skills for Composers

Creating Value

Professional relationships that one hopes to harvest some personal benefit from are transactional in nature – you reap what you sow (sometimes there's a frost, locusts, or some other disaster where you get nothing out of it, but that's life). Systematically building a critical mass of these relationships and creating value for each of your contacts makes an overall net positive more likely. At the foundation of this is a "giving" attitude. Your primary driving force should be "How can I help this person?"

Develop a thoughtful interest in your contact's needs. Ask questions with genuine, humble curiosity and they may surprise you with their honesty. Armed with this information, you have several no-cost options to create value: introduce them to useful companies you've networked with; share informative articles about their industry sector; send them a great deal you found for a Mediterranean vacation because they mentioned wanting to take their kids somewhere special. This behavior makes you much easier to remember than the last 50 composers that sent them a generic cold email, begging for work.

DEMONSTRATING VALUE

Think of creating your value as a never-ending process. In terms of networking, your value is what you can offer your network. However, there are others ways to *demonstrate* value, which we will discuss below. The goal will be to demonstrate your value in such a way that you draw in clients and members of your network, rather than pursuing them yourself. This is ideal because it is self-sustaining. Once you get it started it requires little energy to maintain.

To begin demonstrating your value it is important to consider what your strengths are. What do you have to offer which is unique to you and valuable either to clients or potential members of your network? Then you need to decide how best to demonstrate those strengths. Below are a few of ways of doing this effectively.

The Elevator Pitch

Imagine the typical film scenario where the main character wishes to land a promotion at their company and happens to find herself in an elevator face to face with the C.E.O.

This is where the elevator pitch comes in. It's an introduction that can be communicated within the brief timing of the elevator ride. Regardless if you find yourself in an actual elevator, a meetup, or an interview, it's a great skill to master.

In the Sound Lab (companion site) we offer a step-by-step guide to creating an elevator pitch and additional resources for developing it.

Business Cards

Business cards are a simple way for people you meet to remember your name and what you do. They *are not* a good way for people to get to know you. Sometimes people think that handing out a business card is synonymous with making a meaningful connection. This is demonstrably not the case. Business cards are a way to efficiently exchange information *after* you have made a meaningful connection.

The best business card is one that 1) has all of your information in an easily readable format, and 2) is memorable. So with that, be creative. Some of the best moments we've had at conferences are being given hilarious cards. Just make sure that your card isn't more memorable than you are!

Website

Having an up-to-date portfolio of your work is an essential aspect of demonstrating your value. Demo tapes and CDs have long since given rise to personal websites thanks to cheap and flexible services like Squarespace, Wordpress, and Wix. A personal/professional website is essentially a place to collect samples of your work, client lists, and anything else relevant to your professional life. Websites are great because with one click a potential client can see everything you've ever worked on. It's also easy to tailor your website to any slice of the industry – AAA, indie, mobile, console, etc. simply by choosing what goes on your home page. With that said, there are some important concepts to understand before building your site.

In our view a website is a chance to show off your absolute best work to anyone who clicks onto your page. Don't squander this opportunity. Websites need to be simple, uncluttered, and offer an as easy as humanly possible way to listen to your best work. That said, you don't want to give visitors too much to look at or too much to click on. For this reason we recommend having a very simple and minimal home page with nothing but a handful of audio tracks or videos. Visitors should be able to click onto your site, see the one thing you want them to see (a reel or short video), and hear your work.

Apart from the home page, it's helpful to put a list of clients and projects as a separate page. This is to keep from cluttering up your home page. If you want a page with awards, recognition, or client satisfaction quotes they can go on a separate page as well. Of course, you'll also need a bio page and a contact form. Your bio should tell your story. It should be a picture of yourself and your work. People will want to know what inspires you and what your vision is for doing what you do.

Above all, your website has to define your "brand." This means everything about your site needs to visually point to you and to how you want to be known. If your brand is that you are a dark and broody but visionary composer, then your visuals should look dark and broody. If your brand is a funny, fun-loving sound designer then maybe include some pictures in your bio that show off this side of your personality. A lighter color scheme might be more appropriate here. Keep things clean and visually appealing, but every single thing on the website should point to who you are and what you want your brand to be.

Demo Reel

We've discussed the importance of a well-crafted website. A reel is an important supplement to your website. Demo reels can be presented on your home page, but don't necessarily need to be. More often reels are used after you've made contact with a potential client and they ask for your work. For this reason a demo reel should be crafted at the very beginning of your career, and you should update it as often as necessary. This way when someone asks to hear your stuff, you have something polished and ready to go.

Demo reels are short audio tracks or videos which show *only your very best work*. Some say reels should be a maximum of 60 seconds long, but this also depends on the situation. If you are sending a reel out unsolicited (see the section below on "Finding Work") then 60 seconds is ideal. If a potential client *asks* to hear your work specifically then one to two minutes should be just fine. The bottom line is that a reel is a quick and polished overview of what you can do.

Before crafting your reel (or reels, yes it's possible to have more than one) first ask yourself the following questions.

- Who am I as an artist?
- What are my strengths?
- What am I *passionate* about?

Second, ask yourself:

- Who is the recipient of my art?

The answers to these questions all play an important role in what projects go into your reel. With your reel you want people to see your strengths and what makes your music unique to you. You also, to some degree, want them to see the kind of projects that you are passionate about because that will likely be your focus for future work. It is also important to consider who will be watching this reel. If it is a developer that creates only horror games, then your reel needs to be full of spine-tingling music. For this reason it is common practice to have a few different reels in genres that you are interested in.

Once you decide what tracks you want to put in your reel, you will have a few other technical considerations to decide on. Do you want the reel to be video or audio only? In general, sound design reels should have the visuals with them, so they are usually videos. Music reels can really go either way. The general consensus is that if you have projects that you are proud of and look great, then it's perfectly fine to cut up gameplay videos and splice them into a reel. If you are looking to redo the audio to a gameplay capture or cutscene, then you may be better of just sending the reel as an audio file unless it is absolutely spectacular work. Overdubbing work that is not your own can sometimes show immaturity and lack of experience, which is not what you want potential clients to walk away from your reel thinking. All that being said, it really depends on who is receiving your reel. Many indie developers would accept audio overdubs if it was in the right style. AAA audio directors would be much less likely to do so. Do your research and find out exactly what they are looking for before you send your reel.

Before sending your reel make sure that everything is as polished as possible. Use a professional service like Soundcloud, YouTube, or Vimeo rather than just sending a WAV file or MOV. There are even services like Reelcrafter which allow you to set up a professional portfolio in minutes and track its engagement. Regardless of what you use, make sure your name and information are clearly visible and be sure to link to your website. If your reel is in video format, make sure your name and the project title are written in clear type on the screen *at all times*.

One last demo reel note: Don't let lack of experience hold you back from creating a demo reel or looking for work. At some point every sound designer and composer had zero credits. Be aggressive and create your *own* opportunities for work. Join game jams, create sounds for hypothetical games, go out and meet college students who are studying game design and ask to collaborate. Look for projects on indie game forums in the "Help Wanted" section and make a post to collaborate. There are plenty of ways to self-start, and everybody has to do it at some point. The bottom line is to get out there and *do what you want to do*. You cannot create a demo reel if you haven't created anything.

We recommend checking out resources like Power up Audio's "Reel Talk" stream on Twitch.[2] Every week they review and provide feedback on viewers demo reels. Even if your reel isn't selected for review, you can learn a lot from the information discussed. There are various Facebook groups that also offer demo reel or content review on selected days of the month. There is also the GANG Demo Derby, held annually at GDC which provides individualized feedback on sound design and music reels.

Gina

It's a small victory if a potential employer or client views your demo reel. You will want to be sure the material will hook them right away so they feel more inclined to keep listening. Head over to the Sound Lab for information on building a great demo reel.

Public Speaking Engagements

Public-speaking engagements are a great way to demonstrate your value. This is an example of **inbound networking**. Inbound networking simply means that potential clients or members of your network are *coming to you* rather than you going to them. An example of **outbound networking** would be going door to door in your neighborhood asking if anyone needs audio for their game projects. Both strategies have their merits. Usually outbound networking is necessary when building a foundation for your career. The downside is that outbound networking is high effort with minimal reward. It's more of a numbers game than anything else, and it can get tiresome for you and the people that you are "selling" to. Inbound networking requires patience and persistence, but once you have established yourself it yields moderate to high rewards for minimal effort. In other words, it takes some effort to set up a system of inbound networking, but after that it is mostly an issue of maintenance, and your network will grow almost on its own.

Public-speaking engagements are an ideal form of inbound networking because they give you a forum to share your work with peers. When lecturing at a conference, for 40 or so minutes *you* are the star attraction. Often people will engage with you afterward, asking questions and offering their business cards. It's essentially advertising for free. Another bonus is that almost anyone with a bit of experience can apply to speak at a conference. This isn't to say that it's easy by any means, but it *is* a fair shot if you have some interesting work to present. Public speaking is an opportunity to be creative and share what makes you and your work special.

An important point to remember about these conferences is that speaking engagements are usually very risk free. This may seem absurd considering that public speaking is always listed as a top fear in the general population. But the truth is that when you've adequately prepared, these engagements are usually confidence building and fun. Most conferences have a process that will provide feedback on your talk, and offer supportive ways to improve on it. GameSoundCon is one of the best conferences to attend for this reason. The attendees are all audio professionals as well, so they all have similar interests. If a subject is interesting enough for you to submit a talk, it will likely be interesting to the people in the audience. If you are still terrified of speaking you can start smaller at a local meetup. The Game Audio Network Guild (GANG) Northeast has been holding monthly panels/talks at NYU (in conjunction with NYC Game Audio) for almost two years, and plenty of local audio professionals have taken advantage of this forum to practice public speaking. Many of them have gone on to give these same talks at GSC and even GDC. Beyond that, we would recommend teaming up with some partners. Having a few friends on a panel can make the whole process much less stressful, and it usually offers an opportunity to form deeper connections with your fellow panelists.

Teaching and Mentoring

This overlaps a bit with public-speaking engagements, but there are plenty of ways to employ teaching as an inbound method of networking. YouTube, Facebook, and blogs are perfect avenues to share your knowledge and experience with the world. This can be in the form of tutorials or practical articles – or really anything that you feel good committing your time to. The only caveat is that what you're sharing should be something that people you would want to be in your network will value. The idea here is to provide worthwhile information to people so that they continue to follow you and your work. It places you in the role of the "expert" or "advisor." The goal is *not* to show off all of your projects, which may or may not be helpful to your viewers. Give the viewers what *they* want, and your network will grow.

If you're interested in going this route, organizations like GANG and the Audio Mentoring Project (the AMP) are absolutely the best. There you will have opportunities to either be a mentor, or to mentor others. GANG also has many leadership roles, and is in need of volunteers in local chapters as well. If your town doesn't have a chapter, start one up!

FINDING WORK

We have now arrived at the most commonly asked question in the industry: "Where do I find work?" Unfortunately, there really is no easy answer. Everybody has a different career path, and what works for one person may not work for you. To grossly oversimplify this topic, finding work is about three things:

1. How many people do you know?
2. How many of those people are in a position to hire you?
3. How many of *those* people trust you enough to follow through with a job offer.

This is true of freelancers and in-house professionals alike. Yes, the more people you know the more likely you will know someone who can hire you. That takes care of points #1 and #2. But if you don't have the skills or confidence to convince them that you're hirable you will have exactly zero people in category #3. So it is a balancing act of meeting people, staying on their radar, and making sure that you are the epitome of a reliable professional.

One place to start is by asking people how they got to where they are. It can shed light on some tactics that you may never have thought of otherwise. Just don't expect things to go exactly for you as it did for others. It can also show you strategies that commonly work, and strategies that commonly *don't* work. These two are of equal importance because they'll clue you in on which tricks to use and which to avoid. The act of asking people about their career path alone can lead to a meaningful network connection, so it is always worth the time chatting about it.

Work opportunities (whether for in-house roles or freelance projects) can come from any combination of the following places: solicitation, network relationships, adverts, repeat clients, or professional recognition. Tactics range from outbound (solicitation) to inbound (professional recognition), so the order here is relevant. Outbound tactics are typical of a new career, but as you develop your business the goal is to move further toward inbound tactics. This will allow you to focus more on working and less on finding work. Also note that these are very broad categories. For example, network relationships can encompass so many possible avenues for finding work. And these categories are not mutually exclusive. Any combination of these categories can yield a work opportunity, so it's best to consider all four strategies. Below we'll discuss each of these strategies in detail.

Solicitation

When first starting out, odds are that you'll have to experiment to some degree with solicitation. This just refers to some form of cold-calling. By definition this is an outbound form of networking. You are essentially sending an application that no one asked for ahead of time. So it is important to respect the recipient's time and make things as simple and brief as possible. It used to be easier to get in touch with game studios via phone, but nowadays this is much less likely. Unless you have a personal connection to a larger studio, this usually means that most "cold-calling" will in effect be "cold-emailing" or "cold-social-media-ing." Nevertheless, this is an essential aspect of some people's networking schtick. There are however a few important points to consider when taking this approach.

Before sending out any solicitation it is crucial to observe exactly *who* you are reaching out to. And we mean that literally. Who is the human being that will be on the receiving end of your email? Not only that, but what company does this human being work for and what is the company looking for in a potential applicant? Most of what we discussed in the section on demo reels is directly applicable here. What kind of games do they make? Are they looking for the kind of work that you can produce?

Are they a startup studio looking to take a chance? Or are they an AAA studio looking for top-tier work? These are real issues to consider *before* sending an email. Do your research or you will get nowhere with solicitation.

Another crucial point regarding cold-calling is the development cycle. Again, do your research here. Do they have an audio team already? Are they 90 percent finished with the game? If so, this may be a dead end. It may also come off as stepping on the toes of the current audio team. In short, the earlier they are in the development cycle the better your chances of getting in. Spend some time getting to know the development team and looking into their social media and blog history before reaching out.

Lastly, and this is a big one, be genuine and passionate. It may take more time to craft an email this way, but it is far more likely that a developer will respond if you show genuine interest in them and their projects. This is exactly the same concept as building your network. It *is* a numbers game, but it shouldn't be *just about numbers*. You're looking to make a meaningful and memorable connection here, and you have one short email in which to do it.

> Note the difference in formality during interactions. When speaking informally (Facebook Messenger for example) the goal is to incite an authentic conversation. Ask questions and get to know the person before soliciting work. In an email scenario however, be short and to the point. It is simply an introduction (or a reiteration of how you met that person and why you connected in some cases) and statement of purpose. The conversation will come later.
>
> Spencer

Two final tips regarding all solicitation: *never* open a dialogue by criticizing their current audio, and avoid talking about pricing right away. Pricing comes later during the contract phase, and criticizing audio is a slippery slope that can end in heated resentment. Make the interaction about your passion for their work and your commitment to providing valuable audio to their project. Remember that you are reaching out to them for work. Place the emphasis on what *you can do for them*, not what they can do for you.

Network Relationships

This method of landing gigs is simple – build a network and wait for it to create work opportunities. This can be a mix of inbound and outbound networking depending on how you've chosen to build your network. The opportunities themselves can arise in many ways – referrals, seemingly out-of-the-blue work requests, introductions to potential clients, etc. Due to the unpredictability of the sources of these opportunities the best approach is to follow the advice in the previous section "Networking and Creating Value" to cultivate a lucrative network. All that is needed afterward is patience and persistence.

Applications

Applications for advertised jobs are essentially the means by which all in-house roles will be filled, so it is a very important aspect of finding work. This is not really outbound or inbound networking because it is in effect *the developer's* way of sourcing inbound talent. As a freelancer it is less likely that you will have to make too many applications. Some developers post adverts for contract work, but in these cases there are usually so many applicants that the odds can be stacked heavily against you.

The application process is much less mysterious than other means of finding work because the requirements are all laid out for you to see. You will have a clear idea of what the client is looking for, what skills you must have in order to be a successful applicant, and what your responsibilities will look like if you are accepted. This means that you have a very reliable metric to direct your practical experience. If the job specifications you're looking at are all about Wwise and Unreal, then go start a personal project using Wwise and Unreal. If the specifications are emphasizing field-recording experience then get your gear and start recording some sound libraries. Studying each specification will give you a goal to work toward, and one that you can objectively evaluate.

The key to a successful application is honesty. It is all too common to hear audio directors discussing the slew of unqualified applications they receive for senior-level positions. If you don't have the experience required, then don't apply for the position. Most adverts are crystal clear on what they are looking for in a candidate, so respect the employer's time and yours. It's perfectly acceptable to pass on an application if you aren't qualified. It gives you the opportunity to work on building your skills.

Even if you do have all the skills required, it is important to put your best foot forward. Audio roles are extremely competitive, so make sure your cover letter is concise and polished. There should be no typos or spelling errors, and it should clearly state your skills and why you are the *perfect* candidate for the role you are applying for. If you are applying for a sound design role *do not* discuss how much you love composing music. This is a red flag that you are not interested in the responsibilities of the role. Again, the demo reel advice comes in handy here: Tailor your supporting documentation to the role that you're applying for. With a bit of luck and persistence you will land an interview.

Repeat Clients

"Repeat clients" is technically not its own category because it doesn't tell us any information about how the opportunity to work with said clients arose in the first place. However, the fact is that developers *want* to work with people that they know and trust. So if you have landed a project – *any* project – you have a leg up on the competition. For this reason focusing on repeat clients is by and far the best way to build a sustainable career, which is why it gets its own category. In fact, many composers and sound designers make a very comfortable living off of repeat clients – and you can too! Let's dig into a few ways to take advantage of repeated business.

Repeat clients are the result of an interesting mix of outbound and inbound networking. For the most part it is inbound because often clients will seek you out to work on their latest projects. However, it is outbound as well because it requires some effort on your part to foster a positive, mutually beneficial relationship. In short, clients need to feel that you were easy to work with, reliable, and that you delivered a quality final product.

Being easy to work with can mean many things depending on the situation, so it can be the trickiest of the three elements of positive client relationships. This mostly boils down to communication skills and attitude. In your interactions with your client make sure that your communications are clear and simple so that you eliminate the possibility of misinterpretation. This will save you and your client time. Additionally, you will need to intuit what exactly your client wants from you so that you can deliver it. In this case we aren't referring to what *assets* the client wants. We are referring to what kind of *relationship* your client wants. Some developers are looking for an audio designer who can deliver quality assets from a specified list autonomously and within a specified deadline. Others are looking for more of a collaborator, capable of creative decision making and even giving feedback on the project as a whole. It is your job to determine what your client wants out of this partnership, and to do your best to fill that role. On top of that, an energetic and positive attitude can seal the deal for future business.

Reliability is equally important when nurturing client relationships. Developers are often under tight deadlines, so they need to feel that you will deliver the services you were hired for on time and exactly as specified. Do your best to manage your and your client's expectations. It is always better to be realistic about timeframes and schedules than it is to over-promise, even if it feels awkward at first. If a client asks you when you can finish producing a number of assets, don't beat around the bush and *don't* give them false expectations. Setting realistic expectations might be uncomfortable, but it is far more uncomfortable in the long run to have to ask for an extension. Such a situation will leave your client thinking that you are not reliable and therefore not worthy of repeat business.

Where reliability is concerned, it can be helpful to maintain consistent contact with clients to update them on your progress. This brings up an important issue – asking questions. Many people think that asking questions will show them to be unprofessional or unreliable. This is very clearly not the case. In her book *Dare to Lead*,[3] Brené Brown shows that asking questions actually *increases* supervisors' trust in their workers. Thus, it also increases the perception of reliability. It is impossible to do your job to the best of your ability if you don't understand elements of it. The bottom line is if you need clarification, ask. Give yourself the tools you need to successfully deliver and your clients will be amazed at your reliability.

The final element of a good client relationship is delivering a quality final product. This is actually the easy part. It is a culmination of understanding the client's needs and putting in all of the creative and technical skills we have discussed in Parts I, II, and III. Be passionate, confident, and committed to great audio and you will always deliver a worthy product.

Professional or Popular Recognition

Professional and popular recognition is the only category that is *entirely inbound*. This is the ideal place to be in terms of career development because at this point you will find work based on your reputation alone. Unfortunately it can be difficult to get here, sometimes taking years of work and a fair bit of luck to achieve. As a game audio creator much of your outward success comes from the success of your projects, and that isn't always something within your control.

Some potential places to find work at this stage are things like referrals, industry awards and recognition, soundtrack popularity, and agents. These offer some very powerful opportunities because they come from peers or clients who *already* value your work. You have an edge on other audio professionals, as you do when dealing with repeat clients. Some example scenarios are a developer reaching out to you because they enjoyed a previous game soundtrack you've composed, or a friend referring you to an audio director for a role on her sound team because you have a reputation as a master voice editor. The thing to remember about these types of situation is that they are self-fueling. By doing quality work and making a name for yourself you will receive more opportunities, which in turn allows you to do more work and so on and so forth.

The topic of working with an agent is a very popular one, especially among composers. Sound designers often work on teams, and there are plenty of in-house positions so finding an agent who specializes in representing sound designers is less than common. By contrast there is usually only one composer needed per game, and there are more composers than sound designers in the industry. For this reason agencies are really only relevant to those looking to write music for games specifically. Getting an agent can be tough work. Referrals are possible of course, but they only guarantee you a meeting with an agent. They do not guarantee that you will be represented. You essentially have to already have a prolific career before an agent will even be interested in you. The prevailing advice for those in the industry seeking an agent is this: if an agent hasn't contacted you, then you are not ready for representation.

The role of an agent is to find potential clients and represent you in contract negotiations. Agents get a percentage of what you make for a project, so it is in their best interest to get the best deal possible. This can be hugely helpful on larger contracts, which can be convoluted and confusing. Agents also play a role in bringing in **pitches** for new work (see the section below "The Design Test"). We'll discuss the process of pitching for a project later, but pitches are essentially an audition for large game projects.

There are some helpful things to remember as you build your career that can make representation more likely in the future. For one thing just meeting an agent, even at an early stage of your career, makes it more likely that they will represent you later on. Agents are very well connected, so they make it a point to meet with many composers. But making a good impression on them early in your career can help plant the seed of a relationship. Later on, when you've accrued some larger projects you can check back in and see if their agency is looking for more talent. Additionally, if you're having a hard time finding representation and are looking to make yourself more known to the professional world, it might be a good idea to start out exploring options in public relations before you start soliciting agents. PR professionals are great at finding your strengths and telling your story to the world. This can make a big difference in how valuable an agency perceives you to be.

Finding Work Summary

Remember that the above categories are broad. It would be impossible to detail an exhaustive list of every circumstance leading to work. But if you balance your efforts in each of these categories and have patience, you will find that opportunities for work will start cropping up. Start with solicitation and a heavy emphasis on your network, but keep in mind the goal of creating a slew of inbound clients. It is worth mentioning again the Pyramid of Sustainability from Chapter 10. You can improve your technical skills and your networking skills, but there is always an element of luck that is out of your control. If you are persistent and prioritize overall sustainability, your career path will be surprising and rewarding.

The Design Test

Notice that up to this point we have used the term "work opportunities" rather than "work." This is intentional. Digging up the *opportunity* for work is half the battle. The other half is actually landing the gig. There are a number of things that go into being hired for a position or project. We already mentioned the application process, and this is usually the case for in-house positions. For freelance work, whether it is for a sound design role, music composition role, voiceover work, or anything else it usually includes some kind of design test or pitch. This is an audition, and it is your chance to show the developer what you can do. Ideally, a pitch will be a short sample of music or sound design to video that is memorable and uniquely fits the style and mood of the game in question. To that end, the very first piece of advice we have to offer is to follow directions. Even the boring ones! If an audio director is asking you to pitch along with ten other highly qualified composers, and you are asked to compose music to a short gameplay clip, you can bet that you will be shooting yourself in the foot if you instead deliver a WAV file of a new track you produced. In fact, you may not even have your pitch heard at all. Observe all of the requested technical specifications as well as the creative directions, and deliver on time if not early. Audio directors and developers will commonly send a reference for the style of music or sound design that they want. If this is the case then make sure you are clear on exactly what they are looking for and follow it to the "T." Ask questions if it helps, but don't deliver a hybrid orchestral track if they want orchestra only. These types of oversight can lose you the project and waste time.

An important factor to remember here is that pitches are extremely competitive. You should approach a pitch for a project with excitement and enthusiasm, but also with a sense of boundaries. Try allotting a timeframe to work on your pitch. Choose something realistic, but don't go over. You should be able to make a short, high-quality sample, and then move on to other tasks. Don't let the excitement of a pitch hold you back from delivering on current projects.

It can be helpful to look at the pitching process as if it were test-driving a car. In fact, it's an example of the developer or audio director test-driving *you*. Part of what they may be interested in seeing is how you will work under their direction, so employ all of the techniques we mentioned earlier in the "Repeat Clients" section. Be easy to work with, be reliable, and above all deliver the best final product that you can. Some pitches even allow for revisions, so make it clear when you deliver that you are genuinely interested in hearing what they think of your work. This is a win–win because it shows your professionalism and commitment to quality, and even if you lose the pitch you will have some solid ideas on how to improve.

One final point to observe when pitching is to maintain a positive attitude. Pitches are full of tough competition, and even the best of us lose far more pitches than we win. The upside to this is that the excitement and fun of the pitching process can be a phenomenal motivator. Even if you lose the pitch you will probably walk away with a great demo piece. So many of our pitch samples that did not win us a project have gone on to be part of lectures or conference talks, or even to serve as demo material that *did* get us the opportunity to work on a different game. The bottom line is to take as many pitches as you can, but set appropriate boundaries (especially concerning the amount of time you put in) and expectations. Then move on when you've finished. If you are chosen for the project, then congratulations! But if not, there are other fish in the sea, as they say. Check out the Sound Lab (companion site) for some more information regarding design tests.

BUSINESS AND PRICE CONSIDERATIONS

Once you nail the design test you will then move into a **bidding process**. A bid is just a concise price estimation. Unfortunately, the process of pricing ourselves and charging for our services makes many of us squeamish, especially if asked in regards to a project we like. But pricing our work and trudging through the bidding process is part of the business of game audio, and it's one we need to lean into to be sustainable.

Bids come in all shapes and sizes. Sometimes you'll see a price per asset ($X per sound effect); other times you'll see a price for a given timeframe ($X per minute of music), and still others can be an hourly rate or a monthly retainer. Recently it has become increasingly common to offer lump sums per project. This is a great option for lower-budget indie projects because it makes it simpler for them to estimate the total cost of audio. We recommend this method if you are hired for both sound design and music, or if you have worked with the client previously and are familiar with their development process. Regardless of which pricing model you use, it's important to consider all factors that affect price.

What Determines Your Price?

The price you set for a project can be broken down into a few categories. These categories add up to the price that you will offer on your final bid. We'll look at each category below.

Studio Fees

Composers and sound designers, as well as musicians and voiceover artists, should consider recording studio fees when determining price. If you are working with any kind of talent – musicians, voice artists, etc. – you will have to find a place to record them. If a home studio recording is inadequate for the task, or if distance is a barrier, then the only option is to pay to rent out a studio. Some studios offer very reasonable pricing, but it is important to estimate the length of the recording session, multiply it by the hourly rate, and then add this sum to your bid. Note that it is common practice to overestimate the session length to avoid running out of money for the studio. For these estimations – and any estimations for added cost of a project in the following categories – the best bet is to reach out to professionals that offer the services you need and ask them for a realistic quote. This gives you some real-world data to back your calculations with.

Equipment/Software Fees

As a game audio professional (even as a relative newbie) you have likely spent hundreds if not thousands of dollars building your studio. Your business will eventually have to make this back or risk going under. It is wise to factor in equipment and software fees where appropriate. For example, if you need to rent remote recording gear for a gig, or if you need to buy a shiny new library because a project does not have the budget for live players, then factor it into your bid. We have more than once successfully convinced a developer to spring for a new sound design or sample library by arguing that it would drive up the quality of the work without costing an arm and a leg. The beauty is that after the project ends you can keep the library. It's a win–win.

Talent Fees

This goes without saying, but when working with musicians or voice artists you must pay them. This payment should come from the developer and not out of pocket. The only exception is if a developer is skeptical about the benefits of a live recording vs. a MIDI mockup. In cases like these we have sometimes paid a small sum ($25–$50) out of pocket for a very short speculative ("spec") recording. We then send the developer a comparison between the mockup and the live recording. It is rare that a developer will opt for the mockup over spending a couple hundred dollars on a great recording.

Management/Organizational Fees

Some projects require hours and hours of spreadsheets, Skype calls, and online communications. Most game audio folks are happy to do this, but it should always be factored into the bid. Make sure to clarify that all business interactions including feedback meetings, creative brainstorming sessions, and time spent organizing spreadsheets of data are all billable hours. This has the added benefit of making meetings efficient rather than endless timewasters.

Mixing/Mastering/Outside Collaborator Fees

Toward the end of a project you may end up contracting out miscellaneous tasks including mixing or mastering of assets. These tasks could be things that you're capable of doing, but lack the time to complete. It also could be due to the need for a specialized engineer's touch for a final polish. Either way, if this is the route you're likely to go then make it part of your pricing.

Buffer Budget

Aaron Marks calls this category "The Kicker" in his book *Aaron Marks' Complete Guide to Game Audio*.[4] Essentially it is the extra budget that is commonly added to a bid to account for margin of error or unexpected extra costs. It's good practice to calculate about 5–10 percent of the total bid and leave it open for any extra recording fees, changes in direction, software damage, or other unpredictable elements of production. This is less necessary for smaller projects that won't be a massive time suck, but for larger projects it can save you quite a bit of stress and hassle.

Creative Fees

At last we have come to the creative fee. This will make up the lion's share of your pricing and is a hotly debated topic. Of course, there are ranges you can find online for what you *should* be charging per minute of music. But in reality these ranges are far wider than they are purported to be. This is because the game industry is changing. Mobile and indie developers are making games with budgets under $10,000. Some games are even developed for less, and this is by no means an indication of the quality of the final product. You can't expect a developer to pay you $2,500 per minute of music if the budget of the entire game is $2,500. So how do you set your creativity price? To break this topic down, we must start with some basic numbers.

$1,000 per minute of music and/or ~$150 per sound effect

These are common numbers for the price of the average game audio professional with experience in the industry. These numbers would go down for newer professionals, and go up for professionals with more experience or notoriety. The only problem with these numbers is that they are more or less arbitrary. They don't actually reflect real-world prices because they bear no relationship to a particular project. We don't know what the details are in regards to the pricing categories (recording fees, talent fees, etc.) so we can't know what fraction of those numbers is calculated for the creative fee, or the studio fees, or anything else. What's worse is that these numbers have been floating around for years – possibly decades – and they certainly have not been adjusted for inflation. The myriad changes in the game audio market have not been adequately considered when it comes to the traditional pricing wisdom. According to a 2017 GameSoundCon survey the most common prices per minute of music are $100 and $1,250.[5] This is an astronomical gap, which suggests that developer budgets also have a wide range. As we mentioned earlier, mobile games are a huge portion of the consumer market now and some mobile apps can have a total budget of $1,000. Is it reasonable to charge $1,000 for a minute of music in that scenario? More than that, do you think that it is *profitable* to price your services in a way that does not reflect changes in the market itself?

10 to 20 percent of the overall project budget

Here is another common generalization of game audio pricing. These percentages would cover the total audio budget as compared to the budget of the project as a whole. So 10 to 20 percent of the total project budget would cover music, sound design, implementation, and any voice or musician fees. This estimation does work a lot better because it is more reflective of changes in the market. If project budgets are on the rise, then so is the overall audio budget, which is very reasonable. It is also a very realistic way to share your bid calculations with a developer. The obvious downside is that it offers nothing in the way of business development. For example, with this model an experienced sound designer might take a micro game project on for maybe $50 or $100. For someone with considerable experience, a killer portfolio, and a rock-solid reputation this project will not put food on the table, nor will it be likely to advance her career in any way. The previous model of $1,000/$150 per minute/sound effect actually gave us a stable reference point for career development, but this model does not. Considering this model even further, it is unlikely that you will be able to sniff out the exact budget of every developer that you enter negotiations with, leaving your pricing completely up to chance and guess work. Believe us, this might seem tempting when you are bidding for your first few gigs, but it is far more stress than it is worth! So how do we set our price point for our creative fees to reflect both the market and our career development?

Anchor Value

Anchor value pricing is a method of setting your price outlined by a 2018 GameSoundCon article.[6] In essence, this method allows you to set your price based on the *value* of your overall services. The article asserts that the first number that a client sees will be the value that they attribute to you and you work. This is crucial to your career development. When clients value your work higher, they are willing to pay more, and will respect your technical and creative input more. They will also be more appreciative of your collaboration in general. It affects just about every aspect of your interactions with them in a positive way. You'll find that once you start putting a higher number as your anchor value, you will also feel more confident in yourself and your work as well.

So how should you set the anchor value? The truth is that there is no real answer to this. Basing this number on number of years of experience is common, but also somewhat arbitrary. Years of experience don't necessarily lead to quality. For that matter, quality itself is somewhat arbitrary. You would be surprised how variable the term "quality audio" can be when you ask a game developer to evaluate work. For one client "quality audio" means fantasy-driven sound effects and epic music. For another, a plucky cartoonish aesthetic is what they mean by "quality." Quality really boils down to individual taste, and that is impossible to quantify.

Our favorite method for setting an anchor value is simple. Imagine yourself in a situation where your dream project comes along and the developer asks you for a bid (per hour, or per asset, or for the whole project, it doesn't really matter). Now pick a number in your head that you are fully comfortable and confident in as your price point. Now *double* it. This may seem greedy, but as creative, passionate people we tend to undervalue ourselves. This is especially true when faced with projects that excite and inspire us. This method forces us to push outside our comfort zones and prioritize our career and personal development. For the skeptics out there, sit tight because the process isn't over. The anchor value is not necessarily the price that you will be paid. But it is an important part of your creative fee.

THE FINAL BID

Now that you have all the elements of an effective bid put together, including your anchor value, it's time to send all this information to your potential client. The process of sending over your bid is equally as important as the bid itself, so be fastidious in your communications. The goal here is to set your anchor price and then *leave negotiations open*. This serves multiple purposes. If you are dealing with more of an indie budget it shows that you are an experienced professional, but that you are cognisant of (and sympathetic to) smaller budgets. This also puts the ball in *their court* so that you don't have to waste valuable time trying to guess what their budget is and calculate a percentage off of that.

If you are dealing with an AAA developer this tactic still works, but possibly to a lesser extent. For one thing, AAA studios often tell *you* what they are willing to pay for a project and you can either accept or decline. Most AAA studios are only interested in buyouts (see the next section, "Navigating Contracts") so they are offering more per minute/asset anyway. Additionally, in many AAA deals you will have an agent or manager negotiating the price point for you, so you won't have to deal with too much haggling. If you are stuck negotiating an AAA contract yourself (which is still a common occurrence), then there are usually more data-driven means of determining a realistic anchor price. You might be able to dig up some public salaries or budgets for previous games. You can then extrapolate using the 10–20 percent rule based off of that. You will also want to ask questions to clarify exactly how much work you will be required to produce (how many minutes of music, how many sound assets, etc.) to better understand the scope of the project. Below is a brief list of considerations you should make when trying to dig up an appropriate price for an AAA bid.

Questions to Ask Before Placing Your Bid

Get to know the developer.

- How big is the team? What are the roles?
- Are team members full time or contract?
- How long has the team been in place?

- What are the details of the studio's past projects?
- Will you be working in-house or remotely?

Get to know the game.

- Is there a publisher?
- What is the development cycle length, timeline, and ship date?
- What are the intended release platform(s)?
- What is the story and overall gameplay length?
- What are the characters, levels, and core mechanics?
- What games or other media were an influence for this game?

Understand their audio scope.

- Will you be responsible for music, SFX, VO, or implementation?
- What is the music and SFX style?
- Will you be working with middleware and engines?
- Will there be added animations, cutscenes, trailers, or marketing assets?
- What is the audio budget, and who owns the rights to the final product?
- Do they have budget for live instruments?
- What is the Foley and field recording budget?

These are all important questions to ask your client. The answers they give you will provide an overview of the project *before you submit your bid*. Knowing the scope and responsibilities of the project should suggest how much time you'll be spending on it. For example, knowing whether or not you are responsible for implementation makes a *huge* impact on how to price your work. Sound effects can be effectively priced per asset, or even estimated as a lump sum if you are given all the details of a project. But implementation is often unpredictable and can be more time consuming than you expect. It will depend on the workflow of the developer, which is not under your control. For this reason we've found the most useful form of pricing for sound design and implementation (or implementation alone) to be an hourly rate. Note that anchor pricing works exactly the same for all types of pricing. Go with the option that is most comfortable for you, and makes the most sense for each project.

Rates Sheet

An important step in this process is to have a basic rates sheet. This is just a pdf document that first contains your anchor value (which will be your buyout price), then below it one or two licensing options. Keep it simple and don't offer so many options as to make it confusing. Remember that this is a starting point for your negotiations.

Make sure the sheet is polished with no spelling errors or typos. The goal of the sheet is to show off your anchor value in a visually appealing format. The secondary goal is to keep you from wasting time guessing what to charge. Yes, your rate sheet will slowly change as you increase your anchor value over your career. But from project to project it will largely be the same. If someone asks "What are your rates?" you can casually say "Here's my rates sheet" and send it over without a second thought. If the person asking is an indie developer or a developer you really would like to work with, then follow this up with, "These are my usual studio rates. How do they fit into your budget?"

Creative Payment Options

An important aspect of this method of pricing is the concept of **leverage**. When you set a high anchor value you are amassing leverage, *even if a client can't pay your rates*. This allows you to make strides in your career regardless of your project budgets because you can use this leverage to negotiate other forms of payment. There are in fact many ways to be compensated as a game audio professional, and they are all useful tools for building momentum in your career. We've listed a few payment options below.

Buyout

This is the most basic form of compensation, and is usually the ideal form of payment for game audio professionals. Essentially, the developer is paying you a maximum price so that *they own all of the assets you produce*. Legally, this means that you don't own your sounds, music, or recording. Technically speaking, the developer has the legal right to say that they, as an entity, wrote your music or produced your sound assets since you were under contract (see "Navigating Contracts"). Often this isn't the case as AAA studios are perfectly happy to allow composers the right to use their music on their website and demo reels for promotional purposes. But these points *do* need to be stipulated in the contract; they are not assumed. Nevertheless, these restrictions allow you to command higher prices than the other options. Usually this means that the negotiations end at the buyout, meaning none of the following options can be mixed with a buyout contract.

Licensing

This topic is quite complex, and for the most part requires a fair piece of research to understand fully. To break it down as simply as possible, you can offer an **exclusive license** to a developer and it will amount to virtually the same thing as a buyout in terms of the final product. The asset(s) in question will be permitted for use by the developer in synchronization with the game for a reduced price. In exchange this gives you more leverage to negotiate your terms. For example, a common situation is for composers to offer an exclusive license and retain full rights to the soundtrack. They then release the game soundtrack and retain 100 percent of the sales to compensate for the reduced up-front price. Another example is to negotiate for a backend **revenue share** (revshare), which we'll look at shortly.

Note that these payment options can (usually) be used equally for music and sound design. In the case of a non-exclusive license, sound design assets can be created for a game project and then uploaded to a royalty-free sound-effect website or The Unity Store and sold as an asset pack. Note that this is *only* legal because the license scenario is non-exclusive. This is the difference between the two licensing options. An exclusive license grants use of the assets solely to the developer, and may not be used in synchronization with competing works. A non-exclusive license grants the developer use of the assets for a deeply reduced price. In exchange the licensor (you) has the legal right to sell the assets to other entities. For more information on a deep but important topic, check out Chapter 11 on the Sound Lab (companion site) where you'll find "Music Rights for Game Audio" (a pdf by Materia Collective).

Revenue Share

Revenue share payments are a fantastic way to get paid on a project with a low upfront budget. A revshare payment is money that you will make *after a game is released*. Revshares are negotiated on percentages. For example, a developer might not be able to pay your buyout rate, but will instead offer you 10 percent of the game sales after it is released. On the upside this allows you to get straight to work without haggling over price too much. It also allows you to work on passion projects that will possibly make you a buck or two if you're lucky. This is especially helpful if you have no credits, or are looking for some new experience (i.e. implementation, or a new genre of music you haven't yet explored).

On the down side, in this you will be taking on all of the risk. The unfortunate truth is that games with low budgets tend to bottom out and disappear from the face of the earth, leaving you with no money up front and no portfolio piece. Even if it is released, the game may also tank, leaving you with hours and hours of time spent on audio and no money to show for it. It can even be tough to track down developers after the game release. They are often so busy with the release that it can take a long time to receive payment, not to mention the fact that the payout can be confusing, making it hard to tell what you are owed.

To minimize your risk we recommend ignoring the 10 to 20 percent rule. Since you will be taking on all the risk, shoot for more like 20 to 30 percent of the budget and negotiate down from there. This is especially fair if you are implementing the audio, or if you are experienced and are thus adding some valuable notoriety to the game itself. You should also be sure to scour the contract to determine exactly how the percentages will be calculated. Sometimes they will be calculated based on *total revenue*; other times they will be calculated based on *profit* – two completely different things. Whatever deal you make with the developer, be sure to make it crystal clear in the contract so you know exactly what to expect.

Another way to minimize risk is to mix and match some of these payment options. For example try negotiating for *some* money up front, *plus* a backend revshare percentage. This can at least cover your costs while you wait for the game to release. Above all, if you are working for revshare, always, *always* retain the rights to your assets. This way if the project bottoms out you can still release your audio and possibly make some money from it. You can even add a clause in the contract stipulating a change in terms if the game does not release within a reasonable timeframe.[7]

Units Sold

Similar to a revshare, you can use the number of units sold to add extra income on the backend of the development process. Basically you would negotiate for a bonus if a game sells "X" number of copies. The argument is that it provides an incentive for you as a contractor to do your best work and to aid in the promotion of the product. It is basically the "we're all in this together" philosophy.

Final Thoughts on Pricing

There are clearly dozens of factors that go into the minutiae of a bid. But when it comes to finding your creative fee and setting your anchor value we would encourage you to keep things as simple as possible. Don't overthink or try to guess at a developer's budget *before* you set an anchor value. Trust your rate sheet and be consistent regardless of the project budget. The point of the anchor value is to show potential clients what you're worth, not what you want them to pay you.

The unfortunate truth is that by throwing out an anchor value, some developers will scoff at it and end negotiations. This is not such a pleasant experience, but it serves an important purpose – to weed out low-paying clients that do not value audio as integral to the development process. The bottom line is that if you respect audio, you will pay a fair price for it. In our experience this method teaches clients to value audio *more*, not less. This is good for you and for the industry as a whole.

Finally, we offer a simple word of caution. Do not let your value as a business person influence your self-worth. Remember the Pyramid of Sustainability (Chapter 10) – your personal health and happiness is the foundation of your career. Personal success is under your control while financial success is often influenced by other factors. Keep this in mind and you will find the business side of game audio as rewarding as the technical side.

In the Sound Lab (companion site) you'll find an example of a final bid. Check it out before moving on to the next section.

NAVIGATING CONTRACTS

The final step in landing a project is negotiating the finer points in the contract. It can be intimidating to work through contract negotiations, but you do not have to be a lawyer to negotiate or draw up a contract. The wording in contracts can be sneaky, but there are plenty of tools and references to aid you. The Game Audio Network Guild website [8] is a helpful resource for contracts and negotiations. The "Resources" page contains a sample **work for hire agreement**, license agreement with royalties, and a standard non-disclosure agreement (all of which we will discuss on the companion site). Likewise, Aaron Marks' book *Aaron Marks' Complete Guide to Game Audio*[9] is an indispensable resource for the bidding process and as a reference for sample contracts. Lastly, for even more detailed specifics on contracts and buyouts vs. licenses you can check out the article on game audio contracts by Duane Decker on Gamasutra.[10]

Although it is best to have a lawyer look over or draw up any contract that you will be signing, it is an unreasonably expensive cost for smaller contracts. The reality is that most of the time (especially as a freelancer) you are left on our own to write up and negotiate contracts. With a bit of patience and research this can become a helpful skill to aid your career development.

Before we move on to types of contracts and negotiable contract points, we will impart a word of caution: Do not *ever* work without a contract. Contracts protect both parties. Beware of the client who tries to convince you to work without a contract. These clients are either unprofessional, or they are intentionally trying to pull one over on you. Contracts make professional agreements *clear* and *clear is kind*. Don't work without a contract – ever.

In the Sound Lab (companion site) we'll dive into the types of contracts that you will come across as a game audio professional.

NOTES

1 D. Carnegie, *How to Win Friends and Influence People*.
2 www.twitch.tv/powerupaudio
3 B. Brown, *Dare to Lead*.
4 A. Marks, *Aaron Marks' Complete Guide to Game Audio*.
5 B. Schmidt, "GameSoundCon Game Audio Industry Survey 2017."

6 B. Schmidt, "Becoming a Game Music Composer."
7 It is important to note here that revshares are usually *non-negotiable* for AAA projects, and they are especially uncommon in the case of buyouts. Many of these creative payment choices are aimed more for lower or middle-budget indie projects.
8 www.audiogang.org/
9 A. Marks, *Aaron Marks' Complete Guide to Game Audio.*
10 D. Decker, "Game Audio Contracts."

BIBLIOGRAPHY

Brown, B. (2018). *Dare to Lead*. New York: Random House.

Carnegie, D. (1936/1981). *How to Win Friends and Influence People*, revised edn. New York: Simon & Schuster.

Decker, D. (n.d.). "Game Audio Contracts." Retrieved from www.gamasutra.com/view/feature/4149/game_audio_contracts.php?print=1

Demers, J. (March 8, 2018). "Inbound Vs. Outbound Leads: Which Are Better?" forbes.com forbes.com. Retrieved from www.forbes.com/sites/jaysondemers/2018/03/08/inbound-vs-outbound-leads-which-are-better/#315d86c62392

Marks, A. (2017). *Aaron Marks' Complete Guide to Game Audio: For Composers, Sound Designers, Musicians, and Game Developers*, 3rd edn. Boca Raton, FL: CRC Press.

Schmidt, B. (2018). "Becoming a Game Music Composer: What Should I Charge for Indie Games?" Retrieved from www.gamesoundcon.com/single-post/2016/03/11/Becoming-a-Game-Music-Composer-What-Should-I-charge-for-Indie-Games

Schmidt, B. (2017). "GameSoundCon Game Audio Industry Survey 2017." Retrieved from www.gamesoundcon.com/single-post/2017/10/02/GameSoundCon-Game-Audio-Industry-Survey-2017

12 Your Game Audio Strategy Guide

Now that we have everything we need to dig into the business aspects of game audio, we will outline a game plan (pun intended) for your career. Remember that no two career paths are alike. This is why we have included a few different career paths on the companion site as downloadable templates. Head over to the Sound Lab (companion site) for your individualized five-year plan.

AUDIO CREATOR AS GAME DEVELOPER

Finally, we have come to the end of our journey. This last section is dedicated to our thoughts on game audio and how we as composers, sound designers, musicians, and voice actors fit into the game industry as a whole. We'll discuss our advocacy toward teaching and game audio pedagogy. We'll also take a look at inclusivity and accessibility in games, and explore what our ethical responsibilities might be.

Game Audio Philosophy

Our firm belief is that game audio and audio in general are not synonymous. We have clearly shown that games require audio professionals who understand nonlinearity and have the skills and imagination to devise and implement interactive audio systems. We are not audio specialists contracted out for game work. We are, and should strive to be, game developers ourselves. We need to know the ins and outs of our games, understand how audio tells a story in a nonlinear environment, and be fluent in the other aspects of game development in order to effectively immerse ourselves in the development workflow. Audio creators who work on games are *game developers*.

Not all of us think of ourselves as developers, but we think it is important to cultivate this idea within the game audio community. It is now easier than ever to dabble in game development and design due to the tools and technology that previously were not available to us. Game engines like Twine and GameMaker allow those of us who are inexperienced with coding to focus mostly on the design aspects of the process rather than programming. These tools are a great place to start when you're looking to understand nonlinear media and get a feel for the development process as a whole. This experience and the skills you will learn will absolutely transfer into your audio work and help you communicate more effectively with developers. This is important because we need to begin to see ourselves as game developers so that we can bear more responsibility during the process. Audio might be (technically speaking) "invisible," but the effects of music and sound on a game are not. Often it is a key factor in the relationship between player and game.

As game developers we have a responsibility to the game, to our community of audio creators, and to the world in general. Our responsibility is to integrate the game with sound in a way that supports both gameplay and story as effectively as possible. This can only be done by upholding our responsibility to the game audio community and committing our knowledge and our time to teaching and mentorship opportunities. By sharing our knowledge and teaching others we are creating a truly collaborative community, one whose success or failure depends not on a single project, but on our actions as a whole and on our level of support for fellow audio folks. In our view secrecy benefits no one. Yes, the industry is tough and competitive, but by sharing our experiences and offering support to one another we will all be learning and growing together – which can only result in great audio. This mentality is already present in the game audio community more so than any other niche industry we have seen. We intend for this book to add to and further cultivate that group mentality which is so special and important to our culture.

Lastly, games are powerful. We tend to think of games as something we can casually play to relax and unwind. But games and game technology have been used for decades for education, medicine, and research. Games are extremely powerful tools for learning because they go hand in hand with our natural inclination toward reward-based systems. Game technology is being steadily introduced into public education systems as many of you who are parents might already know. This makes sense because a number of studies have shown that games have the ability to literally shape our brains. One 2017 study in *Frontiers in Human Neuroscience* suggests that games can make the brain regions responsible for attention and visuospatial skills more efficient.[1]

On the player side of things, game technology (especially VR) has been used for decades to study psychology and human behavior. Jim Blascovich and Jeremy Bailenson in their book *Infinite Reality*[2] detail dozens of research projects and medical advances of the past 30 years which have utilized VR technologies. These applications for a technology that is most often associated with gaming are mind blowing, and should not be underestimated. As developers we need to understand the power of games and game technology. With that knowledge it follows that we have an ethical responsibility to the world to be open-minded, inclusive, and to make choices that make the world better for future generations.

Teaching and Game Audio Pedagogy

We've put quite a lot of emphasis in this book on teaching, and it isn't just because it's a great way to earn a stable income while venturing out into the competitive world of game audio. In our view teaching is an essential part of any field, and should be taken very seriously. Unfortunately, there are not many college programs devoted specifically to game music or game audio. A cursory Google search yields a ton of game-design programs, and even a handful of film-scoring programs, but very few game-audio-specific college degrees. Even fewer are present at the graduate level. In a field that demands as much technical prowess and creativity as game audio, we think that this needs to change.

Teaching doesn't just offer freelancers an outlet for their knowledge, it also makes the industry better. By developing a rigorous game audio pedagogy (i.e. teaching methodology) we will be creating generations of game audio professionals who start out with more knowledge than we did. This can only be good for our craft and our industry. Game audio is specific and demanding. By teaching to those demands we will also be bolstering the value of game audio to developers, who can often underestimate audio in games. This will also put game audio on the table as more of a legitimate academic area of research, furthering the pedagogical development.

The need for more practical course materials in game audio was a huge reason we wrote this textbook in the first place. Many people are already doing it on their own, and they are doing it well. But as teachers we still see a growing demand to learn game audio in a way that consolidates best practices and expels the mysteries of the trade. We can't offer an exact template for this proposed pedagogy, but we hope that this book will inspire a serious and continuing discussion on what that would look like. What we *can* offer is something of a call to arms. So find a way to incorporate teaching into your life. Be a mentor, find a mentee, teach a class, create a class, write a book, give a talk, and think hard about how to teach game audio.

Inclusivity and Accessibility in Game Audio

Ethically speaking, we believe our top priority as developers is to create an inclusive and accessible space for game audio professionals. This topic is complex enough that it can and should warrant its own textbook, but here we will attempt to distill some of our thoughts and advice. In the game audio community we have seen more enthusiasm for this topic than in many other industry niches. But the fact still stands that there is a diversity problem in game audio and in game development as a whole. The number of women in game audio is actually on the rise. However, according to the

2017 GameSoundCon Industry Survey,[3] women (despite comprising roughly half of the population) make up only about 12.7 percent of the total composers and sound designers in the industry. What is more disheartening is that women in the industry only earn 83 percent of what men earn on average.

According to the IGDA (International Game Developers Association) 2017 Developer Satisfaction Survey (DSS) Report[4] the vast majority of developers identify as white/Caucasian/European at 68 percent and only 1 percent identified as black/African American. By contrast, according to the U.S. 2016 Census 61 percent of the population was white, and 13 percent was black. Now this survey is voluntary, and international – about 40 percent of respondents are from the U.S. so these numbers are not a perfect representation of the worldwide game development community. Nevertheless, these demographics are skewed and the disparity is pretty clear. What's more alarming is that 14 percent of respondents reported that their company had "no policies whatsoever directed toward diversity or equality. Among those whose company did have some form of policy, most said that their company had a 'general non-discrimination policy' (57%), an 'equal opportunity hiring policy' (49%) or a 'sexual harassment policy' (48%). Only 26% said that their company had a 'formal complaint procedure,' and 21% reported a 'formal disciplinary process' related to equality and diversity policies."

Surveys like this one are important, and despite the fact that they show us how far we have to go, they also show us that most if not all developers consider inclusivity a priority. Of all respondents, 81 percent reported that diversity in the game industry was "very" or "somewhat" important. This was the highest in the history of IGDA's DSS. This shows that inclusivity is already important to people, and we just need to push forward and explore ways of increasing our progress towards diversity.

No discussion of inclusivity is complete without a discussion of sexism and harassment in online gaming. One 2015 study conducted by Michael Kasumovic and Jeffrey Kuznekoff reported the following results: "We show that lower-skilled players were more hostile towards a female-voiced teammate, especially when performing poorly. In contrast, lower-skilled players behave submissively towards a male-voiced player in the identical scenario."

The scenario described above is tragically commonplace. As individuals it is not our fault that this phenomena pervades gaming culture, but as game developers it is our responsibility to do all we can to defend against this toxic behavior. On a societal level we need to reflect on our culture, and the culture of gaming to negate tropes and to promote inclusivity and diversity in games. At the individual level it is vital for us to do what we can to halt this behavior if witnessed, and to hold others and ourselves accountable when it happens. Events like the Games for Change Festival[5] are platforms for social impact games to be seen and played. Social impact games deal with real-world issues like climate change and social equality. They are created with the idea that games are powerful enough to educate players, and to inspire lasting, positive change in the world. While it is not necessary to focus our efforts on a specific genre, we can still take a lesson from Games for Change and social impact games by injecting some of these ideals into our respective companies and game designs. Simply speaking up about an idea that might make your game more inclusive or diverse can make a big difference.

Unfortunately, there is no perfect answer to the question of how to create the perfectly inclusive, safe space for diversity to blossom in the industry. For one thing, just discussing the issue is in itself an important step. Women in Games International (WIGI)[6] regularly hosts roundtables and discussion groups at GameSoundCon and GDC focused on the underrepresentation of women. Blacks in Gaming (BIG)[7] hosts similar events and meetups. There are a number of other advocacy organizations working toward similar goals. Girls Who Code[8] is an NPO devoted to increasing education and opportunities for women in the field of computer science. Kellen Beck's 2018[9] article also outlines a number of developers like Culture Shock Games and Dim Bulb Games looking to make game development more inclusive and diverse.

Anna Anthropy, a game developer and advocate, has a very interesting take on the game industry as a whole. Her book *Rise of the Videogame Zinesters*[10] is essentially a call to arms for hobbyists, amateurs, and basically everyone to start developing and self-publishing games. The hope is that by making game development and publishing possible at the individual level, minority groups will be more adequately represented in the industry. This would certainly go a long way towards making the conversation about diversity and industry ethics a priority. She recommends using the streamlined game development tools mentioned above (see "Game Audio Philosophy," Chapter 12, page 380) to keep the process from being overwhelming. Anthropy suggests that the result of such a personalized indie game development boom would be a striking increase in the diversity of subject matter and game mechanics. The way we interact with games could be positively impacted by simply encouraging more people to make games.

Above all, we would recommend immersing yourself in the events, talks, and round tables available at industry events. Progress on these issues is halted by ignorance and complacency. So don't be shy! These discussion events are safe spaces, and they are vital to progress. Show up, listen, speak up about your experiences, and make friends. The more people who are aware of the issue of diversity, the more allies we have to help defend against it.

Accessibility in games is, of course, its own issue that deserves equal attention. According to Adriane Kuzminski (Smash Clay Audio, owner, designer), sound design is not just a tool to entertain players. It can and should be used to make games more accessible to players with disabilities ranging from blindness and low vision to dyslexia, and even those with autism or sensory processing disorders (SPD).[11] According to Kuzminski, "One of the first things is to use UI sounds as a means to teach the gamer to memorize how the game works. The way to do this is by making 'iconic' UI sounds, which just means that they are consistently tied to an action or outcome. When a sound means the same thing every time, you create an audible user interface."[12]

This concept is taken to the extreme in MMO games like *Paladins* (Hi-Rez Studios). Our work on this project taught us how important it is to create recognizable and predictably timed weapon sounds. For example, creating a reload sound that predictably rose to a specific pitch gave players a competitive edge because it offered valuable information about when the opponent would be ready to fire. According to Kuzminski, design decisions like that also makes games more accessible, which in turn makes them more marketable for developers. In her own words, "We have too much technology and too many people working on this to say that the gaming experience is not meant for certain groups of people."[13]

Making our industry space more inclusive and diverse is not only ethically important, it is also important for games. Look back at Chapter 6 and re-read "Generating Musical Ideas and the Creative Cycle." If creativity is the fire, then the fuel is a diversity of ideas. This does not happen without true diversity at the source of those ideas – people. Diversity in the game industry will yield innovative approaches to development, and truly beautiful games. Not only that, but inclusivity in the industry will increase the likelihood that marginalized groups will be able to relate to the subject material, thereby opening up a wider audience and a more inclusive gaming experience. This, in turn, will encourage people from disabled and marginalized communities to rise up as game developers themselves and share their stories. A focus on accessibility in gaming will ensure that *everyone* is able to enjoy and benefit from those games. Games are powerful, and games are for everyone.

The Possible Future of Game Audio

If it isn't clear already, we have the utmost respect for the craft of creating audio for games. We also have a sincere degree of optimism for its future. Technologically speaking, we have explored some of our thoughts on the possible future of game audio. Machine learning could come into play in a very real way at some point. In many cases it is already being explored in regards to design and development. But what if machine learning was used to create music and sound? Is there a chance that future generations of game audio practitioners may find themselves in the role of aesthetic coordinator rather than content creator? Will they be feeding original content into algorithms that "procedural-ize" audio for them? Will they be *creating* the algorithms themselves? One thing is perfectly clear – composers and sound designers alike are finding themselves in increasingly technical roles in the game industry. It may be that the lines between audio creators and programmers will become more and more blurred.

While we can't predict the stylistic trends in game audio, we can say what we hope to happen. We hope that game audio continues to find a unique and personal niche in the world of art and entertainment. We also hope that the public begins to understand the technical and academic validity of game audio as the singular and exceptional medium that it is. All too often we hear comparisons between film and game sound, or film and game composers. Obviously there are similarities, but we hope to see the differences come into public awareness as well. The difference between film sound and game sound is exactly what makes games special. Regardless of the genre, composers and sound designers are producing audio that is dynamic and interactive. These audio experiences are essentially tailored to each player, and that is something very special in our view. It's common today to hear people leave theaters saying, "Did you hear that theme at the end of the movie? It was awesome." We like to think that in the not-too-distant future it'll be just as commonplace to hear people saying, "Did you hear how the soundscape evolved as I defeated the final boss? I felt like it was designed specifically for me! *Sick*." And then sound designers, composers, Foley artists, voice actors, implementers, musicians, mix engineers, audio programmers, and audio directors can all smile knowing that our impact as game developers is as indispensable as the games themselves.

We will leave you with one final thought. *Game audio is for everyone.* With enough passion and persistence anyone and everyone can learn it. As sophisticated as the topic of game audio is, the essence of it is still passionate people making sounds for games that they love. That's it! It is our sincere hope that this book demystifies the process and makes the methods and technology clearer and more accessible for those who want to learn. If you finish this book with just one piece of advice, let it be this: *just make sounds.* Go out and record some sounds, or make some music, or make a game. Whatever you do, do it now. Good luck!

NOTES

1 U. Pfeiffer *et al.*, "Towards a Neuroscience of Social Interaction."
2 J. Blascovich and J. Bailenson, *Infinite Reality.*
3 B. Schmidt, "GameSoundCon Game Audio Industry Survey 2017."
4 J. Weststar, V. O'Meara, and M.-J. Legault, *IGDA Development Satisfaction Survey 2017.*
5 www.gamesforchange.org
6 https://getwigi.com
7 www.facebook.com/pages/category/Organization/Blacks-in-Gaming-142639569140671/?
8 https://girlswhocode.com
9 K. Beck, "Diversity in the Video Game Industry Is (Surprise) Not Good."
10 A. Anthropy, *Rise of the Videogame Zinesters.*
11 T. Chopp, "Why Now is the Time."
12 Ibid.
13 For more information on accessibility, join the IGDA GA-SIG (Game Accessibility Special Interest Group): https://igda-gasig.org/oer/

BIBLIOGRAPHY

Anthropy, A. (2012). *Rise of the Videogame Zinesters: How Freaks, Normals, Amateurs, Artists, Dreamers, Drop-outs, Queers, Housewives, and People Like You Are Taking Back an Art Form*. New York: Seven Stories Press.

Beck, K. (2018). "Diversity in the Video Game Industry Is (Surprise) Not Good." Retrieved from https://mashable.com/2018/01/09/video-game-diversity/#VNSRiqhXhOq6

Blascovich, J. and Bailenson, J. (2011). Infinite Reality: Avatars, Eternal Life, New Worlds, and the Dawn of the Virtual Revolution. New York: William Morrow.

Chopp, T. (2017). "Why Now is the Time to Create Accessible Video Games." Retrieved from www.voices.com/blog/why-now-is-the-time-to-create-accessible-video-games/

Pfeiffer, U., Timmermans, B., Vogeley, K., Frith, C., and Schillbach, L. (eds.) (2013). "Towards a Neuroscience of Social Interaction." Retrieved from www.frontiersin.org/research-topics/211/towards-a-neuroscience-of-social-interaction

Schmidt, B. (2017). "GameSoundCon Game Audio Industry Survey 2017." Retrieved from www.gamesoundcon.com/single-post/2017/10/02/GameSoundCon-Game-Audio-Industry-Survey-2017

Weststar, J., O'Meara, V., and Legault, M.-J. (2018). *IGDA Development Satisfaction Survey 2017*. Retrieved from https://cdn.ymaws.com/www.igda.org/resource/resmgr/2017_DSS_/!IGDA_DSS_2017_SummaryReport.pdf

Index